W0079421

Lecture Notes in Biomathematics

Lecture Notes in Biomathematics

Managing Editor: S. Levin

12

Arun V. Holden

Models of the Stochastic Activity of Neurones

Springer-Verlag
Berlin · Heidelberg · New York 1976

Editorial Board

W. Bossert · H. J. Bremermann · J. D. Cowan · W. Hirsch
S. Karlin · J. B. Keller · M. Kimura · S. Levin (Managing Editor)
R. C. Lewontin · G. F. Oster · L. A. Segel

Author

Dr. Arun Vivian Holden
Department of Physiology
School of Medicine
University of Leeds
Leeds LS2 9JT/Great Britain

Library of Congress Cataloging in Publication Data

Holden, Arun V   1947-
  Models of the stochastic activity of neurones.

  (Lecture notes in biomathematics ; 12)
  Includes bibliographies and index.
  1. Neurons--Mathematical models. 2. Action potentials (Electrophysiology)--Mathematical models.
3. Neural transmission--Mathematical models. I. Title.
II. Series.
QP363.H64   591.1'88   76-41774

AMS Subject Classifications (1970): 60-02, 60 G 35, 60 J 70, 60 K 10, 92-02, 92 A 05, 94-02, 94 A 35

ISBN-13: 978-3-540-07983-5   e-ISBN-13: 978-3-642-46345-7
DOI: 10.1007/978-3-642-46345-7

This work is subject to copyright. All rights are reserved, whether the whole or part of the material is concerned, specifically those of translation, reprinting, re-use of illustrations, broadcasting, reproduction by photocopying machine or similar means, and storage in data banks.

Under § 54 of the German Copyright Law where copies are made for other than private use, a fee is payable to the publisher, the amount of the fee to be determined by agreement with the publisher.

© by Springer-Verlag Berlin · Heidelberg 1976

CONTENTS

PREFACE

These notes have grown from a series of seminars given at
Leeds between 1972 and 1975. They represent an attempt to gather
together the different kinds of model which have been proposed to
account for the stochastic activity of neurones, and to provide an
introduction to this area of mathematical biology. A striking
feature of the electrical activity of the nervous system is that
it appears stochastic: this is apparent at all levels of recording,
ranging from intracellular recordings to the electroencephalogram.
The chapters start with fluctuations in membrane potential,
proceed through single unit and synaptic activity and end with
the behaviour of large aggregates of neurones: I have chosen this
sequence is chosen to suggest that the interesting behaviour of the nervous
system - its individuality, variability and dynamic forms - may
in part result from the stochastic behaviour of its components.

I would like to thank Dr. Julio Rubio for reading and
commenting on the drafts, Mrs. Doris Beighton for producing the
final typescript and Mr. Peter Hargreaves for preparing the
figures.

Arun Holden

June 1976   Leeds

1. STOCHASTIC FLUCTUATIONS IN MEMBRANE POTENTIAL

The resting membrane potential V, which is the potential difference across the membrane separating the cytoplasm and the extracellular
fluid, is negative and is due to the selective ionic permeability of
the membrane and the difference in ionic composition between the cytoplasm and the extracellular fluid. These differences in ionic composition are maintained by an active sodium-potassium exchange pump. The
ionic permeability of the membrane is a measure of the ease with which
ions move through the membrane, and so will be dependent on the ionic
mobility within the membrane, the membrane thickness and the partition
coefficient between the aqueous extra- and intra-cellular fluids and
the membrane. The membrane permeability will not depend on the ionic
concentration or membrane potential unless these factors affect the
ionic mobility or partition coefficient.

Since it is easier to measure the current carried by an ion than
the ionic flux it is useful to consider the membrane conductance rather
than permeability. The membrane conductance g_m will depend on the
concentration of the ion within the membrane (and hence on the extra-
and intra-cellular ion concentrations and the partition coefficient)
and the membrane permeability.

If I assume a homogenous membrane with a thickness D, ionic mobility μ, partition coefficient β and a constant potential field V/D
the permeability P_a of a univalent ion $\underline{a}$ is defined by

$$P_a = \mu \beta RT/DF$$

where R is the universal gas constant

T is absolute temperature

F is Faraday's constant.

The only forces producing ionic movement through the membrane, or an
ionic current, I_a, are the concentration $d[a]/dx$ and electrical dV/dx
gradients, where $[.]_i$ denotes concentration inside, and $[.]_o$ outside
the cell.

Thus

$$-I_a = RT\mu d[a]/dx + [a]\mu FV/D$$

with boundary conditions

$$x = 0 \text{ (inside surface)} \quad [a] = \beta [a]_i$$

$$x = D \text{ (outside surface)} \quad [a] = \beta [a]_o$$

Integration and substitution gives

$$I_a = P_a \frac{F^2 V}{RT} \left\{ \frac{[a]_o - [a]_i \exp(-VF/RT)}{1 - \exp(-VF/RT)} \right\}$$

but $g_a = I_a / (V - V_a)$

where V_a is the reversal potential for the ion $\underline{a}$ and

$$V_a = \frac{RT}{F} \ln([a]_i / [a]_o)$$

Thus

$$g_a = \frac{P_a F^2 V [a]_o}{RT} \left\{ \frac{1 - \exp((V_a - V)/RT)}{1 - \exp(-VF/RT)} \right\} \frac{1}{V - RT \ln([a]_i / [a]_o)}$$

when $V = V_a$ or $I_a = 0$ this simplifies to

$$g_a = \frac{P_a F^3 V_a}{(RT)^2} \left\{ \frac{[a]_o [a]_i}{[a]_i - [a]_o} \right\}$$

Thus when the intra- and extra-cellular ion concentrations are constant the constant field assumption gives $g_a \propto P_a$.

A simple circuit model of the passive electrical properties of a membrane is that of a resistance (the reciprocal of conductance) and capacitance in parallel (see Figure 1.1(a)). Any change in V_m means that current is flowing across the membrane. This membrane current is the sum of ionic and capacitance membrane currents.

$$I_m = I_a + C_m dV/dt$$

where C_m is the membrane capacity/unit area. C_m is approximately $1 \mu F/cm^2$ in all cells, and this capacity is due to the high dielectric constant of the lipoid membrane material and the membrane thinness. C_m may be taken as a constant, as there is only a 2% change during the 100 mV depolarization of an action potential (Cole, 1968).

Thus, any fluctuations in V will be due to fluctuations in membrane
ionic currents, which will be due to changes in membrane ionic conductances.

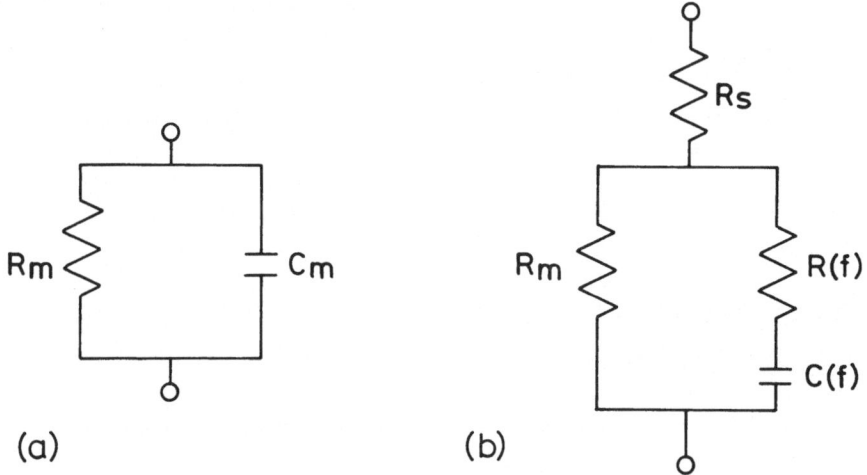

(a)   (b)

Figure 1.1. Equivalent circuit models of the passive electrical properties of a

membrane. (a) Simple RC model. (b) Modification of simple RC model suggested by

AC analysis, with a nonideal, frequency sensitive capacitance.

It appears that ions, which are hydrophilic, can only pass
through the hydrophobic lipoid membrane at specializations which are
sparsely distributed - these specializations are called ionic channels.
The low density of ionic channels corresponds to the low ionic conductance of the membrane. Thus stochastic fluctuations of V might
provide insights into the opening and closing kinetics of ionic
channels.

The principal ionic species of interest are sodium, potassium
and chloride ions. The resting membrane potential depends on their
concentrations and permeability by

$$V = \frac{RT}{F} \ln \left\{ \frac{P_{Na}[Na^+]_o + P_K[K^+]_o + P_{Cl}[Cl^-]_i}{P_{Na}[Na^+]_i + P_K[K^+]_i + P_{Cl}[Cl^-]_o} \right\}$$

Good accounts of the biophysics and electrophysiology of membranes are found in Katz (1966), Cole (1968) and Jack, Noble and Tsien (1974): here I will only be concerned with some models of stochastic fluctuations in membrane potential and will assume familiarity with the basic facts of membrane biophysics.

The types of fluctuation in membrane potential can be classified as:

a)   thermal noise
b)   shot noise
c)   flicker noise
d)   conductance fluctuations.

Stevens (1972) has reviewed these types of membrane potential fluctuation and discussed their implications.

1.1   Thermal or Johnson-Nyquist Noise

Fatt and Katz (1950) suggested that voltage fluctuations due to thermal ionic movement might contribute to detectable fluctuations in the resting membrane potential. In a later paper on spontaneous sub-threshold fluctuations in membrane potential at the frog neuromuscular junction (see section 11) Fatt and Katz (1952) estimated the r.m.s. amplitude of thermal noise fluctuations. For an ideal resistance R at an absolute temperature T the r.m.s. noise value V_t is given by (Nyquist 1928)

$$V_t^2 = 4 \ kT \int_o^{f_1} R \ df \tag{1.1.1}$$

where k is Boltzman's constant (1.38×10^{-23} J/°C) and f_1 is the upper frequency (Hz) of the frequency band over which noise is distributed.

$$V_t^2 = \frac{8kT}{\pi} \ \sqrt{R_a R_m} \ (2r)^{-1.5} \int_o^{f_1} A \ df \tag{1.1.2}$$

where $A = \sqrt{\{ \frac{1}{2}[1/(1 + 4\pi^2 f^2 C_m^2 R_m^2) + 1/\sqrt{(1 + 4\pi^2 f^2 C_m^2 R_m^2)}]\}}$

and so V_t increases as fibre radius decreases. Reasonable values of

R_m, R_a and C_m give V_t of $2\mu V$ for an axonal diameter of 100 and 0.5 mV for an axonal diameter of 0.1 μ.

However, the membrane impedance is not an ideal resistance. If the spectral density of the thermal noise is $V_t(f)$, equation (1.1.1) may be rewritten as

$$V_t(f) \;=\; 4\,k\,T\,\mathrm{Re}\,Z(f) \tag{1.1.3}$$

where $Z(f)$ is the complex impedance of the membrane. Thus for the electric circuit model of Figure 1.1(a)

$$V_t(f) \;=\; 4\,k\,T\,R_m \,/(1 + 4\pi^2\,(R_m C_m f)^2) \tag{1.1.4}$$

and hence an experimental estimate of $V_t(f)$ could give support to the parallel $R - C$ circuit model and give estimates of R_m and C_m. The imaginary part of the impedance is simply the Hilbert transform of $\mathrm{Re}\,Z(f)$.

Sine wave analysis has given a more accurate equivalent circuit for an axonal membrane where an ideal resistance is in parallel with a frequency dependent series resistance and capacitance (Cole 1968) - see Figure 1-1b. For small fluctuations in voltage which do not influence the Hodgkin-Huxley rate coefficients (see section 3) the membrane impedance is well described if (Stevens 1972)

$$R(f) \;=\; R_o\,(2\pi f\tau)^{\alpha-1}\sin\,\alpha\pi/2$$

$$C(f) \;=\; C_o/\{(2\pi f\tau)^{\alpha}\,\cos\,\alpha\pi/2\}$$

where R_o, C_o and τ are constants. If

$$a \;=\; C_o/\cos(\alpha\pi/2)$$

$$b \;=\; R_o\,\sin\,\alpha\pi/2$$

$$V_t(f) \;=\; 4kTR_m\left\{\frac{1 + (ab/\tau)^2 + (a^2 bR_m/\tau^2)\,(2\pi f\tau)^{1-\alpha}}{1 +\{(ab/\tau) + (aR_m/\tau)\,(2\pi f\tau)^{1-\alpha})\}^2} + \frac{R_s}{R_m}\right\} \tag{1.1.4}$$

For the squid giant axon membrane $\alpha \sim 0.1$ and so for high frequencies

$$V_t(f) \sim 4kT(b/(2\pi f\tau)^{1-\alpha} + R_s)$$

1.2 Shot noise

The ionic membrane current is carried by individual ions moving through the membrane. Since the membrane resistance is high the probability of an ion transit is low, and I will assume that in any small period of time Δt there is a small constant probability P of an ion transit, so that the rate of ion transit is $p = P\Delta t$. Thus the probability of m ions crossing the membrane in a period $\tau = N\Delta t$ is

$$P(m, \tau) = \binom{N}{m} p^m (1 - p)^{N-m}$$

and so

$$<m> = \sum_{m=0}^{N} mP(m,\tau) = Np$$

and $<\Delta m^2> = <m>$

i.e. ion transit is a Poisson process.

Since the ionic channels are sparse ion movement is likely to be independent. If I assume that all monovalent ions cross the membrane by a similar mechanism, say by passage through an aqueous 'pore', the transit time for all univalent ions will be similar and the current i(t) carried by a <u>single</u> ion will be similar.

With these three assumptions the spectral density of the ionic current at a constant voltage will be that of shot noise (Rice 1954)

$$I_s(f) = p \left| \frac{1}{\sqrt{2\pi}} \int_{-\infty}^{\infty} i(t) \exp(-2 fjt)dt \right|^2 \tag{1.2.2}$$

where $j = \sqrt{-1}$

Thus the voltage shot noise spectral density will be

$$V_s(f) = I_s(f)|z(f)|^2 \tag{1.2.3}$$

where Z(f) is the membrane complex impedance which is filtering the
current noise.

The average transit time for an ion through the thin (approx
10 nm) membrane is short - if I assume the ionic mobility in the membrane to be the same as that in free aqueous solution the average
transit time will be less than 1 µsec. For frequencies much less than
the average transit time the voltage shot noise spectrum will be flat.
Thus the shot noise process will have a flat spectrum at low frequencies, and to demonstrate interactions between ion movements the spectral
density of the shot noise process would have to be accurately obtained
at arbitrarily high frequencies.

1.3 Flicker Noise

The voltage noise generated by electronic devices at high frequencies (> than 10^5 Hz for a vacuum tube device) has a flat spectral density and can be quantitatively accounted for by Johnson-Nyquist and
shot noise processes. For lower frequencies, the r.m.s. noise level is
often greater than would be expected and often has a spectral density
which is inversely proportional to the frequency.   This low frequency
noise is called 'excess', 'flicker' or '1/f' noise in the electronic
engineering context: see Robinson (1974) or van der Ziel (1970). In
membrane biophysics '1/f' noise is the preferred term (Verveen and
Derksen, 1968; Stevens, 1972); however, I will use the term flicker
noise as the term '1/f' noise has unfortunate consequences as f → 0.

If a system is in thermodynamic equilibrium the only possible
noise source is Johnson-Nyquist noise, and so flicker noise must be
associated with a system which is not at equilibrium. If the noise is
generated by a stochastic process characterized by a relaxation time
τ_r the correlation function will have the form

$$c(t) \quad = \quad A_r \quad \exp(-t/\tau_r) \tag{1.3.1.}$$

which gives by the Wiener-Khintchine theorem (Lee, 1967 - see chapter 2)
a spectral density

$$V(f) \quad = \quad 4 \int_0^\infty A_r \quad \exp(-t/\tau_r) / \cos 2\pi ft \quad dt$$

$$= \quad 4 A_r \tau_r / (1 + (2\pi f\tau_r)^2) \tag{1.3.2}$$

Thus at low frequencies a process with a single relaxation time gives $V(f)$ varying inversely with f^2. van der Ziel (1959) has proposed a simple model for flicker noise in which the stochastic relaxation time τ depends on the activation energy E

$$\tau = \tau_o \exp(E/kT)$$

where k is Boltzman's constant and T is the absolute temperature. Then if activation energies between E_1 (giving τ_1) and E_2 (giving τ_2) are equiprobably distributed then the normalized probability distribution of relaxation times $g(\tau)d\tau$ is given by

$$g(\tau)d\tau = d\tau/(\tau \ln \tau_2/\tau_1)$$

which gives

$$V(f) = \frac{A (\tan^{-1} 2\pi f\tau_2 - \tan^{-1} 2\pi f\tau_1)}{2\pi \ln\tau_2/\tau_1 \cdot f} \qquad (1.3.3)$$

and so flicker noise is obtained when $f\tau_2$ is large and $f\tau_1$ is small.

The amplitude distribution of flicker noise is Gaussian (Hooge and Hoppenbrouwers, 1969) and cannot be used to distinguish between linear and non-linear models. A non-Gaussian amplitude distribution would impose constraints on the type of model used to account for flicker noise.

The occurrence of flicker noise in a wide range of electronic devices which are not in an equilibrium state but are subject to a direct current bias has generated a large number of specific models: see du Pre, 1950; van der Ziel, 1950; 1954; 1959; Richardson, 1950; McFarlane, 1950; Petritz, 1952; Bess, 1953; Morrison, 1956; McWorter, 1957; Bell, 1958; Sautter, 1960. These models may be considered as specific examples of the type given in the derivation of equation (1.3.3).

Although surface effects in semiconductors influence the magnitude of flicker noise, Hooge (1969) has shown that the magnitude of flicker noise is proportional to the reciprocal of the total number of mobile charge carriers in samples of different semiconductors. Evidence that flicker noise is a bulk effect, and not a surface effect,

is given in Hooge and Hoppenbrouwers (1969), Hooge, van Dijk and
Hoppenbrouwers (1970), Hooge (1970) and Hooge and Gaal (1971), in
which flicker noise is given by the empirical relation

$$V(f) \; = \; \frac{\alpha \, I^2}{N_T \, f}$$

(1.3.4)

where N_T is the total number of charge carriers, α is a constant
ranging from 2.10^{-3} for electrons in metals and 1 for ions in solution
(but depends on concentration for electrolyte solutions).

Verveen and Derksen have extensively investigated voltage
fluctuations across a single node of Ranvier in single frog axons
(Derksen, 1966; Siebenga and Verveen, 1970; 1971a; 1971b; Verveen,
Derksen and Schick, 1967). The properties of these spontaneous voltage fluctuations may be summarized as

a) the power spectral density varies inversely with frequency
over a frequency range from about 1 to 1500 Hz.

b) at low frequencies there is a decrease in the slope of the
log noise power-log frequency plot.

c) at high frequencies the spectral density approaches that of
white noise.

d) interference with metabolism or sodium ion transport did not
influence the flicker noise component.

e) the magnitude of the flicker noise spectral density varies
with passive potassium ion flux.

f) the flicker noise component vanishes in thermal noise at the
potassium equilibrium potential given by the Nernst equation

$$V_K \; = \; \frac{RT}{F} \; \ln \, [K]_o / [K]_i$$

g) the amplitude distribution of the noise is Gaussian for membrane potentials above about -70 mV and is positively skewed for membrane potentials which are more negative. The skewness of the amplitude distribution increases with further hyperpolarization. The
Gaussian and flicker noise depend on passive K^+ movements; the skewed
noise is associated with irregular depolarizing noise bursts associated
with a TTX insensitive Na^+ current.

Poussart (1969, 1971) has investigated spontaneous fluctuations in membrane current in giant axons of the lobster under voltage clamp. Flicker noise current was observed for frequencies from about 3 to 1000 rad/sec. By changing the external potassium ion concentration, and using lanthanum substitution for calcium and blocking regenerative sodium conductance with TTX he showed that the intensity of the flicker noise current was proportional to the mean magnitude of the K^+ current. This in both frog and lobster axons the process generating flicker noise appears to be passive K^+ movement.

Stevens (1972) has interpreted Poussart's 1971 results, by using equation (1.3.4) together with

$$g \propto I$$

$$g = q \, \mu \, N_T$$

where q is the charge/ion and μ the average ionic mobility in a channel, to estimate $\mu(g)$. Except at very low g_k, $\mu(g)$ is approximately constant which suggests that an increase in conductance is due to increasing the number of channels rather than to increasing the mobility and hence the ionic flux within a channel.

The problem is to find a biophysically plausible model for K^+ flux that will generate flicker noise. Offner (1970; 1971a, b; 1972) has proposed an attractively simple model based on a random walk with drift. A simple random walk would generate a spectral density that varied inversely with f^2; however when there is a drift back towards the resting membrane potential with a time constant $\tau = 1/f_c$, Offner (1970, 1971a) apparently generated flicker noise in computer simulations. However Hawkins (1972), Alberding (1973); and Bird (1973, 1974) have shown by computer simulations and analytically that this apparent flicker noise is a numerical artefact, and that Offner's models generate relaxation spectra proportional to $(1/f^2 + f_c^2)$.

Lundstrom and McQueen (1974) have pointed out that the interior of the cell membrane might be similar to the structure of a liquid crystal, and if vibration of the hydrocarbon chains in the membrane influences g_K flicker noise could be generated. Thus they propose that the flicker noise spectrum is determined by the normal modes of vibration in a liquid crystal.

An alternative physical model has been proposed by Schick and

Verveen (1974). They noticed flicker noise in a model system which was the flow of small grains through a long, thin glass tube. The power spectral density was white at low frequencies and changed to approximately 1/f as frequency increased. Thus we have a simple physical system which generates flicker noise and is well behaved as $f \to 0$, and so there is no need for a mathematical model to deal with the 'infra-red catastrophe' or explosion in power at $f = 0+$ by concepts such as the conditional spectrum introduced by Mandlebrot (1967). This model system suggests that flicker noise may be generated by the analogous process of ion passage through long pore-like channels in the membrane. Thus the noise current will consist of a series of pulses, the pulses corresponding to ionic current flow through single ionic channels and having a variable shape. This gives a doubly stochastic process, with one set of parameters determining the shape of the pulse and another set of parameters determining the interpulse interval probability density functions.

Stepanescu (1974) has described a two parameter stochastic process model for flicker noise in semi-conductors in which

(a) each conductivity change (or current pulse) has a rectangular shape with duration τ which has a distribution

$$g(\tau) = \frac{1}{\tau_o} \exp (-\tau/\tau_o)$$

where τ_o is the average duration

(b) the p.d.f. $p(x)$ of the inter-pulse intervals x is Poisson with parameter r

$$p(x) = r \exp (-rx)$$

(c) each trap (channel in our context) acts independently, and the trapping and releasing of a carrier (opening and closing of channels) are processes whose activation energy are distributed uniformly over a wide range.

This model, with the two parameters r and τ, generates a 1/f spectrum within frequency limits determined by the range over which the activation energies E are distributed, since the pulse duration and inter-pulse interval are related to the activation energies.

Schick (1974) has developed this kind of model with reference to membrane potential fluctuations, and has treated Poisson and non-Poisson pulse sequences and different pulse shapes. If each pulse has an amplitude scaling factor a, time constant τ and shape $s(t, \tau)$ and the pulse train has an interpulse distribution function $p(x)$ the current will be

$$I(t) = \cdots + a_{n-1} s_{n-1} (t - (x_1 + x_2 + \cdots + x_{n-2}) \tau_{n-1}$$
$$+ a_n s_n (t - (x_1 + x_2 + \cdots + x_{n-1}), \tau_n) + \cdots \qquad (1.3.5)$$

which, when x, τ and a are independent, has a power spectrum

$$I(f) = r < a^2 > < |F(f, \tau)|^2 > .$$
$$\left\{ 1 + \frac{2 <a>^2 \cdot |<F(f, \tau)>|^2 \cdot \text{Re } (\phi/1-\phi)}{<a^2> \cdot <|F(f, \tau)|^2>} \right\} \qquad (1.3.6)$$

where $<\ >$ denotes average value, r is the number of pulses/unit time and

$$F(f, \tau) = \int_\infty^\infty s(t, \tau) \exp (-2\pi jft) \, dt \qquad (1.3.7)$$

$$\phi = \int_0^\infty p(x) \exp (2\pi jfx) \, dx \qquad (1.3.8)$$

for a Poisson distribution $\text{Re } (\phi/1-\phi) = 0$ and so equation (1.3.6) simplifies to

$$I(f) = r < a^2 > < |F(f, \tau)|^2 > \qquad (1.3.9)$$

If the time constants τ have a distribution $f(\tau)$ there must be a maximum τ, τ_{max}, such that $f(\tau_{max}) \to 0$ and so

$$<F(f, \tau)> \simeq \int_0^{\tau \, max} F(f, \tau) \, f(\tau) \, d\tau$$

and so for frequencies such that $f\tau_{max}$ is small $<F(f, \tau)>$ and hence $<|F(f, \tau)|>$ is flat. Similarly, there is an x_{max} such that $p(x) \to 0$ and so when fx_{max} is small $\text{Re}(\phi/1-\phi)$ is flat. Thus $I(f)$ is flat at low frequencies if there is no coupling between the pulses.

Equation (1.3.6) shows that there are two possible ways in which

I(f) might have a range of 1/f behaviour

  a) there is a class of distribution functions $f(\tau)$ which will
generate a range of 1/f behaviour of I(f): for the low frequency end
of this range to be approximately 1 rad/sec as in nerve membranes τ
must be of the order of seconds. Such long time constants seem unlikely: however the conductance changes produced by the action of
cyclic antibiotics on artificial membranes have long time constants
(Hladky and Haydon, 1970).

  b) A class of non-Poisson distributions p(x) can generate a
range of 1/f behaviour of I(f). Thus the individual pulses can be
generated by processes with short time constants, and $I(f) \to 0$ as
$<a> \to 0$ or as V tends to the Nernst equilibrium potential for the ion
generating the flicker noise. Such a non-Poisson p(x) implies an
interaction between the processes generating the current pulses (see
section 8.4).

1.4 Conductance fluctuations

  If ion flow is taken to be through membrane specializations or
channels, the two ways in which membrane conductance can fluctuate are

  a) if a channel can only be in two states, conducting and non-
conducting, then the membrane conductance will fluctuate as the number
of open channels fluctuates

  b) if a given channel has a conductance which can very continuously over some range the membrane conductance will fluctuate as the individual channel conductances fluctuate.

  Different models of membrane conductance mechanism will give
different types of membrane conductance fluctuation, and so a study of
conductance fluctuations might give insight into the mechanisms of ionic conductance. If the membrane conductance is thought of simply as
the reciprocal of the membrane resistance the membrane conductance
would give the relationship between the total membrane ionic current
and the membrane potential. However, the relationship between membrane
ionic current and potential is non-linear, and the total ionic current
is the sum of different ionic currents passing through membrane channels
having different properties. Further a given ionic current mechanism
will only pass current when there is an electrochemical gradient driving permeable ions through the mechanism. Thus, instead of considering
membrane conductance it is more appropriate to consider the current-
voltage relationships of membrane channels. A given channel may be

specified in terms of its ionic selectivity and kinetics.

The general relationship between the ionic current through a channel and the membrane potential may be written as

$$I_a = F(t, V) \qquad\qquad (1.4.1)$$

where a is the ionic species which can pass through the channel and $F(.)$ is a non-linear function. The ionic channel is characterized by a reversal potential, the potential at which the electrochemical gradient driving a ions through the channel is zero. If only one ionic species can pass through the channel this is the Nernst potential for that ion.

If I consider a single type of membrane channel which is permeable only to say potassium ions, then

$$I_K = f(t, V)$$
$$f(t, V = V_K) = 0$$

The chord conductance is defined by

$$g_K = I/(V-V_K) = f(V)/(V-V_K) \qquad\qquad (1.4.2)$$

and so g_K is a non-linear function of V. The slope conductance is defined by

$$\frac{\partial I}{\partial V} = \frac{\partial F(V)}{\partial V} \qquad\qquad (1.4.3)$$

and so in general the slope and chord conductances are non-linear functions of voltage and are not equal - see Fig. 1-2. The chord conductance can be used to calculate absolute current values at a given potential, and the slope conductance can be used to calculate how the membrane will react to small changes in current or voltage.

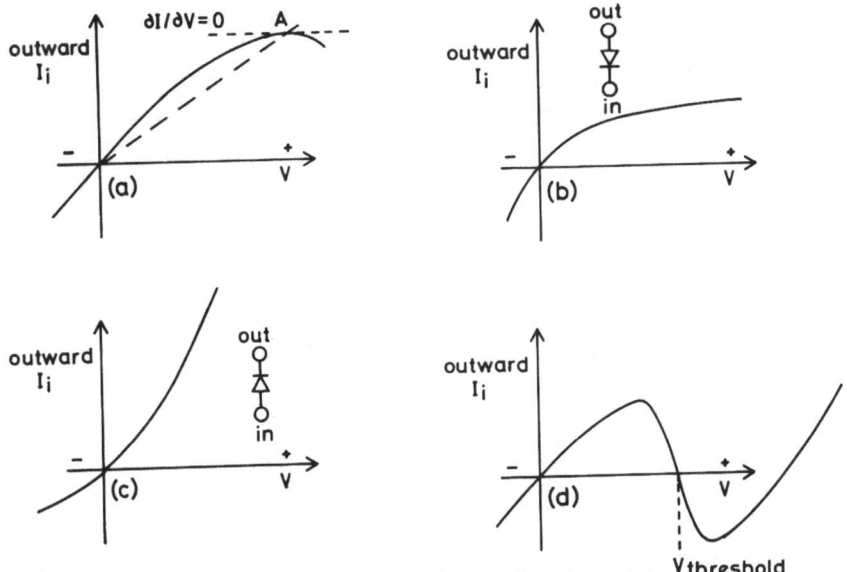

Figure 1.2. Some nonlinear membrane current-voltage relations. (a) At A the slope conductance $\partial I/\partial V$ is zero while the chord conductance I_i/V is positive. (b) inward-going rectification. (c) outward-going rectification. (d) N-shaped I-V relation characteristic of excitable membranes. The relation intersects the voltage axis at 3 points: the intersection with the negative slope conductance is the threshold.

The current-voltage relationship also changes with time, and so the instantaneous and steady-state current-voltage relations are different e.g. for the membrane of the giant axons of the squid, the instantaneous I-V relation is linear whereas the 'steady state' (or more accurately, at some time t) relation is non-linear. I will not be concerned here with the mechanisms responsible for the non-linear I-V relations - a good discussion is given in Jack, Noble and Tsien 1975, but only with the difference between regenerative and non-regenerative I-V relations.

The general form of the I-V relation of an excitable membrane (a membrane capable of generating action potentials) is N-shaped - see Fig. 1-2 d. The positive slope intercepts of the zero current axis are stable points - a small change in potential will give rise to a current tending to restore the membrane potential to its stable value e.g. the resting membrane potential. The negative slope intercept of the zero current axis is an unstable point or the threshold - a small depolarization will produce an inward current producing a further depolarization i.e. a regenerative action potential.

Thus if the I-V relation has a negative slope intercept there are regenerative ionic currents, if the I-V relation is non-linear but does not have a negative slope intercept there are only non-regenerative ionic currents.

The effects of conductance fluctuations in a regenerative system will be discussed in section 3. Two non-regenerative systems will be considered: discrete fluctuations in receptor potentials in section 2 and quantal fluctuations in end plate potential in section 11.

1.5 References

Alberding, N.: Random walk models of semi-conductor noise. J. Appl. Physics 44 1911-2 (1973).

Bell, D.A.: Semiconductor noise as a queuing problem. Proc. Phys. Soc. 72 27 (1958).

Bess, L.: A possible mechanism for 1/f - noise generation in semiconductor filaments. Phys. Rev. 91 1569 (1953).

Bird, J.F.: Neural 1/f noise and membrane models. Biophys. J. 14 563-565 (1974a).

Bird, J.F.: Noise spectrum analysis of a Markov process vs random walk computer solutions simulating 1/f noise spectra. J. Appl. Physics 45 499-500 (1974b).

Cole, K.S.: Membranes, Ions and Impulses. 569 pp. Univ. of California Press, Los Angeles (1968).

Derksen, H.E.: Axon membrane voltage fluctuations. Acta Physiol. Pharmacol. Neerl. 13 373 (1965).

Derksen, H.E. & Verveen, A.A.: Fluctuations of resting neural membrane potential. Science 151 1388 (1966).

Fatt, P. and Katz, B.: Some observations on biological noise. Nature 166 597-8 (1950).

Fatt, P. and Katz, B.: Spontaneous subthreshold activity at motor nerve endings. J. Physiol. 117 109-28 (1952).

Hawkins, R.J.: Modified random-walk model for 1/f noise. J. Appl. Physics 43 1276-7 (1972).

Hladky, S.B. & Haydon, D.A.: Discreteness of conductance change in bimolecular lipid membranes in presence of certain antibiotics. Nature 225 451-3 (1970).

Hooge, F.N.: 1/f noise is no surface effect. Physics Letters 29 A 139-40 (1969).

Hooge, F.N.: 1/f noise in the conductance of ions in aqueous solution Physics Letters 33 A 169-70 (1970).

Hooge, F.N., van Dijk, H.J.A. & Hoppenbrouwers, A.M.H.: 1/f noise in epitaxial silicon. Philips Res. Rept. 25 81-86 (1970).

Hooge, F.N. & Gaal, J.L.M.: Fluctuations with a 1/f spectrum in the conductance of ionic solutions and in the voltage of concentration cells. Philips Res. Rept. 26 77-90 (1971).

Hooge, F.N. & Hoppenbrouwers, A.M.H.: Amplitude distribution of 1/f noise. Physica 42 331-9 (1969).

Jack, J.J.B., Noble, D. & Tsien, R.W.: Electric current flow in ex- citable cells. 500 pp. Oxford University Press, New York (1975).

Katz, B.: Nerve, Muscle and Synapse. 193 pp. McGraw-Hill, N.Y. (1966).

Lee, Y.W.: Statistical Theory of Communication. 509 pp. Wiley, N.Y. (1967).

Lundstrom, I. & McQueen, D.: A proposed 1/f noise mechanism in nerve cell membranes. J. theoret. biol. 45 405-9 (1974).

Mandlebrot, B.: Some noises with 1/f spectrum, a bridge between direct current and white noise. I.E.E.E. Trans. Information Theory. IT-13 289-98 (1967).

McFarlane, G.C.: A theory of contact noise in semiconductors. Proc. Roy. Soc. 363 807-14 (1950).

McWorter, A.A.: in semiconductor Surface Physics. Univ. of Pennsyl- vania Press, Pa. (1957).

Morrison, S.R.: Recombination of electrons and holes at dislocations. Phys. Rev. 104 619 (1956).

Nyquist, H.: Thermal agitation of electric charge in conductors. Phys. Rev. 32 110-113 (1928).

Offner, F.F.: 1/f noise in semiconductors. J. Appl. Physics 41 5033-4 (1970).

Offner, F.F.: 1/f fluctuation in membrane potential as related to mem- brane theory. Biophys. J. 11 123-4 (1971a).

Offner, F.F.: Quantitative measurement of 1/f noise and membrane theory. Biophys. J. 11 969-971 (1971b).

Offner, F.F.: The excitable membrane - a physico-chemical model. Biophys. J. 12 1583-629 (1972a).

Offner, F.F.: Comments on 'modified random walk model of 1/f noise'. J. Appl. Physics 43 1277-8 (1972b).

Petritz, R.L.: A theory of contact noise. Phys. Rev. 87 535 (1952).

Poussart, D.J.M.: Nerve membrane current noise: direct measurements under voltage clamp. Proc. Nat. Acad. Sci. U.S.A. 57 1350 (1969).

Poussart, D.J.M.: Membrane current noise in lobster axon under voltage clamp. Biophys. J. 11 211-234 (1971).

du Pre, F.K.: A suggestion regarding the spectral density of flicker noise. Phys. Rev. 78 615 (1950).

Rice, S.O.: Mathematic analysis of random noise; in Selected papers on noise and stochastic processes, ed. N. Wax. Dover, N.Y. (1954).

Richardson, J.M.: The linear theory of fluctuations arising from diffusional mechanisms - an attempt at a theory of contact noise. Bell sys. Tech. J. 29 119 (1950).

Robinson, F.N.H.: Noise and fluctuations in electronic devices and circuits. 246 pp. Clarendon Press, Oxford (1974).

Sautter, D.: Noise in semiconductors. In Progress in semiconductors. vol. 4 ed. A.F. Gibson (1960).

Schick, K.L.: Power Spectra of pulse sequences and implications for membrane fluctuations. Acta Biotheoretica 23 1-17 (1974).

Schick, K.L. & Verveen, A.A.: 1/f noise with a low frequency white noise limit. Nature 251 599-601 (1974).

Siebenga, E. & Verveen, A.A.: Noise voltage of axonal membrane. Pflugers Archiv. 318 267 (1970).

Siebenga, E. & Vereen, A.A.: The dependence of 1/f noise intensity of the node of Ranvier on membrane potential. Proc. First European Bio-physics Congress V 219-223 (1971a).

Siebenga, E. & Verveen, A.A.: Membrane noise and ion transport in the node of Ranvier. Biomembranes 3 473-482 (1971b).

Stepanescu, A.: 1/f noise as a two-parameter stochastic process. Il nuovo cimento 23 B 356-364 (1974).

Stevens, C.F.: Inferences about membrane properties from electrical noise measurements. Biophys. J. 12 1028-1047 (1972).

Verveen, A.A. & Derksen, H.E.: Fluctuations in membrane potential and the problem of coding. Kybernetik 2 152-160 (1965).

Verveen, A.A. & Derksen, H.E.: Fluctuation phenomena in nerve membrane. Proc. I.E.E.E. 56 906-916 (1968).

Verveen, A.A. & Derksen, H.E.: Amplitude distribution of axon membrane noise voltage. Acta Physiol. Pharmacol. Neerl. 15 353 (1969).

Verveen A.A., Derksen, H.E. & Schick, K.L.: Voltage fluctuations of neural membrane. Nature 216 588-9 (1967).

van der Ziel, A.: On the noise spectra of semiconductor noise and of flicker effect. Physica 16 359-72 (1950).

van der Ziel, A.: Noise. 450 pp. Prentice-Hall, N.J. (1954).

van der Ziel, A.: Fluctuation phenomena in semiconductors. 168 pp. Butterworth. London (1959).

van der Ziel, A.:Noise: sources, characterization, measurement. 184pp.
Prentice-Hall, N.J. (1970).

Updates:

Two reviews of membrane noise have appeared:
Verveen, A.A. and DeFelice, L.J.: Membrane Noise. Progress in Biophysics
and Molecular Biology 28 189-265 (1974)

Conti, F. and Wanke, E.: Channel noise in nerve membranes and lipid
bilayers. Quarterly Review of Biophysics 8 451-506 (1975)

A microscopic model based on the macroscopic model suggested by
Lundstrom and McQueen (1974) has appeared:
Clay, J.R. and Schlesinger, M.F.: Theoretical model of the ionic
mechanism of 1/f noise in nerve membrane. Biophysical J. 16 121-
136 (1976).

The idea that interactions between sparse channels generates
flicker noise (see section 8.4) has been further developed:
Holden, A.V.: Flicker noise and structural changes in nerve membrane.
J. Theoretical Biology 57   243-246 (1976).

Holden, A.V. and Rubio, J.E.: A model for flicker noise in nerve
membranes. (1976) submitted for publication.

2. QUANTAL FLUCTUATIONS IN GENERATOR POTENTIAL

A single primary sensory neurone or receptor transmits information about the intensity and time course of an applied stimulus to
second order sensory neurones. This process may be idealized as a
sequence:

a)   The stimulus directly (as in mechanoreceptors) or indirectly
(as in photoreceptors) alters the properties of a specialized area of
membrane, the receptor membrane, so that the receptor membrane has a
changed ionic permeability and hence changed conductance. The receptor
membrane has a non-regenerative I-V relation.

b)   This change in ionic chord conductance, together with the
electrochemical gradient across the membrane, causes a current to flow
across the membrane.

c)   This receptor current produces a change in membrane potential,
the receptor potential.

d)   The receptor potential spreads electrotonically through the
receptor terminals. The size and time course of the receptor potential
depend on the size and time course of the receptor current, the membrane
I-V relations and the geometry of the receptor terminals.

e)   The receptor potential spreads to an area of membrane which
has a regenerative I-V relation - this is the spike initiating region.
The electrotonic spread of the receptor potential into this region is
called the generator potential.

f)   If the potential in the spike initiating region reaches a
value where the I-V relation has a negative slope conductance intercept
with the zero current axis (the threshold potential) an action potential
is generated.

Many receptor neurones have a resting or background discharge of
action potentials in the absence of an externally applied stimulus.
This background discharge is stochastic, in the sense that the interspike
interval is variable. The response of most receptors to a constant
applied stimulus is also stochastic. This variability in the background
and evoked discharges imposes restrictions on the amount of information

a single neurone can transmit in a given time (see section 13) and may
be due to

a)   stochastic fluctuations in the threshold potential causing a stochastic change in excitability. This possibility is discussed in section 4.

b)   spontaneous fluctuations in the generator potential due to the
mechanisms discussed in section 1.

c)   fluctuations in receptor potential due to either random fluctuations
in the receptor membrane properties or the intrinsic stochastic nature
of the physical stimulus. These possibilities will be discussed with
special reference to photoreceptors.

2.1 Psychophysical evidence for quantal sensitivity in vision

The human visual system is sensitive to light over about 10 decades of intensity: in the dark adapted state it is possible to estimate
the absolute threshold for visual perception. Hecht et al (1942) have
shown that under the most favourable conditions approximately 50 photons
incident on the eye can produce a visual sensation. Of these photons
only about 5 are absorbed by the photolabile pigment of the retinal
rods, rhodopsin. Since these photons are falling on a retinal area
encompassing perhaps 500 rods it is unlikely that more than one photon
is absorbed by rhodopsin in any one rod. Thus the absorption of a
single photon by a rod produces some kind of retinal response, and when
several such responses occur in a short period of time in a small area
of retina a visual sensation can be evoked. The sensitivity of dark
adapted rods to single photons means that the receptor can operate at
the maximal theoretical limit of sensitivity. This extreme sensitivity
of the visual receptors has been confirmed by many workers - Baumgardt
1972 has reviewed psychophysical evidence and theories for absolute
visual thresholds.

At very low levels of illuminances, the light energy incident on
a given area of retina should show random fluctuations in time. These
fluctuations are due to the statistical fluctuations in the number of
photons, which has a Poisson distribution, the probability of there
being k quanta in a given sample being

$$P(k) \; = \; \exp{(-\bar{n})} \, . \; \bar{n}^{\,k}/k! \qquad\qquad (2.1.1)$$

where n̄ is the average number of quanta/sample. Thus one would expect
the absolute visual threshold to be a variable, not constant, energy
value. The absolute visual threshold does fluctuate, but it is unlikely that this is only due to fluctuations in the number of photons.
Falk and Fatt (1974) have discussed synaptic noise in the retina as a
further source of stochastic fluctuations in absolute visual threshold.

If there is a minimum number s of photon absorptions that must
occur for a visual sensation to be produced, the probability P(s) that
s absorptions occur in a test flash of an intensity that has an
average number n̄ of quanta is

$$P(s) = 1 - \exp(-f\bar{n}) \sum_{0}^{s-1} f\bar{n}^{p}/p! \qquad (2.1.2)$$

where (1-f) is the fraction of incident quanta that are lost by reflection and absorption by non-photoreceptor pigment between the
cornea and receptor absorption. Equation (2.1.2) should describe the
frequency-of-seeing curve if   a)   the probability of perception is
one if n > s absorptions occur in the test area of retina and
b) s is a constant. The experimental frequency-of-seeing curves obtained can be fitted by equation (2.1.2) with 2 < s < 12. This extreme sensitivity of the human visual system suggests that the absolute
threshold might be limited by internal retinal noise due to thermal
decomposition of rhodopsin.   de Vries (1948) estimated that about
10^{-9} to 10^{-6} thermal decompositions of rhodopsin occur/rod/second.
Thus for a brief flash (about 0.1 sec) over a retinal area about 0.5°
in diameter the retinal threshold is about three orders of magnitude
greater than the number of spontaneous thermal decompositions of
rhodopsin.

The psychophysics of intensity or contrast discrimination may
also be accounted for by the quantal nature of light. The incremental
intensity threshold ΔI on a background intensity I is, over six
decades of I, proportional to $\sqrt{I}$. de Vries (1943) and Rose (1948)
accounted for this by assuming that ΔI was suprathreshold when ΔI
was greater than the statistical fluctuations in photon flux I, which
are proportional to $\sqrt{I}$.

This psychophysical evidence suggests two approaches: models of
the psychophysical laws by a neural machine with a quantal input

(section 2.2) and models of the effects of quantal conductance
changes on receptor potential (section 2.3).

2.2   Some neural machines

The behaviour of the visual system at the absolute threshold may
be considered as a problem in coincidence-detection, in which the
criterion for perception is that at least K photons must be absorbed
within a retinal area A and time T.   K is lowest for short (< 0.1 sec:
Bloch's time) and small (10 min of arc: Ricco's area, A) flashes.
Thus a natural neural machine is a coincidence-detector defined over
a sampling unit (A,T).   This machine is the KC-scaler of Bouman (1964),
Bouman and Ampt (1966) and van der Grind and Bouman (1968) or the
summation pool of Rushton (1965).

Ricco's area encompasses many, say u, receptors.   The input to
each receptor is Poisson with mean rate $\bar{n}$, thus the input to the KC-
scaler is the summation in space of these inputs, and so has a mean
rate $\bar{r} = \bar{n}.u$ and the superposition in time of the receptor inputs,
and so has a Poisson distribution with parameter r (see section 8).

The perception criterion is exceeded when rt > K, for 0 < t < T
and so the interval distribution between outputs from the KC-scaler
is

$$P_k \; (t) \quad = \quad \exp \; (-\bar{r}t) \; \bar{r}^k \; t^{k-1} / \; (k-1)! \qquad\qquad (2.2.1)$$

with a mean interval $\bar{t}$ and standard deviation σ given by

$$\bar{t} \; = \; (K - 1) \;\; / \; \bar{r} \; ; \quad \sigma \; = \; (K - 1)^{\frac{1}{2}} / \; \bar{r} \qquad\qquad (2.2.2)$$

Thus the coefficient of variation decreases for any r with increasing
K (Bouman and Ampt (1966)).

A simpler neural machine is the K-scaler of van der Grind et al
(1971), which generates an output event on the arrival of the Kth input event since the last output event.   Thus the restriction that the
input events occur in Bloch's time has been removed, or alternatively,
for the KC-machine, T → ∞.   This K-scaler is identical to the perfect
integrator model neurone subject to a purely excitatory quantal input -

see sections 3, 6 and 7. If the K-scaler is taken as a neural model
it may be considered to operate in two different modes: a decision or
switching mode at small values of r and an encoding or integrating
mode at high values of r (van der Grind et al, 1971, and Jenik, 1962,
1964). In the decision mode the question is whether or not threshold
has been reached; in the encoding mode the question is the form of
the relation between the output and input sequence of events.

A more general form of the K-scaler in the KT-scaler, or leaky
integrator, which generates an output event when g(t) < K, where g(t)
is the output from a leaky integrator:

$$g(t) \quad = \quad \sum_{i=1}^{m} h_i \int_{-\infty}^{t} \exp \left(-(t-s)/\tau \right) f_i (s) \, ds \qquad (2.2.3)$$

where $f_i (s)$ is the input signal, in this case a train of events. As
$T \to \infty$ the KT-scaler degenerates into a K-scaler.

For both the K-scaler and KT-scaler machines the output event
rate increases at any K as the input event rate increases: since the
rate of action potentials in neurones can only vary over approximately
three decades these machines would not cover the 10 decade range of
light intensity the visual system is sensitive to, unless different
machines had different sensitivity ranges. This suggests a modification where K increases as I(t) (or $\bar{n}$) increases, and so there is some
form of adaptation. Van der Grind et al have proposed two types of
adapting neural machines - the de Vries-Rose machine (van der Grind
et al 1970) and Weber machines (Koenderink et al 1970). The application of these machines designed to model psychophysical phenomena to neural models is discussed in van der Grind (1971). The de
Vries-Rose machine (VR machine) has a scaling factor K(t) which depends on the light intensity by

$$\dot{K}(t) \quad = \quad K_o + CI_B(t) \qquad (2.2.4)$$

where $I_B(t)$ is the predicted, expected number of output events to
come in the next time period. Time is divided into discrete clock
periods T. If $I_A(t)$, the actual number of events occurring in a time
period, is greater than $I_B(t)$ there is an increase in $I_B(t)$ and hence
K(t). K_o is the minimum scaling factor, and C is a constant. If a

VR machine is started from rest, and there are n_1 input events in the first clock period with $n_1 \gg K_o$ and m_1 output events,

$$n_1 \simeq m_1 K_o + 0.5 \ a \ C \ m_1 (m_1 - 1) \tag{2.2.5}$$

where a change $\underline{a}$ of $I_B(t)$ causes a change $a.C$ of $K(t)$.

Thus $m_1 \simeq (2 / (2 \ a \ C)^{\frac{1}{2}}) \ n_1^{\frac{1}{2}}$

and $\qquad K_1 \simeq (2 \ a \ C)^{\frac{1}{2}} \ n_1^{\frac{1}{2}} \tag{2.2.6}$

or the scaling factor is proportional to the square root of the number of input events, which gives the de Vries-Rose law. In the steady state, in any given period of time the average scaling factor is proportional to the average number of output events. Thus, any feedback scaling machine with this property, produced by a KC or KT machine in the feedback loop, will act as a VR-machine and model the de Vries-Rose law if it is coupled to a detector which can detect a fixed change in the number of output events. The continuous parameter neural model of Siebert and Gambardella (1968) is a form of VR-machine, and successfully mimics post-stimulus time histograms of primary auditory fibres in the cat. van der Grind et al (1970) have shown that the clock time period VR-machine may be characterized by a differential equation

$$\frac{1}{\bar{m}} \frac{d\bar{m}}{dt} - \frac{1}{\bar{n}} \frac{d\bar{n}}{dt} = \frac{1}{2T} - \frac{\bar{Cm}^2}{2\bar{n}} \tag{2.2.7}$$

from which the flash response and sine wave responses may be obtained.

A different type of adapting scaler is the Weber machine (Koenderink et al, 1970, 1971). If the scaling factor is w, adaptation is produced by the rule

$$w(t + \Delta t) = z.w(t) \tag{2.2.8}$$

where Δt is sufficiently small so no output pulses occur between t and $t + \Delta t$ and $z > 1$. The Weber machine (W-machine) generates a logarithmic relation between the number of input events n and the number of output events m in any given time period

$$m = \log ((z - 1)n + 1) \simeq \ln (z - 1)n / \ln (z)$$

when z is an integer. Thus adaptation produces a geometric series, whereas for the VR-machine adaptation produces an arithmetic series. The W-machine will adapt completely, so that in the steady state $m = 0$. This may be overcome by introducing dark adaptation

$$w(t_2) = w(t_1) \exp (- (t_2 - t_1) / \tau_d) \; ; \; w(t) > 1 \qquad (2.2.9)$$

where τ_d is the time constant of dark adaptation.

Equations (2.2.8) and (2.2.9) give the stationary behaviour for a mean input rate r

$$\overline{T} = \tau_d \ln (z)$$

$$\overline{w} = \overline{T} r$$

where $\overline{T}$ is the mean output inter-event interval. Thus in the station- ary case the output is independent of the input intensity when the input intensity is high enough for light adaptation to occur.

If the detection criterion is taken as a fixed increment ℓ in the number of output events in a given interval, the threshold proper- ties in response to brief flashes, and the dynamic behaviour of the W machine + detector, may be investigated. When quantal fluctuations in n are neglected or $(rT)^{\frac{1}{2}} \ll w\ell$ Webers law and the slope of the frequency of seeing curve can be obtained. The change from de Vries to Weber dominated behaviour occurs at $n \simeq 25$. The dynamic behaviour of the W-machine mimics that of on-units, and off-unit behaviour may be produced by introducing a bias input. Thus the W-machine can be used as a model of retinal ganglion cell discharge.

2.3 Fluctuations in the Limulus eccentric cell generator potential

The success of the neural machines described in section 2.2 in accounting for psychophysical aspects of the sensitivity of the visual system suggest a) identification of the machines with single retinal

neurones or nets and b) the identification of electrophysiological
processes corresponding to the effects of photon absorption. Vertebrate retinal neurones are small and so intracellular recording from
rods is difficult; however, the lateral eye of the horse-shoe crab
Limulus is a preparation eminently suitable for electrophysiological
studies. Knight (1973) has given a review of the structure and dynamics of the lateral eye of Limulus.

The lateral eye of Limulus is a compound eye containing a few
thousand ommatidia, each of which contributes one active axon to the
optic nerve. Thus each ommatidium comprises a single functional unit.
Each ommatidium contains about eleven retinula cells arranged around
the distal process of single eccentric cell. Although both retinula
and eccentric cell axons form the optic nerve only the axons of eccentric cells propagate action potentials. The photopigment, which
is similar to rhodopsin, is in the rhabdome which is a structure formed by microvilli of adjacent retinula cells.

The eccentric cell axon responds to a step increase in light intensity by a burst of impulses, which adapts to a steady state discharge rate which is proportional to the logarithm of the light intensity over a four decade range of intensities. In the steady state
there is a linear, inhibitory interaction between neighbouring ommatidia which may be quantitatively described by the Hartline-Ratcliff
model (Hartline and Ratliff, 1957, 1958; Ratliff and Hartline, 1959,
Ratliff, Hartline and Miller, 1963)

$$r_p = e_p - \sum_{\substack{j=1 \\ j \neq p}}^{n} k_{p,j} (r_j - r^o_{p,j}) \qquad (2.3.1)$$

where r is the steady state spike rate under diffuse illumination, e_p
is the steady state spike rate when only ommatidium p is illuminated,
$k_{p,j}$ is the inhibitory coefficient of unit j on p and r^o the inhibitory
threshold. In this form of the equations e_p includes the effect of
self-inhibition. $k_{p,j}$ is independent of e_p, but varies with the distance separating the ommatidia. Kirschfeld and Reichardt (1964) have
shown that a Gaussian function with a half width of about 8 ommatidia
adequately describes the decay of average inhibitory coefficient with
distance. The lateral inhibition in Limulus eye is recurrent, and in
the steady state acts to enhance spatial contrast. The Hartline-

Ratliff equation may be generalized to a frequency-dependent form
(Knight et al 1970)

$$r_p = G(f) I - k_{p,p} T_s(f) r_p - T_1(f) \sum_{p \neq j} k_{p,j} \left[f_j - r^o_{p,j} \right]$$

(2.3.2)

where the self-inhibition is now explicit with a self-inhibitory coefficient $k_{p,p}$; and $G(f)$ is the light to voltage, $T_s(f)$ the self-
inhibitory and $T_1(f)$ the lateral inhibitory frequency response functions which have been experimentally determined. I is the light intensity.

Intracellular recording from the eccentric cell soma shows a
resting membrane potential of approximately - 50 mV. In response to
illumination there is a slow depolarization, the generator potential,
upon which is superimposed action potentials. The slow depolarization
is a generator potential, not a receptor potential, since the photolabile pigment containing receptor cells are the retinula cells. The
steady state spike rate is proportional to the generator potential
depolarization and injected current intensity (Fuortes, 1959). Thus
the nonlinear transduction between steady state spike rate and light
intensity must depend on the relation between the generator potential
and light intensity. If spike rate is a single valued function of
potential at the impulse initiation site, the effect of light on the
generator potential could be modelled by a photosensitive, variable
conductance channel, with an equilibrium potential about 50 mV +ve
to the resting membrane potential, shunting the resting conductance
of the membrane (Rushton, 1959). The photosensitive conductance
channel is located in the distal dendrite, and so the somatic generator potential is due to electrotonic spread from the distal dendrite.

Lateral inhibition producing a decrease in steady state spike
rate does not produce a change in generator potential (Purple and
Dodge 1965). This does not contradict the assumption that the spike
rate is a single valued function of V(t) as the inhibitory synapses
and spike initiating region are on the eccentric cell axon, separated
from the soma by a length of cable with a non-regenerative I-V relation.

Yeandle (1957, 1958) observed discrete irregular waves in the somatic potential of dark adapted eccentric cells. These waves were present in the dark. If the occurrence of at least one discrete wave in response to a brief test flash depends on at least q photons being absorbed

$$P(q) = \sum_{r=q}^{\infty} \bar{q}^r \exp(-\bar{q}) / r! \qquad (2.3.3)$$

where $\bar{q}$ is the mean number of photons absorbed / flash. Experimentally $q = 1$ fits the probability of discrete wave occurrence - light intensity curves, which is consistent with the hypothesis that a discrete wave is the response to a single photon. However, an alternative model is that the discrete waves are the eccentric cell responses to molecules or quantal packets of transmitter released by the rhabdome in response to photon absorption.

These discrete waves were studied in greater detail by Fuortes and Yeandle (1964), who found at low illuminances a linear relation between the rate of discrete waves n and light intensity I

$$n = b (I + I_0) \qquad (2.3.4)$$

where b is a constant, and I_0 a constant 'equivalent light' generating the mean spontaneous rate in the dark. The interval between discrete waves at a constant (low) light intensity had a Poisson distribution. The response to flashes had a long variable latency which increased with hyperpolarization and decreasing temperature, and the number of waves evoked by a given number flashes was proportional to the flash intensity. The probability of evoking n waves by a flash intensity which evoked a mean $\bar{n}$ waves was in good agreement with the Poisson series $P(n) = \bar{n}^n / n! \exp(-\bar{n})$.

The Poisson nature of the statistics of the discrete waves suggests a correspondence with the Poisson nature of photon absorption at low luminances. The question is whether this is a one-to-one correspondence, or whether one discrete wave is generated by q photon absorptions.

If $P(r,q)$ denotes the probability that r waves are produced by a flash, q absorptions being necessary (and sufficient) for the production of a wave

$$P(r,1) = \exp(-\bar{m}_1)\,\bar{m}_1^{\,r} / r! \tag{2.3.5}$$

$$P(r,2) = \exp(-\bar{m}_2)\left[\bar{m}_2^{\,2r} / 2r! + \bar{m}_2^{\,(2r+1)} / (2r+1)!\right] \tag{2.3.6}$$

$$P(r,3) = \exp(-\bar{m}_3)\left[\bar{m}_3^{\,3r} / 3r! + \bar{m}_3^{\,(3r+1)} / (3r+1)! \right.$$
$$\left. + \bar{m}_3^{\,(3r+2)} / (3r+2)!\right] \tag{2.3.7}$$

and the relation between average number $\bar{n}$ of waves produced and the average number of absorbed quanta $\bar{m}_1$, $\bar{m}_2$ and $\bar{m}_3$

$$\bar{n}_{(q=1)} = \bar{m}_1 \tag{2.3.8}$$

$$\bar{n}_{(q=2)} = \sum_{r=0}^{\infty} r\left[\bar{m}_2^{\,2r} / 2r! + \bar{m}_2^{\,(2r+1)} / (2r+1)!\right]\exp(-\bar{m}_2) \tag{2.3.9}$$

$$\bar{n}_{(q=3)} = \sum_{r=0}^{\infty} r\left[\bar{m}_3^{\,3r} / 3r! + \bar{m}_3^{\,(3r+1)} / (3r+1)! \right.$$
$$\left. + \bar{m}_3^{\,(3r+2)} / (3r+2)!\right]\exp(-\bar{m}_3) \tag{2.3.10}$$

The experimental relation between the average number of waves n and the intensity of a light flash could be fitted by relations (2.3.8) - (2.3.10) on the assumptions of different proportionality constants b between the mean number of photon absorptions and the mean number of photons in a flash, and the value of the equivalent light I_o of equation (2.3.4). The fit is best for the one-photon hypothesis, but the two and perhaps even the three-photon hypotheses are not invalidated.

Adolph (1964) investigated the statistics and time-course of the discrete waves, which he called slow potential fluctuations or SPF's, and a description of the behaviour of the SPF's was given in terms of the theory of Poisson shot noise (Rice, 1954; also see section 1.2). If f(t) is the shot shape, with a mean shot amplitude $\bar{a}$ and mean shot rate v the mean μ of the Poisson shot process is

$$\mu = v\,\bar{a}\int_0^{\infty} f(t)\,dt \tag{2.3.11}$$

and variance σ^2

$$\sigma^2 = v \; \overline{a^2} \int_0^\infty f^2 (t) \; dt \qquad\qquad (2.3.12)$$

where $\overline{a^2}$ is the second moment of the shot amplitude distribution. If the somatic or excitatory potential is the linear sum of a Poisson shot process, its mean and variance will be given by equations (2.3.11) and (2.3.12). Experimentally, the mean and variance of the somatic potential increase linearly over a 30-fold range of light intensity; above this range the deviation from the extrapolated linear relation is unequal for μ and σ^2 and may be due to changes in the shot shape f(t), the amplitude distribution of the SPF's, h(a), or a non-Poisson inter-shot interval distribution.

A detailed examination of the SPF interval distributions showed (a) they were Poisson in the dark (b) under steady illuminance, the distributions were non-Poisson, there being significantly fewer short intervals than a Poisson distribution. A non-Poisson distribution would decrease the variance, not mean, of the excitatory potential.

The amplitude distribution of the SPF's, h(a) changes with the state of dark adaptation of the ommatidium, the $\bar{a}$ being larger in the dark adapted state. Light adaptation produces a decrease in both the mean and the variance of h(a). A light-adapted unit returned to the dark has a higher mean rate v of SPF's than a dark adapted unit: thus light adaptation is associated with a facilitation of the SPF mechanism.

There was little consistent change in the shape f(t) of SPF's: however f(t) fell into two classes, a slow class in which both the rising and falling phases were slow, and a fast class which had a fast rising and falling component superimposed on a slow SPF. This suggested two possibilities - there are two distinct independent classes of SPF, or alternatively the fast class of SPF is the result of the superposition of a transient on some slow SPF's. The distribution of the peak amplitudes of the SPF's showed a clear separation into the slow and fast classes, the amplitude distribution of the fast SPF's having a larger variance than the distribution of the slow SPF's. A small regenerative potential trnasient is seen on rapid release from hyperpolarized potentials, and Fuortes and Poggio (1963) have observed larger depolarizing transients with a similar time course. Thus the fast class of SPF may be due to a regenerative mechanism.

At low light intensities, the decrease in membrane resistance is proportional to the depolarizing change in membrane potential. If a constant light intensity produces a conductance change with a constant mean and variance, the mean and standard deviation of the light-produced excitatory potential should vary linearly with the intensity of an injected hyperpolarizing current. This is observed for small injected currents ($<$ 1 x 10^{-9} amp); at greater current intensities there is a consistent deviation from this linear relation suggestive of a regenerative mechanism.

The existence of two types of SPF complicates the stochastic modelling of the responses to small numbers of photons. The fairly smooth excitatory potential at low light intensities suggests that at low intensities slow SPF's predominate, and summate to give a smooth potential, and so the slow SPF's are the more direct response to photon absorption in the rhabdome. If this is true one would expect a difference in the average latency of the two classes of SPF in response to light flashes.

Borsellino and Fuortes (1968 a, b) have investigated the latency of the SPF's evoked by flashes and analysed their results in terms of a model proposed by Fuortes and Hodgkin (1964). Fuortes and Hodgkin showed that the generator potential response to a brief flash could be fitted by

$$V(t) \quad = \quad K \ (\lambda t)^d \ \exp \ (-\lambda t) \qquad\qquad (2.3.13)$$

where d varied from 7 to 13 for different cells. This may be consider-ed as the response of a network of d stages arranged sequentially in a cascade, each stage having a gain and time constant. An electric circuit model of this system is given in Figure 2. 1.a - the response of this circuit to a brief voltage pulse $\bar{v}_o \ \Delta t$ is

$$v_d(t) \quad = \quad \lambda_2 \ \bar{v}_o \ \Delta t \ (\lambda_2 t)^{d-1} \ \exp \ (-\lambda_1 t) \ / \ (d - 1)! \qquad (2.3.14)$$

where $\lambda_1 = 1 \ / \ RC_o$ and $\lambda_2 = \mu \ / \ C_o$. An alternative interpretation of the model in terms of a sequence of compartments is shown in Figure 2. 1.b; this compartmental formulation facilitates the derivat-ion of the response of the model to a stochastic input.

(a)

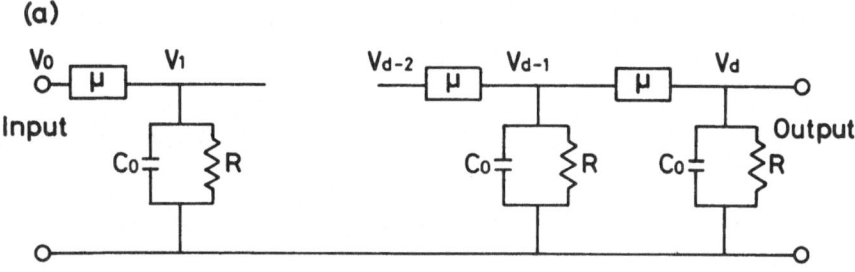

(b)

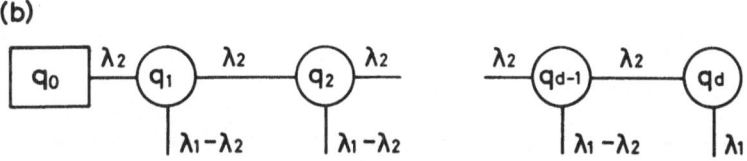

Figure 2.1. Two equivalent representations of Fuortes and Hodgkin (1964) model for evoked SPFs in Limulus. The amplifiers μ of the electrical model correspond to the negative parameters $-\lambda_2$.

The zero compartment represents photon absorption and the rapid light reactions: absorption of n photons generates m particles, each of which has a short life span τ_o

$$m = \alpha n \qquad (2.3.15)$$

and so if $q_r(t)$ is now the number of particles in compartment r, the average number of particles during a flash of duration Δt in the zero compartment is

$$\bar{q}_o = \alpha n \tau_o / \Delta t \qquad (2.3.16)$$

Particles in any compartment may be lost or gained from the outside or lost to the subsequent compartment:

$$dq_r / dt = -(\lambda_1 - \lambda_2) q_r - \lambda_2 q_r + \lambda_2 q_{r-1}$$

$$= -\lambda_1 q_r + \lambda_2 q_{r-1} \qquad (2.3.17)$$

A brief flash introduces $\lambda_2 \bar{q}_a$ Δt particles instantaneously into compartment 1; with this as an initial condition

$$q_d(t) = \lambda_2 \bar{q}_0 \Delta t (\lambda_2 t)^{d-1} \exp(-\lambda_1 t) / (d-1)! \quad (2.3.18)$$

which if $\bar{v} = V / q_d$ and $\alpha = \bar{q}_0 / I \tau_0$ gives equation (2.3.14) when the q_r are large.

When the q_r are small q_r (t) is a random function, say X_r (t). If

$$P_{r,x}(t) = \text{Prob} \{ X_r(t) = x \} \quad (2.3.18)$$

denotes the probability that there are x particles in compartment r at time t, Borsellino and Fuortes (1968 b) derived

$$\frac{d}{dt} P_{r,x}(t) = P_{r,x}(t) \left[-\lambda_1 x + \lambda_2 q_{r-1}(t) \right]$$
$$+ P_{r,x+1}(t) \lambda_1 (x+1) + P_{r,x-1}(t) \lambda_2 q_{r-1}$$
$$(2.3.19)$$

where $q_r(t) = \sum_{x=0}^{\infty} x P_{r,x}(t)$

Introducing $\lambda_2 q_0 \Delta t$ particles instantaneously into compartment one and solving (2.3.19) gives

$$P_{r,x}(t) = \frac{1}{x!} \left[q_r(t) \right]^x \exp(-q_r(t)) \quad (2.3.20)$$

which is a time dependent Poisson process.If a SPF is generated when a particle reaches the dth compartment the latency distribution predicted by the model can be obtained.

The probability of no arrivals at d in a time t is

$$P_{d,o}(t) = \exp(-\lambda_2 \int_0^t q_{d-1}(t) \, dt) \quad (2.3.21)$$

and so the probability density for the arrival of the *first* particle
at a time t, or the latency distribution is

$$\frac{-d}{dt}\ P_{d,o}(t)\ =\ \lambda_2\ q_{d-1}\ (t)\ exp\ \left[-\ \lambda_2 \int_0^t q_{d-1}(t)\ dt\right]\ (2.3.22)$$

The multiplication factor F of the chain of compartments is the ratio
of the number of particles introduced in the final dth compartment to
the number introduced in the zero compartment

$$F\ =\ \tau_0\ \lambda_1\ (\lambda_2\ /\ \lambda_1)^d \qquad\qquad (2.3.23)$$

The evoked SPF's could be classified in two populations; the
small S waves with a peak amplitude < 6 mV and a rise time > 60 msec
(corresponding to Adolph's slow class of SPF) and the large L waves
with a peak amplitude > 6mV and a rise time < 60 msec (Adolph's fast
SPF's). The S waves had a shorter latency with a mean latency of
about 60 msec; the mean L wave latency was about 140 msec. Analysis
of the experimental latency distribution in terms of equations (2.3.22)
and (2.3.23) gave the multiplication factor F a value of about 25 for
the S waves for a chain with about 10 compartments. The temperature
dependence of the peak voltage and time to peak supported the identification of the compartmental chain with a cascade of chemical reactions. The effect of the metabolic poison di-nitrophenol suggests
that at least half of the compartments correspond to metabolic reactions, the lack of effect of osmotically induced cell volume
chances suggest that these reactions are not in the cytoplasm but
in the membrane.

The magnification factor of this biochemical cascade means
that a detectable membrane response may be produced by a single
photon absorption without having to postulate a regenerative membrane
I-V relation. An alternative model assuming a regenerative membrane
I-V relation has been given by Bass and Moore (1970) in which a single
photon absorption opens a low resistance channel in the microvilli of
the rhabdome; this could produce a depolarization of several mV.

Dodge, Knight and Toyoda (1968 a, b) have used a different
analytical approach to show that the generator potential is the sum

or superposition of shot or bump voltage changes produced by photon absorption. The stationary shot noise process is characterized by its autocovariance function

$$C(\tau) = \overline{[g(t) - \overline{g}] \; [g(t + \tau) - \overline{g}]}$$
(2.3.24)

which is averaged over time t and g is membrane conductance. For an inhomogenous Poisson process the autocovariance of the steady-state response and the average response to small cosinusoidal fluctuations in the mean rate are related

$$C(\tau) = A \int_0^\infty |r(t)|^2 \cos(2\pi \; f\tau) \; df$$
(2.3.25)

where A is a constant and r(f) is the frequency response function. (Rice, 1954). The close agreement between estimates of C (τ) by equations (2.4.24) and (2.4.25) supports the idea that the generator potential is produced by a superposition of shot events. From Campbells theorem (Rice, 1954),

$$\overline{g} = r \, T\alpha$$
$$\overline{(g - \overline{g})^2} = r \, T\alpha^2$$
(2.3.26)

(where r is the shot rate, T the shot duration and α the shot amplitude), the dependence of the mean and variance of the generator potential on light intensity can be used to obtain the dependence of r, T and α on light intensity I. Over 5 decades of light intensity T decreases 4-fold ; r deviates from a linear relation at higher intensities, showing a reduced photon efficiency and $\alpha \simeq 1 / \sqrt{I}$.

Thus these different experimental and analytic approaches all demonstrate that fluctuations in the generator potential of Limulus can be accounted for by a shot process of conductance changes produced by photon absorption. Whether or not these fluctuations in generator potential completely account for the variability of the interspike interval will be discussed in section 3.2.

2.4 References

Adolph, A.R. : Spontaneous slow potential fluctuations in the Limulus photoreceptor. J. Gen. Physiol. 48 297-321 (1964).

Bass, L. and Moore, M.J. : An electrochemical model for depolarization of a retinula cell of Limulus by a single photon. Biophys. J. 10 1-19 (1970).

Baumgardt, E. : Threshold quantal problems. In: Handbook of Sensory Physiology Vol. VII - 4 pp 29-55 eds. D. Jameson & L.M. Hurrich (1972).

Borsellino, A. and Fuortes, M.G.F. : Responses to single photons in visual cells of Limulus. J. Physiol. 196 507-539 (1968 a).

Borsellino, A. and Fuortes, M.G.F. : Interpretation of responses of visual cells of Limulus. Proc. I.E.E.E. 56 1024-1032 (1968 b).

Bouman, M.A. : Efficiency and economy in impulse transmission in the visual system. Proc. of the 27th International Congress of Psychology 'Psychophysics and the ideal observer' Acta Psychol (Amst.) 23 239 (1964).

Bouman, M.A. and Ampt, C.G.F. : Fluctuation theory in vision and its mechanistic model. In: Performance of the eye at low luminances, eds. M.A. Bouman, J.J. Vos. Excerpta Medica International Congress Series No. 125, Amsterdam (1966).

Dodge, F.A., Knight, B.W. and Toyoda, J. How the horseshoe crab eye processes optical data. I.B.M. R.C. 2248 (No. 11117), I.B.M. York-town Heights, New York (1968 a).

Dodge, F.A., Knight, B.W. and Toyoda, J. : Voltage noise in Limulus visual cells. Science 160 88-90 (1968 b).

Falk, G. and Fatt, P. : Limitations to single-photon sensitivity in vision. In 'Physics and Mathematics of the Nervous System' eds. M. Conrad, W. Guttinger and M. Dal Cin. Lecture Notes in Biomathematics 4 Springer-Verlag, Berlin (1974).

Fuortes, M.G.F. : Initiation of impulses in visual cells of Limulus. J. Physiol. 148 14-28 (1959).

Fuortes, M.G.F. and Hodgkin, A.L. : Changes in time scale and sensitivity in the ommatidia of Limulus. J. Physiol. 172 239-263 (1964).

Fuortes, M.G.F. and Poggio, G.F. : Transient responses to sudden illumination in the cells of the eye of Limulus. J. Gen. Physiol. 46 435-452 (1963).

Fuortes, M.G.F. and Yeandle, S. : Probability of occurrence of discrete potential waves in the eye of Limulus. J. Gen. Physiol. 47 443-463 (1964).

van der Grind, W.A., Koenderink, J.J., van der Heyde, G.L. and Bouman, M.A. Adapting coincidence scalers and neural modelling studies of vision. Kybernetik 8 85-122 (1971).

van der Grind, W.A. and Bouman, M.A. : A model of a retinal sampling unit based on fluctuation theory. Kybernetik 4 136-141 (1968).

van der Grind, W.A., Koenderink, J.J. and Bouman, M.A. : Models of the processing quantum signals by the human peripheral retina. Kybernetik 6 213-227 (1970).

Hartline, H.K. and Ratliff, F. : Inhibitory interaction of receptor units in the eye of Limulus. J. Gen. Physiol 40 357-376 (1957).

Hartline, H.K. and Ratliff, F. : Spatial summation of inhibitory influences in the eye of Limulus, and the mutual interaction of receptor units. J. Gen. Physiol 41 1049-1066 (1958).

Hecht, S., Shlaer, S. and Pirenne, M.H. : Energy, quanta and vision. J. Gen. Physiol 25 819-840 (1942).

Jenik, F. : Electronic neuron models as an aid to neurophysiological research. Ergebn. Biol. 25 206-245 (1962).

Jenik, F. : Pulse processing by neuron models: In: Neural theory and modelling p. 190 ed. R.F. Reiss, Stanford Univ. Press, Stanford (1964).

Kirschfeld, K. and Reichardt, W. : Die Verarbeitung stationärer optischer Nachrichten im Komplexange von Limulus. Kybernetik 2 43-61 (1964).

Knight, B. : The horseshoe crab eye : a little nervous system that is solvable. In: Lectures on Mathematics in the Life Sciences 5 Some Mathematical Questions in Biology IV ed. J.D. Cowan, American Mathematical Society (1973).

Knight, B.W., Toyoda, J.I. and Dodge, F.A. : A quantitative description of the dynamics of excitation and inhibition in the eye of Limulus. J. Gen. Physiol 56 421-437 (1970).

Koenderink, J.J., van der Grind, W.A. and Bouman, M.A. : Models of retinal signal processing at high luminances. Kybernetik 6 227-237 (1970).

Koenderink, J.J., van der Grind, W.A. and Bouman, M.A. : Foveal information processing at photopic luminances. Kybernetik 8 128-144 (1971).

Purple, R.L. and Dodge, F.A. : Interaction of Excitation and inhibition in the eccentric cell of the eye of Limulus. Cold Spring Harbor Symposia on Quantitative Biology 30 529-538 (1965).

Ratliff, F. and Hartline, H.K. : The responses of Limulus optic nerve fibers to patterns of illumination of the receptor mosaic. J. Gen. Physiol 42 1241-1255 (1959).

Ratliff, F., Hartline, H.K. and Miller, W.H. : Spatial and temporal aspects of retinal inhibitory interaction. J. Opt. Soc. Amer. 53 110 (1963).

Rice, S.O. : Mathematical analysis of random noise in N. Wax. ed. Selected papers on noise and stochastic processes. Dover, New York (1954).

Rose, A. : The sensitivity performance of the human eye on an absolute scale. J. Opt. Soc. Amer. 38 196-208 (1948).

Rushton, W.A.H. : A theoretical treatment of Fuortes' observations upon eccentric cell activity in Limulus. J. Physiol 148 29.38 (1959).

Rushton, W.A.H. : Visual adaptation. The Ferrier Lecture 1962. Proc. Roy. Soc. B 162 20-46 (1965).

Siebert, W.M. and Gambardella, G. : Phenomenological model for a form of adaptation in primary auditory nerve fibres. M.I.T. - Q.P.R. No. 88 330-334 (1968).

de Vries, H. : The quantum nature of light and its bearing upon the threshold of vision, the differential sensitivity, and the visual acuity of the eye. Physica 10 553-564 (1943).

de Vries, H. : Die Reizschwelle der Sinnesorgane als physikalisches problem. Experientia 4 205 (1948).

Yeandle, S. : Studies on the slow potential and the effects of cations on the electrical responses of the Limulus ommatidium. Ph.D. Thesis, the John Hopkins University (1957).

Yeandle, S. : Electrophysiology of the visual system. Discussion. Amer. J. Ophthalmol. 46 82.87 (1958).

3   MODELS OF ACTION POTENTIAL INITIATION

If a nonmyelinated axon is represented by an infinitely long,
thin cable with an axoplasmic conduction core with specific resistivity
ρ ohm-cms, and a surface membrane with capacitance C_M μF / cm^2 and
resistance R_M ohm-cm^2 in parallel, the spread of voltage with
distance and time is given by the partial differential equation

$$\lambda^2 \partial^2 V / \partial x^2 - \tau_M \partial V / \partial t - V = 0 \qquad (3.0.1)$$

where x is the distance along the cable, D the cable diameter and λ
and τ_M the space and time constants :

$$\lambda = (R_M D / 4 \rho)^{\frac{1}{2}} \qquad (3.0.2)$$

$$\tau_M = R_M C_M \qquad (3.0.3)$$

The derivation and solutions of this equation are given in Jack,
Noble and Tsien (1974), who also treat different geometries and
non-linear I-V relations. The steady state solution of equation
(3.0.1) to a constant voltage V_o at x = 0 is

$$V(x) = V_o \exp (- x / \lambda) \qquad (3.0.4)$$

and so the space constant λ gives the distance in which the voltage
will fall to 1 / e of its value at x = 0. The space constant for
most axons is of the order of a few mm, and so electrical signals
may not be passively transmitted over distance greater than a few
mm. In the nervous system, information transmission is by the prop-

agation of action potentials generated by the non-linear I-V relation
of excitable membranes.

The action potentials in any axon have a constant amplitude
and shape, and so information is coded only in the time sequence of
the action potential train. Neurones can generate trains of action
potentials over a rate range of about 3 decades, and the discharge
rate can be modulated by sensory or synaptic inputs or by injected
currents.

In response to a constant input, the discharge of most neurones
is stochastic in the sense that the interspike-interval is a random
variable. Brink et al (1946) published the first description of this
variability as an interspike interval histogram, and the first analysis
of interspike interval fluctuations was given in a series of papers
by Hagiwara (1949, 1950, 1954). Variability in the discharge rate is
also seen in the absence of an experimentally controlled input, in the
background or 'spontaneous' discharge.

This variability in response to a constant input suggests two
kinds of model : deterministic models of spike generation driven by
a stochastic generator potential, or stochastic models of spike generation in which there are fluctuations in threshold. In this section
I will consider some of the deterministic mathematical models which
have been proposed for action potential generation; fluctuations in
excitability will be considered in section 4.

3.1 The Logical Neurone

McCulloch and Pitts (1943) introduced a simple neural model
which could be used either as a model neurone or as an element in a
model neural network. The characteristics of this logical neurone are

a) A logical neurone is in either one of two states, state 0
(inactive) or state 1 (active).

b) Each logical neurone has a single output which is connected
simultaneously to an arbitrary number of logical neurones (which may
include itself). These connections can be either excitatory or inhibitory.

c) There are N_e excitatory inputs and N_i inhibitory inputs
to a given logical neurone where N_i and N_e are non-negative integers.

d) Time in a network of logical neurones is synchronously
quantized, in that the logical neurones can only change their states

at a discrete sequence of times

$$t = t_o + m \Delta t$$

where m is an integer. Without any loss of generality t_o can be set to 0 and Δt to 1. In the interval

$$t_o + m \Delta t < t < t_o + (m + 1) \Delta t$$

there can be no change in the state of any logical neurone.

e) Each logical neurone has a threshold k, which is a positive integer.

f) A given input line is active at a time $t_o + (m + 1) \Delta t$ if and only if its neurone of origin was active at a time $t_o + m \Delta t$, or its neurone of origin fired at time $t_o + m \Delta t$.

g) A neurone is active at a time $t_o + m \Delta t$ if and only if $n_e - b n_i > k$ at that time, where n_e and n_i are the number of active excitatory and inhibitory inputs, b is a positive real number, $n_e \leqslant N_e$ and $n_i \leqslant N_i$.

The logical neurone so defined generates an output action potential when a threshold is exceeded, however its range of behaviour is restricted by the synchronous, quantized time axis. Small networks of logical neurones can perform any logical operation and so a universal computer or Turing machine can be built from logical neurones – these applications are discussed in Minsky (1967). Winograd and Cowan (1963) have shown that reliable automata may be constructed from logical neurones even when the connections between the logical neurones are unreliable. The behaviour of randomly connected nets of logical neurones is discussed in section 12.1.

The behaviour of a logical neurone may be represented by

$$y(m) = \begin{cases} 1 \\ 0 \end{cases} \text{ if } n_e (m) - b\, n_i (m) \quad \begin{array}{l} \geqslant k \\ < k \end{array} \qquad (3.1.1)$$

when y(.) is the state of the neurone. Thus there is no integration over time, the output depending only on the inputs at that time, and so the neurone does not have any memory. Caianiello (1961) has modified the McCulloch-Pitts formal neurone in that the instantaneous behaviour is described by the neuronic or decision equation

$$y \, (m + 1) \quad = \quad 1 \, [\, A_m \, - \, \alpha \, \sum_{r=0}^{m} \, \beta^{-r} \, y \, (m - r) - k] \qquad (3.1.2)$$

where $A_m = n_e \, (m) - n_i \, (m)$

$$1 \, [x] \quad = \quad 1, \; x > 0 \; ; \quad 0, \; x \leqslant 0 \qquad (3.1.3)$$

$$\alpha \; > \; 0 \; , \quad \beta \; > \; 1$$

and so there is an exponentially decaying refractory period. However there is still no direct dependence on previous input, only a memory of the output y(.) and so no possibility of subthreshold temporal integration. Nagumo (1974) has investigated the discharge character-istics of this logical neurone with a refractory period.

If the refractory period is ignored, the logical neurone can be represented by

$$y \, (m + 1) \quad = \quad 1 \, [\, E(m) - k] \qquad (3.1.4)$$

where $E(m)$, the net excitation, is

$$E \, (m) \quad = \quad \sum_{j=1}^{n} \, a_j \, x_j \, (m) \qquad (3.1.5)$$

and a_j is the coupling coefficient which can be positive for excitation and negative for inhibition, and x_j (m) is the state (0 or 1) of the jth input line.

de Luca and Ricciardi (1968) have obtained asymptotic theorems for the interspike interval probability density function when n, the number of input lines, becomes large. The input activity in any line j is a two-valued, stationary random variable

$$x_j \quad = \begin{cases} 1, \text{ with Prob. } p_j \\ 0, \text{ with Prob. } q_j \end{cases} \qquad (j = 1, 2, \cdots, n) \qquad (3.1.6)$$

$$p_j + q_j \quad = \quad 1$$

Introducing the random variable e_j

$$e_j = \begin{cases} a_j, \text{ with Prob. } p_j \\ 0 \ , \text{ with Prob. } q_j \end{cases} \qquad (j = 1,2,\cdots,n) \qquad (3.1.7)$$

with mean $< e_j > = p_j a_j$, variance $\sigma^2_j = p_j q_j a^2_j$,

which describes the effect of the random variable input activity x_j on the excitation, the firing probability P_f can be written in terms of the density function $f_j (x_j)$

$$f_j(x_j) \ dx_j \ \equiv \ P \{ x_j < e_j < x_j + dx_j \} \qquad (3.1.8)$$

This assumes the e_j's are continuous random variables, which with using the Dirac delta function includes the discrete case of this model.

$$F (\chi) = \int_{-\infty}^{+\infty} dx_1 \ \int_{-\infty}^{+\infty} dx_2 \cdots \int_{-\infty}^{+\infty} dx_{n-1} \ \prod_{j=1}^{n-1} f_j (x_j) \ f_n (\chi - \sum_{j=1}^{n-1} x_j)$$

$$(3.1.9)$$

$$P_f = \int_{k}^{\infty} F (\chi) \ d\chi \ \equiv \ 1 - \phi (k) \qquad (3.1.10)$$

where $\phi (x)$ is the distribution function of the net excitation E with mean

$$< E > = \sum_{j=1}^{n} p_j a_j$$

standard deviation μ given by

$$\mu^2 = \sum_{j=1}^{n} p_j q_j a^2_j$$

$$\phi (x) \equiv \int_{\infty}^{x} d\chi \ F (\chi) \ \equiv \ P \{ -\infty < \chi < x \} \qquad (3.1.11)$$

The above relations hold for any finite n; if $n \to \infty$ use of the central limit theorem gives the result (see de Luca and Ricciardi)

$$P_f \sim \frac{1}{\sqrt{2\pi}} \int_z^\infty \exp\left(- t^2 / 2\right) dt = 1 - G(z) \qquad (3.1.12)$$

where $z = (k - <E>) / \mu_n$ and $G(z)$ is the Gaussian distribution
function. Using a theorem of Liapounoff it was shown that when n is
large but finite (say 10^5) the deviation of the probability of firing
distribution from the Gaussian distribution given by equation (3.1.12)
is < 11%. Thus although the lack of memory in logical neurones means
that the behaviour is only interesting when networks are considered,
an analysis of a logical neurone can give bounds on the error introduced by using the central limit theorem to approximate the input as a
Gaussian distributed random variable.

3.2   The Perfect Integrator Model

A serious deficiency of the models considered in section 3.1
is that there is no memory of previous inputs :   the opposite extreme
is the perfect integrator, or integrate and fire model, which integrates
inputs without any decay until a threshold voltage V_o is reached.   An
action potential is generated and the membrane potential reset instantaneously to its resting value. Without any loss of generality the resting membrane potential may be set to 0 and the threshold potential to 1.

If $x(\cdot)$ is the input as a function of time the subthreshold
generator potential is given by

$$V(t) = \int_o^t x(u) \, du \qquad (3.2.1)$$

if an action potential occurred at t = 0.   The time t_i to reach a
constant threshold V_o is the solution of

$$V_o = \int_o^{t_i} x(u) \, du \qquad (3.2.2)$$

If the input is a constant c, the firing rate is $r = 1/t_i$
$= c/V_o$, and so the discharge rate is proportional to the input.
Often there is such a linear relationship between the firing rate of
a neurone and the measured generator potential.

If there is a cosinusoidal modulation of a constant input, and so

$$x(t) = \alpha \exp(j(\omega t + \phi)) + c \qquad (3.2.3)$$

with an angular frequency ω and phase ϕ at $t = 0$, substitution into equation 3.2.2 and integration gives

$$V(t, \phi) = \frac{\alpha}{j\omega} \exp(j\phi)(\exp(j\omega t - 1)) + ct \qquad (3.2.4)$$

and an action potential is generated when $V_o = 1$ is reached. The rate of discharge determined by a constant input c is the carrier discharge rate.

For $\alpha \leqslant c$, the expected density of action potentials $n(\phi)$ averaged over a number of cycles is

$$n(\phi) = \alpha \exp(j\phi) + c \qquad (3.2.5)$$

this is derived in Knight, 1969; Rescigno, et al, 1970; Stein, French and Holden, 1972; Knight, 1972, and Knox, 1974. Thus the density of pulses is sinusoidally modulated with no change of phase and a frequency-independent gain. The spectral density will have components at harmonics of ω and many other frequencies (see section 5.3).

The perfect integrator with constant threshold can be used as a simple analytically tractable model for the generation of stochastic spike trains by making the input function, $x(t)$, a random variable. Then the behaviour of the model depends on the type of input.

If the input function is a series of positive impulses, $V(t)$ will increase in a staircase fashion until V_o is reached. This is the behaviour of the K-scaler discussed in section 2.2. If the input process is Poisson with parameter r the probability density function of the output intervals has a gamma density of order $= V_o / a$, where a is the amplitude of the change in $V(t)$ produced by a single input impulse, that is

$$p(t) = r^k t^{k-1} \exp(-rt / (k-1)! \tag{3.2.6}$$

with mean $\mu = k/r$ and variance $\sigma^2 = k/r^2$. Stein, French and Holden (1972) have obtained the frequency response function and coherence function in this case by :

The input spectral density $S_x(\omega)$ of a Poisson process of parameter r is

$$S_x(\omega) = r / 2\pi \tag{3.2.7}$$

The output spectral density $S_y(\omega)$ is

$$S_y(\omega) = \frac{1}{2\pi\mu} \left\{ 1 + \frac{P(j\omega)}{1 - P(j\omega)} + \frac{P(-j\omega)}{1 - P(-j\omega)} \right\} \tag{3.2.8}$$

where

$$P(j\omega) = \int_0^\infty p(t) \exp(-j\omega t)\, dt$$

$$= r^k / (r + j\omega)^k$$

which is the one sided Fourier transform of $p(t)$. Equation 3.2.8 assumes that the output is a renewal process and is derived in Cox and Miller (1965) p. 359. The frequency response function is a constant, as shown above for sinusoidal inputs, and for a Poisson input $G(\omega) = 1/k$. The cross spectral density $S_{xy}(\omega) = G(\omega) S_x(\omega) = r/2\pi k$. Thus the coherence $\gamma^2(\omega)$ is given by :

$$\gamma^2(\omega) = \frac{|S_{xy}(\omega)|^2}{S_x(\omega) S_y(\omega)}$$

$$= \frac{[(1 + j\omega / r)^k - 1]\ [(1 - j\omega / r)^k - 1]}{k\ [1 + (\omega / r)^2)k - 1]} \tag{3.2.9}$$

$$\approx 1\ ,\ \omega \ll r$$

$$1/r,\ \omega \gg r$$

These functions are illustrated in Figure (3.1)

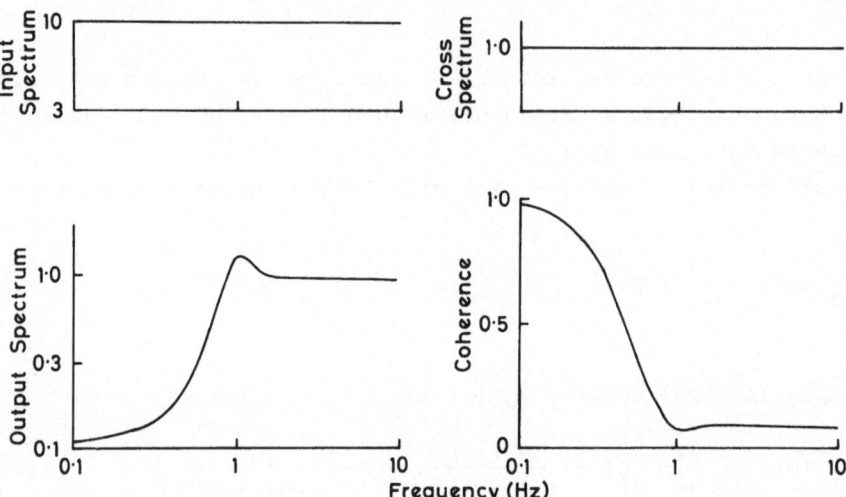

Figure 3.1. The input, output and cross-spectra and coherence function of a
perfect integrator model subject to a Poisson-distributed excitatory pulse train
input. Values calculated from equations 3.2.7 - 3.2.9 with r.= k = 10, which
generates a mean discharge rate of 1 impulse/sec.

In section 2.4 I showed that the changes in generator potential
in Limulus could be accounted for by a shot process of conductance
changes produced by photon absorption. The impulse firing mechanism
of Limulus has been modelled by a perfect integrator (Dodge, 1968;
Knight et al (1970), and so it is interesting to consider if fluctuations in impulse discharge rate in Limulus can be completely accounted for by the shot process underlying fluctuations in generator
potential. Ratliff et al (1968) showed that the variance of the
steady-state firing rate produced by current injection was much less
than the variance of the firing rate produced by light, which depended
on the light intensity and state of adaptation. Shapley (1969, 1971a,
b) has shown that the fluctuations in impulse rate can be quantitatively
accounted for by the generator potential fluctuations. The model is
that of a linear filter with a stochastic input

$$S_i(\omega) = |G(\omega)|^2 S_g(\omega)$$

where G (ω) is the frequency response function of the current to firing
rate process which is a perfect integrator coupled to self inhibition,

S_g (ω) the spectral density of the generator potential and S_i (ω) the spectral density of the instantaneous impulse rate. These functions can all be estimated, and any one computed from the other two. The computed and measured spectral densities and autocorrelation functions agreed within two standard errors over most of the frequency range. The probability density function, under steady state conditions, of both the generator potential and impulse rate were Gaussian under light adaptation and gamma under dark adaptation. Thus the fluctuations in spike rate could be accounted for by the fluctuations in generator potential.

When the input x(t) to the perfect integrator is a series of positive and negative impulses, a random walk results - see section 6. If the input is white noise, or if the rate of the positive and negative input impulses → ∞ as their amplitude a → 0+, a Wiener process is generated and a diffusion results - see section 7.

3.3   The Leaky Integrator Model

In the leaky integrator model the effects of inputs sum linearly and decay exponentially, with a single time constant τ, for subthreshold changes in generator potential. If the voltage reaches a constant threshold V_o an action potential is generated and the voltage is reset instantaneously to its resting value, say 0 and $V_o = 1$. Thus

$$V(t) = \int_0^t x(u) \exp(-(t-u)/\tau) \, du \qquad (3.3.1)$$

for $0 < V(t) < 1$

$$\text{and} \quad V_o = \int_0^{t_i} x(u) \exp(-(t-u)/\tau) \, du \qquad (3.3.2)$$

defines the interspike interval t_i. The properties of this simple neural model have been extensively investigated : Lapique, 1907; Stein, 1965; Gluss, 1967; Segundo et al 1968; Johannesma, 1968; Siebert, 1969; Rescigno et al 1970; Sugiyama et al 1970; Stein et al 1972; Knight, 1972.

The intervals in response to a constant input are given by the solution of equation (3.3.2)

$$t_i = -\tau \ln(1 - 1/c\tau) \tag{3.3.3}$$

where $c\tau > 1$

Thus the rate of firing $1/t_i$ is not proportional to the input, there being an increased sensitivity to small inputs. The response to a periodic input given by equation (3.2.3) gives

$$V(t, \phi) = \frac{\alpha\tau \exp(j\phi) [\exp(j\omega t) - \exp(-t/\tau)]}{(1 + j\omega t)}$$

$$+ c\tau (1 - \exp(-t/\tau)) \tag{3.3.4}$$

for $0 \leqslant V(t, \phi) \leqslant 1$

If an impulse is generated at a phase ϕ_1, the next impulse will occur at a phase $\phi_1 + \omega t_i$ where t_i is the solution when $\mathrm{Re}[V(t, \phi)] = 1$. It can be shown (Rescigno et al, 1970; Stein et al, 1972) that the response will be phase locked to the input function $x(t)$

$$y(t) = \sum_{i=1}^{k} \delta(t - t_i) = \sum_{i=1}^{k} \delta(\phi - \phi_i) \tag{3.3.5}$$

with k impulses occurring at fixed phases every n cycles.

When the input to the leaky integrator model is a train of excitatory impulses, the behaviour is that of the KT-scaler machine discussed in section 2.2. When both excitatory and inhibitory input impulses are permitted a random walk is obtained (see section 6), which in the limit as the amplitude of the input impulses decreases and their rate increases gives a diffusion approximation model (see section 7). When the input $x(t)$ is Gaussian white noise equation (3.3.1) describes a Ornstein - Uhlenbeck process. The distribution of the first passage time to an absorbing barrier of a Ornstein - Uhlenbeck process, which in this case is the interspike interval distribution, is not available in closed form. The problem of this distribution is discussed in section 7.

3.4 Two-time Constant Models

The strength-duration curve of the leaky integrator model obtained
by using square-wave current pulses of duration t_i is

$$i = i_{rh} / 1 - \exp(- t_i / \tau) \qquad\qquad (3.4.1)$$

which, for short t_i, gives

$i = i_{rh} \tau / t_i$ or $it_i = i_{rh}\tau = $ constant where i is the
threshold current intensity for a pulse of duration t_i. Experimentally,
this relationship is approximately true; however, accurate measurements
of the strength-duration curve of axons at weak current intensities
show deviations from equation (3.4.1): Monnier (1934), Hill (1936)
and Rashevsky (1938).

If the input stimulus rises in a linear ramp, the time to excitation and the stimulus level at which excitation occurs both increase
as the slope of the stimulus decreases. This is the phenomenon of
accommodation, which was explained in terms of a two-factor or two
time-constant model independently by Monnier, Hill and Rashevsky. A
more satisfactory explanation is given by Hodgkin-Huxley theory (see
section 3.5).

If I use Hill's 1936 notation and identify the variable V with
generator potential and U with voltage threshold

$$-dV / dt = (V - V_r) / k$$

$$\qquad\qquad\qquad\qquad\qquad\qquad\qquad (3.4.2)$$

$$-dU / dt = (U - U_r) / \lambda$$

where V_r and U_r are the resting potential and threshold, k is now the
time constant of excitation and λ the time constant of accommodation,
the strength-duration curve is given by

$$i / i_{rh} = \frac{(1 - k/ \lambda) / (\lambda/ k)^{\;1/ (\lambda / k - 1)}}{\exp(- t_i / \lambda) - \exp(- t_i / k)} \qquad (3.4.3)$$

which, in the limit $\lambda/k \to \infty$ reduces to equation (3.4.1) with $k = \tau$.

This two-time constant model is attractive in that its behaviour is more realistic than that of the better studied perfect and leaky integrators, and its formulation is far simpler than the Hodgkin-Huxley equations. Noble and Stein (1966) have shown that the strength-duration curve of a point-polarized Hodgkin-Huxley axon is close to the range of curves obtained by the two-time constant model for different ratios of λ/k. It would be interesting to investigate analytically phase locking in the Hill model for different λ/k, and to numerically evaluate the response to Poisson excitatory impulse inputs and white noise inputs.

3.5   The Hodgkin-Huxley Equations

The Hodgkin-Huxley equations (Hodgkin and Huxley (1952 a-d)) describe the electrical properties of the membrane of the squid giant axon, and when coupled with the partial differential cable equation for local circuit current flow can describe excitation and propagation of action potentials under a wide range of experimental conditions. The H-H theory has also been applied to a variety of other excitable tissues – see the review by Noble (1966) and the monograph by Jack, Noble and Tsien (1975).

I will not be concerned here with the extensive body of experimental evidence which supports the assumptions implicit in the H-H analysis, but will just consider the properties of the H-H equations as a mathematical model. If an axon is treated as an infinite cable of diameter D, axoplasmic resistivity ρ, the membrane having resistive and capacitative properties, the general cable equation describes the pattern of voltage V with time and distance x

$$\frac{D}{4\rho} \frac{\partial^2 V}{\partial x^2} = Cm \frac{\partial V}{\partial t} + J_i + J \qquad (3.5.1)$$

where J_i (V, x, t) is the membrane ionic current density and J (x, t) the current density from an externally applied source such as a stimulating electrode. The spread of voltage with distance can be eliminated by short circuiting the axoplasmic resistance with an internal wire electrode, so $\partial V / \partial x = 0$. This is a space clamp, and under these conditions equation (3.5.1) reduces to

$$Cm \frac{\partial V}{\partial t} = J_i (V, t) \qquad (3.5.2)$$

when there is no applied current. H-H theory assumes that any active
ionic pumping processes present are electrically neutral, and so J_i
is produced by the electrochemical gradient driving ions through the
membrane. A further assumption is that different ionic species cross
the membrane independently through specializations having different
properties. Thus

$$J_i = i_{Na} + i_K + i_{Cl}$$

$$= f_{Na} (V_r - V_{Na}) + f_K (V_r - V_K) + f_L (V_r - V_L) \qquad (3.5.3)$$

where V_r is the resting membrane potential, and V_{Na}, V_K and V_L the
Nernst equilibrium potentials or reversal potentials for Na^+, K^+ and
leakage current as defined in section 1. The functions f_{Na}, f_K and
f_L are linear functions of the electrochemical gradients for _instantan-_
eous changes in voltage, but the transient and steady state behaviour
is nonlinear. The form of the functions was determined empirically
using the voltage clamp technique and ion substitution experiments.

K^+ movement through the membrane depends on the state of some
membrane specialization (the gate of the K^+ channel) which can exist
in two states, α and β. If the fraction of the membrane specializations in the α state is n, and if a first order kinetic reaction is
assumed

$$dn/dt = \alpha_n (1 - n) - \beta_n n$$

$$n_\infty = \alpha_n / (\alpha_n + \beta_n) \qquad (3.5.4)$$

$$\tau_n = 1 / (\alpha_n + \beta_n)$$

where α_n and β_n are voltage dependent rate constants for the conversion

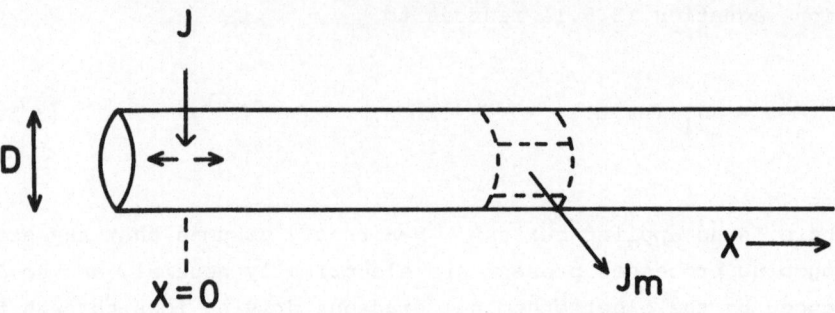

If a current J is injected at a point x=0, the axial current I_a is

$$I_a = - \frac{\partial V}{r_a \partial x}$$

by Kirchoff's law

$$\frac{\partial I_a}{\partial x} = -j_m, \text{ the membrane current density}$$

Thus:

$$J_m = \frac{\partial^2 V}{r_a \partial x^2} = \frac{D}{\partial x^2}\frac{\partial^2 V}{}$$

but this is the sum of the capacitative, ionic and injected current densities

$$\frac{D}{4\rho} = \frac{\partial^2 V}{\partial x^2} = C_m \frac{\partial V}{\partial t} + J_i + J$$

The ionic current is given by the Hodgkin-Huxley equations:

$$J_i = G_{Na}(V - V_{Na}) + G_K(V - V_K) + G_L(V - V_L)$$

where

$$G_{Na} = 120 \, m^3 h, \quad G_K = 36n^4, \quad G_L = 0.3$$

$$dm/dt = \alpha_m(1 - m) - \beta_m m$$

$$\alpha_m = 0.1(25-V)/(exp\{(25-V)/10\}-1)$$

$$\beta_m = 4exp(-V/18)$$

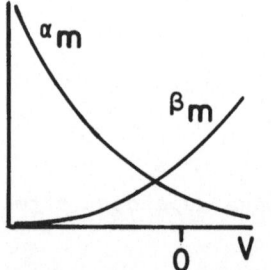

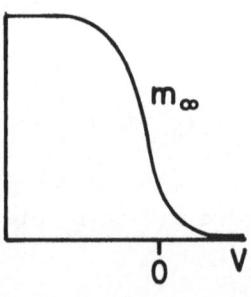

$$dh/dt = \alpha_h(1-h) - \beta_h h$$

$$\alpha_h = 0.07\exp(V/20)$$

$$\beta_h = 1/(\exp\{(30-V)/10\}+1)$$

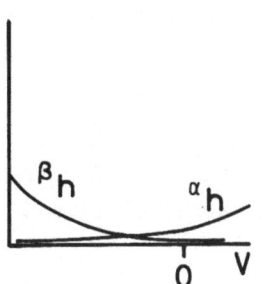
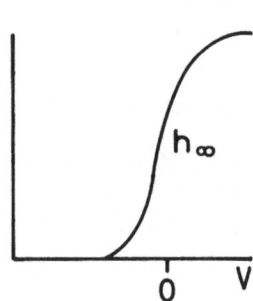

$$dn/dt = \alpha_n(1-n) - \beta_n n$$

$$\alpha_n = 0.01(10-V)/\exp\{(10-V)/10\}-1)$$

$$\beta_n = 0.125\exp(-V/80)$$

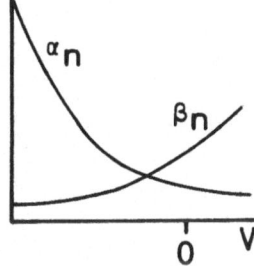
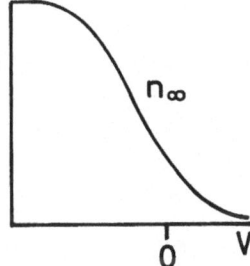

J_i: membrane ionic current density/unit area $(\mu A/cm^2)$

V_{Na}, V_K, V_L: Nernst potentials for sodium, potassium and leakage currents (mV)

m,n,h: the sodium activation, inactivation and potassium activation variables

G_{Na}, G_K, G_L: sodium, potassium and leakage membrane conductances
ρ: axoplasmic resistivity, r_a: axoplasmic resistance/unit length

Figure 3.2.   The Hodgkin-Huxley equations.

from state β to state α (the forward rate constant α_n) and from state α to β (the backward rate constant β_n). n_∞ is the steady-state value of n and τ_n the time constant of the process. The potassium conductance depends on n : $g_K = n^4 \bar{g}_K$ gives the sigmoid delayed rise in i_K seen experimentally under voltage clamp conditions, where $\bar{g}_K$ is the maximum potassium conductance.

The change in g_{Na} under voltage clamp is transient , and its maximal value depends on the preceeding holding potential. These effects are accounted for by two systems, an activation system (m) and an inactivation system (h)

$$dm/dt = \alpha_m (1 - m) - \beta_m m$$

$$m_\infty = \alpha_m / (\alpha_m + \beta_m) \qquad\qquad (3.5.5)$$

$$\tau_m = 1 / (\alpha_m + \beta_m)$$

$$dh/dt = \alpha_h (1 - h) - \beta_h h$$

$$h_\infty = \alpha_h / (\alpha_h + \beta_h) \qquad\qquad (3.5.6)$$

$$\tau_h = 1 / (\alpha_h + \beta_h)$$

where the rate constants α_m and β_h increase and β_m and α_h decrease on depolarization. The sodium conductance is determined by $g_{Na} = m^3 h \, \bar{g}_{Na}$.

The leakage conductance g_L is voltage insensitive. The full set of H-H equations which have been extensively used for numerical investigations is given in Figure (3.2). The H-H system of equations are analytically intractable and so their behaviour may be investigated either by numerical computation or by reducing the H-H equations to a simpler form (section 3.6).

A number of numerical methods have been used to obtain solutions for the H-H membrane equations. Hodgkin and Huxley (1952 d) solved the four simultaneous first-order differential equations using a hand calculator by the method of Hartree (1932-3). Cole et al (1955) and Nemoto et al (1975) used the Runge-Kutta method of integration, and Cooley and Dodge (1966) used closed integration formulas obtained from the trapezoidal rule of integration, in order to achieve numerical

stability without a very small integration step δt. Noble and
Stein (1966), Stein (1967), Moore and Ramon (1974) and Holden (1976)
used the simple Euler method.

The numerical investigations of the H-H equations show:

a)   a strength-duration curve similar to that of the two-
time constant model (Noble and Stein, 1966);

b)   repetitive discharge in response to constant currents,
with a discharge rate limited to 50-150 impulses/sec having an
approximately logarithmic relation to the injected current density
(Agin, 1964;   Stein, 1967);

c)   phase locked action potentials in response to a cosinusoidally
modulated injected current (Holden, 1976;   Nemoto et al, 1975).
The pattern of phase locking is similar to that observed for the
leaky integrator model.

Thus although the H-H equations reproduce the waveform of the
action potential the qualitative aspects of their behaviour is similar
to that of simpler neural models, and so the H-H equations have not
been extensively used as a model of the spike generating mechanism
in models of stochastic spike train generation. However, Stein (1967)
has subjected the H-H equations to unit permeability changes occurring
at random, and has shown that the negative feedback inherent in the
H-H equations tends to produce a small negative serial correlation
between adjacent intervals.

An interesting statistical mechanical interpretation of the
parameters of the H-H equations has been given by Agin (1963). If the
α_n term of Figure (3.2) is rewritten as

$$\alpha_n = \frac{(V + a)/a^2}{\exp ((V+a)/a - 1)}$$

$$= \frac{\phi/a}{\exp (\phi - 1)} \qquad (3.5.7)$$

where $\phi = (V+a)/a$; $a = 10$

Agin showed (his equations 16-20) that

$$\alpha_n = \frac{\frac{1}{a} \sum_{j=0}^{\infty} \phi_j \exp{(-\phi j)}}{\sum_{j=0}^{\infty} \exp{(-\phi j)}} \qquad (3.5.8)$$

which is isomorphic with the mean value function of statistical mechanics. A similar algebraic treatment demonstrated a correspondence between other empirical terms of the H-H equations and other well known functions of statistical mechanics. This suggested that an abstract model based on the framework of statistical mechanics could be used to produce a set of equations analogous to the H-H system of equations.

3.6 Reduced Hodgkin-Huxley Equations

Since the full system of H-H equations are analytically intractable a number of approaches have been made to reduce the H-H equations to a simpler structure.

Knight (1973) has derived a reduction of the H-H equations to that of a leaky integrator model. For a space clamped patch of membrane

$$C_M \, dV/dt = \bar{g}_{Na} \, m^3 h(V - V_{Na}) + i_K + g_L(V - V_L) + i\,(t)$$
$$(3.6.1)$$

where $i\,(t)$ is the source current. For slow changes of $i\,(t)$, and if i_{Na} and i_K are small

$$V \simeq \frac{1}{g_L} \, i\,(t) + V_L \qquad (3.6.2)$$

as long as $m < m_c$, a critical value of m at which m^3 becomes large and hence i_{Na} dominates and produces an action potential. Now from equation 3.5.5,

$$dm/dt = (m_\infty\,(V) - m))/\tau_m(V) \qquad (3.6.3)$$

and $\tau_m(V)$ is approximately constant and $m_\infty\,(V)$ an approximately linear function of V in the range for slow changes of potentials. In this slow range, if

$$\tau_m(V) = \tau_o$$

$$m_\infty(V) = m_o + m'(V-V_c) \quad , \quad V_c \text{ is constant}$$

equations (3.6.2 - 3) give

$$dm/dt = \frac{1}{\tau_o}[m_o + m'(\frac{1}{g_L} i(t) + V_L - V_c) - m] \qquad (3.6.4)$$

or $dm/dt = -\gamma m + s(t) \quad , m < m_c$ \hfill (3.6.5)

where $\gamma = 1/\tau_o$ and

$$s(t) = \frac{m' i(t)}{\tau_o g_L} + \frac{m_o + m'(V_L - V_c)}{\tau_o}$$

the solution of (3.6.5) is

$$m(t) = \int_0^t \exp(-\gamma(t-\tau)) \; s(\tau) \, d\tau \qquad (3.6.6)$$

which is the leaky integrator model. Note that the time constant of this leaky integrator model is determined by $\tau_m(V)$, not τ_M, the membrane time constant. If $\gamma \to 0$ we have the perfect integrator. This reduction of the H-H equations to that of a leaky integrator accounts for the similarity between the frequency response function of a leaky integrator and the frequency response functions of the H-H equations at low frequencies computed using periodic current inputs (Holden, 1976).

A less drastic reduction of the H-H membrane equations is given by the approach of Fitzhugh (1960, 1961), which utilizes the mathematical techniques of non-linear mechanics (Minorsky, 1947) or the geometric theory of differential equations (Lefschetz, 1957). The solution of the H-H equations may be considered to follow paths in a Euclidean phase space whose coordinates are the dependent variables (V, m, n, h). The topological behaviour of all paths in the phase space is determined by studying the singular points (equilibrium points in the phase space), separatrices (paths along which solutions approach a saddle point or unstable singular point) and limit cycles.

The phase space of the H-H equations is four dimensional:
in order to obtain insights into the behaviour of the system it
is useful to reduce the dimensionality of the phase space to two or
three dimensions, which be readily visualized.

The four dependent variables of the H-H system can be
divided into those with 'fast' time constants, (V, m), and those
with 'slow' time constants (n, h). Thus, in a short time interval,
V and m can change appreciably whereas n and h are approximately
constant.
If

$$dh/dt = 0, \quad dn/dt = 0$$

h, n are constant

the H-H system of equations reduces to

$$I = C \, dV/dt + m^3 h \, \bar{g}_{Na} \, (V + 115)$$
$$+ n^4 \bar{g}_K \, (V - 12) + \bar{g}_L \, (V + 10.5989) \qquad (3.6.6)$$
$$dm/dt = \alpha_m \, (V) \, (1 - m) + \beta_m \, (V) \, m$$

and so the state of this reduced system is completely defined by
the point (V, m) in a two-dimensional phase space.

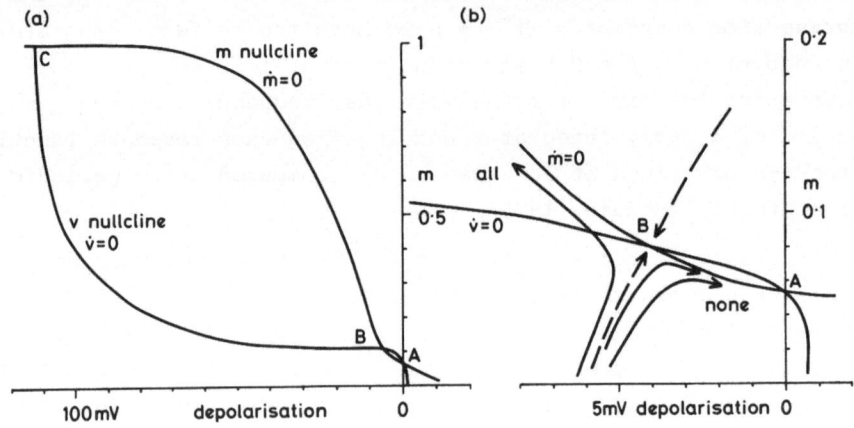

Figure 3.3. Nullclines of V-m reduced system representing the fast behaviour of
the Hodgkin-Huxley membrane equations. The two paths leading to the unstable saddle
point B are the threshold separatrices forming the boundary between all or none
responses. See text.

This reduced system has two interesting nullclines, the
m-nullcline, where $dm/dt = 0$, which is the m_∞ (V) curve, and the
V-nullcline, where $dV/dt = 0$. There are three intersections of the
nullclines (schematically shown in Figure 3.3) where $dV/dt =$
$dm/dt = 0$, of which A, the resting state, and C, the excited state,
are stable and B is an unstable saddle point. The threshold
separatrices, the two paths leading to B, form the boundary between
'all' and 'none' responses. Note that there is no recovery to the
resting state, and so the V - m reduced system can only be used to
examine the threshold behaviour of the system, not spike train
generation or limit cycle behaviour, which requires a restoring
variable.

Lecar and Nossal (1974) have investigated the response of the
V-m reduced system to fluctuations in voltage. If the rate constants
of the V-m system are defined as

$$\gamma_1 = (\bar{g}_K \, n_\infty^4 \, (V_r) + g_L)/C \qquad\qquad (3.6.7)$$

$$\gamma_2 = (g_{Na} \, h_\infty(V_r) \,)/C \qquad\qquad (3.6.8)$$

$$\lambda(V) = 1/\tau_m(V) \qquad\qquad (3.6.9)$$

with V_r denoting the stable resting potential (point C of Figure
3.6), the effective emf's are

$$V_1 = (g_K n_\infty^4 \, (V_r) \, V_K + g_L V_L)/\gamma_1$$
$$\qquad\qquad (3.6.10)$$
$$V_2 = V_{Na}$$

and the external driving force is

$$J = I/C$$

Introduction of a new dimensionless conductance variable $\sigma \equiv m^3$
gives in place of equation (3.6.6)

$$dV/dt = J - \gamma_1 \, (V-V_1) - \gamma_2 \, \sigma(V-V_2) \qquad\qquad (3.6.11)$$

$$d\sigma/dt = \phi(V,\sigma)$$

The use of the dimensionless conductance variable permits equation (3.6.10) to be used also for the toad node of Ranvier membrane, if in this case, where the membrane properties are described by the Frankenhauser-Huxley (1964) equations, σ is now given by $\sigma = m^2$. $\phi(V,\sigma)$ represents the sodium activation kinetics, and can either be obtained by substitution from the H-H equations, which gives

$$\phi(V,\sigma) = 3\lambda(V)\sigma^{2/3}(\sigma_\infty(V)^{1/3} - \sigma^{1/3})) \tag{3.6.12}$$

or left unspecified, to be derived from particular models of the g_{Na} process.

In the presence of noise the paths in the $(V - \sigma)$ phase plane are now time varying probability distributions. The effects of noise may be represented by equivalent random forces or Langevin forces (see section 7) F_V and F_σ

$$dV/dt = J - \gamma_o(V - V_1) - \gamma_2\sigma(V - V_2) + F_V(< \sigma >, t) \tag{3.6.13}$$

$$d\sigma/dt = \phi(V, \sigma) + F_\sigma(< V >, t) \tag{3.6.14}$$

The effect of a brief stimulus is to move the state of the system parallel to the V axis into the vicinity of the threshold separatrix. Here the state point will randomly move back and forth across the threshold separatrix while drifting towards the unstable saddle point B. In the vicinity of the saddle point the paths are repelled away from B towards A (no action potential) or towards C (action potential initiated). Thus the probability of initiating an action potential is the probability that after a long time the state point is on the suprathreshold side of the threshold separatrix.

Lecar and Nossal (1971a) obtained the probability of initiating an action potential as an integrated Gaussian distribution function of the stimulus strength, which agrees with the experimental curves (see section 4). Further, the coefficient of variation does not depend on the stimulus duration, and the probability of firing - stimulus duration and latency distribution curves were obtained.

In a separate paper, Lecar and Nossal (1971b) derived the dependence of the width of the probability of firing curve on different types of noise source.

$$P(\text{fire}|I) = \tfrac{1}{2}[1 + \text{erf} \ (I - I_\theta)/RI_\theta] \tag{3.6.15}$$

where I = stimulating current

I_θ = threshold current

R   is the relative spread of the distribution, the width of the region where $P(\text{fire}|I)$ rises from 0 to 1.

$$R = (\sqrt{2} \ \sqrt{D} \ V_1) \ (Z_{11} \ V_\theta) \tag{3.6.16}$$

V_θ = threshold depolarization as a deviation from V_r.

Z_{11} depends on the geometry of the V-σ phase plane and hence on the membrane variables

$$D = \int_0^\infty \int_0^\infty \exp \ (-p_1 \ (y + z) \ ) < X \ (y) \ X \ (z) > dydz \tag{3.6.17}$$

p_1 is a rate constant associated with movement of the state point away from the threshold separatrix.

X (t) is associated with the random Langevin force.

Thermal (Johnson-Nyquist) noise, flicker (1/f) noise and fluctuations in g_{Na} were the noise sources considered - the effect of conductance fluctuations are discussed in section 4.3. Using experimentally estimated values for membrane parameters the effect of conductance fluctuations on threshold fluctuations was an order of magnitude greater than the effects of thermal or flicker noise.

3.7 References

Bayly, E.J.: Spectral analysis of pulse frequency modulation in the nervous system. I.E.E.E. Trans. Bio-Med. Engng. 15 257-265 (1968).

Brink, F., Bronk, D.W. & Larrabee, M.G.: Chemical excitation of nerve. Ann. N.Y. Acad. Sci. 47 457-85 (1946).

Caianiello, E.R.: Outline of a theory of thought processes and thinking machines. J. theor. Biol. 1 201-235 (1961).

Cole, K.S., Antosiewicz, H.A. & Rabinowitz, P.: Automatic computation of nerve excitation. J. Soc. Indust. Appl. Math. 3 153-172 (1955).

Cooley, J.W. & Dodge, F.A.: Digital computer solutions for excitation and propagation of the nerve impulse. Biophys. J. 6 583-599 (1966).

Cox, D.R. & Miller, H.D.: The Theory of Stochastic Processes. 398 pp. Methuen, London (1965).

Dodge, F.A.: Excitation and inhibition in the eye of Limulus. In: Optical Data Processing by Organisms and Machines. ed. W. Reichardt, Academic Press, N.Y. (1968).

Fitzhugh, R.: Thresholds and plateaux in the Hodgkin-Huxley nerve equations. J. Gen. Physiol. 43 867-896 (1960).

Fitzhugh, R.: Impulses and physiological states in theoretical models of nerve membrane. Biophys. J. 1 445-466 (1961).

Frankenhauser, B. & Huxley, A.F.: The action potential in the myelinated nerve fibre of Xenopus laevis as computed on the basis of voltage clamp data. J. Physiol. 171 302 (1964).

Gluss, B.: A model for neuron firing with exponential decay of potential resulting in diffusion equations for probability density. Bull. Math. Biophys. 29 233-243 (1967).

Hagiwara, S.: On the fluctuation of the interval of the rhythmic excitation. I. The efferent impulse of the human motor unit during the voluntary contraction. Report. Physiograph. Sci. Instit. Tokyo Univ. 3 19-24 (1949).

Hagiwara. S.: On the fluctuation of the interval of the rhythmic excitation. II. The afferent impulse from the tension receptor of the skeletal muscle. Report. Physiograph. Sci. Instit. Tokyo Univ. 4 28-35 (1950).

Hagiwara, S.: Analysis of interval fluctuations of the sensory nerve impulse. Japanese J. of Physiol. 4 234-40 (1954).

Hartree, D.R.: A practical method for the numerical solution of differential equations. Mem. Manchr. lit. phil. soc. 77 91-107 (1932-3).

Hill, A.V.: Excitation and accommodation in nerve. Proc. Roy. Soc. B 119 305-355 (1930).

Hodgkin, A.L. & Huxley, A.F.: Currents carried by sodium and potassium ions through the membrane of the giant axon of Loligo. J. Physiol. 116 449-472 (1952a).

Hodgkin, A.L. & Huxley, A.F.: The components of membrane conductance in the giant axon of Loligo. J. Physiol. 116 473-496 (1952b).

Hodgkin, A.L. & Huxley, A.F.: The dual effect of membrane potential on sodium onductance in the giant axon of Loligo. J. Physiol. 116 497-506 (1952c).

Hodgkin, A.L. & Huxley, A.F.: A quantitative description of membrane current and its application to conduction and excitation in nerve. J. Physiol. 117 500-544 (1952d).

Holden, A.V.: The response of excitable membrane models to a cyclic input. Biol. Cybernetics 21 1-8 (1976).

Jack, J.J.B., Noble, D. & Tsien, R.W.: Electric current flow in excitable cells. 500 pp. Oxford University Press, New York (1975).

Johannesma, P.I.M.: Stochastic neural activity - a theoretical investigation. Ph.D. thesis, Catholic Univ. Nijmegen, The Netherlands (1969).

Knight, B.W.: Dynamics of encoding in a population of neurones. J. of Gen. Physiol. 59 734-766 (1972).

Knight, B.W.: Some questions concerning the encoding dynamics of neuron populations. 4th International Biophysics Congress, 422-436, Puschino USSR (1973).

Knight, W.: Frequency response for sampling integrator and for voltage to frequency converter, in: Systems Analysis in Neurophysiology 61-72, ed. C.A. Terzuolo, Univ. of Minnesota Press, Minneapolis (1969).

Knight, B., Toyoda, J. & Dodge, F.A.: A quantitative description of the dynamics of excitation and inhibition in the eye of Limulus . J. Gen. Physiol. 56 421-437 (1970).

Knox, C.K.: Cross-correlation functions for a neuronal model. Biophys. J. 14 567-582 (1974).

Lapique, L.: Recherches quantitatives sur l'excitation electrique des nerfs traitee comme une polarization. J. Physiol. (Paris) 9 622-635 (1907).

Lecar, H. & Nossal, R.: Theory of threshold fluctuations in nerves. I. Relationships between electrical noise and fluctuations in axon firing. Biophys. J. 11 1048-1067 (1971a).

Lecar, H. & Nossal, R.: Theory of threshold fluctuations in nerves. II. Analysis of various sources of membrane noise. Biophys. J. 11 1068-1084 (1971b).

Lefschetz, S.: Differential Equations: Geometric Theory. Interscience, N.Y. (1957).

de Luca, A. & Ricciardi, L.M.: Formalized neuron: probabilistic description and asymptotic theorems. J. Theoret. Biol. 14 206-17 (1967).

de Luca, A. & Ricciardi, L.M.: Probabilistic description of neurons in 'Neural Networks', ed. E.R. Caianiello, 100-109, Springer-Verlag, Berlin (1968).

McCulloch, W.S. & Pitts, W.H.: A logical calculus of ideas immanent in nervous activity. Bull. of Math. Biophys. 5 115-133 (1943).

Minorsky, N.: Introduction to Non-Linear Mechanics. J.W. Edwards, Ann Arbor (1947).

Minsky, M.: Computation: Finite and Infinite Machines. 317 pp. Prentice-Hall, Englewood Cliffs, N.J. (1967).

Monnier, A.M.: L'Excitation Electrique des Tissues. 326 pp Herman et Cie. Paris (1934).

Moore, J.W. & Ramon, F.: On numerical integration of the Hodgkin and Huxley equations for a membrane action potential. J. Theor. Biol. 45 249-273 (1974).

Nagumo, J.: Response characteristics of a mathematical neuron model. Advances in Biophysics. 6 41-73 (1974).

Nemoto, I., Migazaki, S., Saito, M. & Utsunomiya: Behaviour of solutions of the Hodgkin-Huxley equations and its relation to properties of mechanoreceptors. Biophys. J. 15 469-479 (1975).

Noble, D.: Applications of Hodgkin-Huxley equations to excitable tissues. Physiol. Rev. 46 1-50 (1966).

Noble, D. & Stein, R.B.: The threshold conditions for initiation of action potentials by excitable cells. J. Physiol. 187 129-1962 (1966).

Rashevsky, N.: Mathematical Biophysics 380 pp. Univ. of Chicago Press. Chicago (1938). reprinted by Dover Publications, New York in two volumes (1960).

Ratliff, F., Hartline, H.K. & Lange, D.: Variability of interspike intervals in optic nerve fibres of Limulus. Proc. Nat. Acad. Sci. U.S.A. 60 392 (1968).

Rescigno, A., Stein, R.B., Purple, R.L. & Poppele, R.E.: A neuronal model for the discharge patterns produced by cyclic inputs. Bull. Math. Biophys. 32 337-353 (1970).

Segundo, J.P., Perkel, D.H., Wyman, H., Hegsted, H. & Moore, G.P.: Input-output relations in computer-simulated nerve cells - influence of the statistical properties, strength, number and inter-dependence of excitatory pre-synaptic terminals. Kybernetik 4 157-171 (1968).

Shapley, R.: Fluctuations in the response to light of visual neurones in Limulus. Nature 223 437-440 (1969).

Shapley, R.: Fluctuations in the impulse rate in Limulus eccentric cells. J. Gen. Physiol. 57 539-556 (1971a).

Shapley, R.: Effects of lateral inhibition on fluctuations of the impulse rate. J. Gen. Physiol. 57 557-575 (1971b).

Stein, R.B.: A theoretical analysis of neuronal variability. Biophys. J. 5 173-195 (1965).

Stein, R.B.: The frequency of nerve action potentials generated by applied currents. Proc. Roy. Soc. B 167 64-86 (1967).

Stein, R.B.: Some models of neuronal variability. Biophys. J. 7 37-68 (1967).

Stein, R.B., French, A.S. & Holden, A.V.: The frequency response, coherence and information capacity of two neuronal models. Biophys. J. 12 295-322 (1972).

Sugiyama, H., Moore, G.P. & Perkel, D.H.: Solutions for a stochastic model of neuronal spike production. Math. Biosci. 8 323-341 (1970).

Winograd, S. & Cowan, J.D.: Reliable computation in the Presence of Noise. 95 pp M.I.T. Press. Cambridge, Mass. (1963).

4   FLUCTUATIONS IN EXCITABILITY

The models of section 3 are deterministic, and only produce a
stochastic spike train when there are stochastic fluctuations in the
input or in the generator potential. An alternative approach is to
consider models in which the excitability, or threshold voltage,
undergoes spontaneous fluctuations.

If a single axon is electrically stimulated with a current
intensity which is close to threshold, an action potential is generated
in a fraction of the trials (Blair and Erlanger, 1932; Pecher, 1939),
and this phenomenon has been further investigated by Verveen (1960
1961).

That this variability of the response is due to a fluctuating
threshold, and not uncontrolled fluctuations in stimulus parameters, was
shown in experiments when the responses of two different fibres,
distinguishable by their different latencies, were recorded. At a
given stimulus intensity all possible combinations of response occur,
either fibre responding alone, or both, or neither. The probability
that both responded was the product of the probabilities of each of the
two fibres responding separately: this supports the hypothesis that the
fluctuations in excitability in the two fibres are independent. When
the stimulus repetition rate is low (< 1 stimulus/2 sec), so there is
no interaction between the refractory period and the probability of
response, there is a fluctuation in latency and the relation between
probability of response and stimulus intensity is well fitted by the
Gaussian distribution function, with a mean and standard deviation
dependent on the stimulus duration. The coefficient of variation for
a given axon is approximately constant for stimulus durations from
0.25 - 2.5 msec. Since there is an intrinsic fluctuation in threshold,
the threshold intensity has to be defined as the stimulus intensity
which, in a series of repeated, independent trials, produces an action

potential in a given percentage of trials.

Pecher (1939) argued that fluctuations in threshold could be
accounted for by rapid, localized fluctuations in ion concentration.
From kinetic theory, the coefficient of variation is $1/\sqrt{m}$, where m is
the number of ions necessary for excitation. For a node of Ranvier
of width $d(10^{-4}$ cms), fibre radius r $(10^{-4}$ cms) and distance of
influence of ions D ($\sim 10^{-7}$ cms) the total number of ions m which
can influence excitability is 2π d r D C N, where C is the molar ionic
concentration and N Avogadro's number. This gives a coefficient of
variation of about 0.01, and would generate a Gaussian distribution of
threshold if the amplitude of the fluctuations is much less than the
threshold, i.e. action potentials are not initiated spontaneously due
to local fluctuations in ionic concentration.

Using this theory to generate a Gaussian distribution of
threshold fluctuation with a small coefficient of variation, Landahl
(1941), Householder and Landahl (1945) and Rashevsky (1948) used a two
time-constant model to derive
    a) a probability of response-stimulus intensity curve with a
Gaussian distribution and
    b) the latency of response, and its standard derivation,
decreases with stimulus intensity.
These relations fitted Pecher's data.

4.1 Single time constant models with fluctuating threshold

If the threshold $V_o(t)$ is taken to be

$$V_o(t) = k + n(t) \tag{4.1.1}$$

where n(t) is a stationary random variable with zero mean, and k is
a constant, a simple model is to let the threshold undergo a random
walk (see section 6) with depolarizing and hyperpolarizing steps having
the same amplitude a and equal probability, i.e. zero drift.

In the absence of an input, when the threshold undergoes a
random walk, both the perfect integrator and leaky integrator neurone
models are spontaneously active, as if the threshold is undergoing a
random walk with zero drift the probability that the threshold $V_o(t)$
crosses the constant resting potential tends to one as $t \to \infty$. This is
equivalent to the situation considered in section 6.3 when V(t) undergoes
a random walk with zero drift, an action potential being initiated when

$V(t)$ reaches a constant threshold V_o. Thus the spontaneous interspike interval density will be

$$f(t) = \frac{k}{\sqrt{4\pi b}} \; t^{-3/2} \exp\left(-k^2/4bt\right) \qquad (4.1.2)$$

where, as in equation (6.4.6)

$$b = \lim_{\Delta t, a \to o} \left(a^2/2\Delta t\right)$$

Gestri (1971) has investigated the properties of the perfect integrator model with a random threshold subject to deterministic inputs, as it is not possible to obtain analytic expressions when the somatic potential is a decreasing function of time, i.e. when the time constant of the leaky integrator is finite. If the threshold $n(t)$ is a random variable defined by

$$n(t) = n_o \qquad\qquad 0 < t < t_i \; ,$$

$$n(t) = n_i \qquad\qquad t_i < t < t_{i+1} \; , \; i = 1, 2, 3, \cdots \; ,$$

where

$\{n_m\}$, $m = 0, 1, 2, \cdots$, is a succesion of finite valued random variables $n_m \geqslant 0$ which need not be independent

$\{t_i\}$, $i = 1, 2, 3, \cdots$, is the output stochastic point process from the perfect integrator
the threshold is synchronized with the output, and at the time of spike generation $V(t)$ is reset to zero and $n(t)$ takes a random value which it holds until time t_{i+1}. This model acts as a good pulse frequency modulator, the information carrying parameter being the spike rate and the random carrier rate (the discharge rate produced by a constant bias input) improves the information transmission efficiency of the model. However, although choice of a suitable $n(t)$ can generate an integral Gaussian probability of firing distribution, the model has either a zero or infinite latency and so cannot be used to model the latency fluctuations.

4.2   Two time constant models with fluctuating threshold

Ten Hoopen et al (1963) have investigated the effect of threshold fluctuations on Rashevsky's 1948 two-time constant model. If the response to a step stimulus of duration T and intensity I is

$$
V(t) = \begin{cases} 0 & ; \quad t < 0 \\ a\ I\ f(t) & ; \quad 0 < t < T \\ a\ I\ (f(t) - f(t-T)) & ; \quad t > T \end{cases} \qquad (4.2.1)
$$

where $f(t) = \exp(-t/\tau_1) - \exp(-t/\tau_2)$ the problem is to find the time when V(t) reaches the threshold $V_0(t) = k + n(t)$, where n(t) is a zero-mean, stationary random function with variance σ^2. The closer V(t) is to k, or $<V_0(t)>$, the greater the probability of V(t) crossing $V_0(t)$:   thus it is useful to introduce

$$
X(t) = V(t) - k
$$

If n(t) is a random step function, with step amplitudes having a Gaussian distribution with variance σ^2 and step times having a Poisson distribution with parameter λ, and c(t) Δt is the conditional probability of a response in the interval (t, t + Δt) given no response in the interval [0, t], c(t) is given by:

$$
c(t) = \lim_{\Delta t \to 0} \frac{\text{Prob}\{n(t+\Delta t) \leqslant X(t+\Delta t) \mid n(u) > X(u);\ u \in [0,t]\}}{\Delta t}
$$

$$
= \lim_{\Delta t \to 0} \frac{1}{\Delta t}\ \lambda\ \Delta t\ \Phi\{(X(t+\Delta t))/\sigma + (1 - \lambda \Delta t)\}
$$

$$
\frac{dX(t)/dt.\phi(X(t)/\sigma).\ \Delta t}{\sigma(1- \Phi(X(t)/\sigma))} \qquad (4.2.2)
$$

where $\quad \Phi(y) = \displaystyle\int_{-\infty}^{y} \frac{1}{\sqrt{2\pi}}\ \exp(-z^2/2)\ dz$

$$
\phi(y) = d\Phi(y)/dy
$$

is the Gaussian distribution and density and $dX(t)/dt \geqslant 0$. Thus

$$c(t) = \lambda\Phi(X(t)/\sigma) - \frac{d}{dt}\ln[1 - \Phi(X(t)/\sigma)] \qquad (4.2.2)$$

The probability density of a response at time t, p(t) Δt, is given by

$$p(t)\,\Delta t = [1 - \Phi(X(0)/\sigma)]\,[1-c(\Delta t)\,\Delta t]\,.$$

$$[1 - c(2\Delta t)\Delta t]\,\ldots\,[1 - c(t-\Delta t)\Delta t]\,c(t)\Delta t$$

and letting $\Delta t \to 0$

$$p(t) = [1 - \Phi(X(0)/\sigma)]\;c(t)\;\exp\left(-\int_0^t c(u)\,du\right) \qquad (4.2.3)$$

Substituting equation (4.2.2) in (4.3.3) gives

$$p(t) = \frac{X(t)\;\phi(X(t)/\sigma)}{\sigma} \qquad (4.2.4a)$$

when the mean rate of steps λ is small and dX(t) is large, and

$$p(t) = \{1 - \phi[X(0)/\sigma]\}\lambda\;\phi[X(t)/\sigma]$$

$$\exp\left(-\int_o^t \lambda\phi[X(u)/\sigma]\,du\right) \qquad (4.2.4b)$$

when λ is large and dX(t)/dt is small. ten Hoopen et al obtained the latency distribution p(t) either numerically or by Monte Carlo simulation, and showed

    a)   as the intensity I increases the latency decreases

    b)   as the intensity increases the standard deviation of the latency decreases

    c)   at high intensities p(t) is skewed

    d)   the earliest responses occurred at times for which V(t) was approximately (k-3σ) - this is the onset of the 'effective time' during which V(t) > (d-3σ).

Use of a different random function, a regular train of steps with a Gaussian amplitude distribution, gave quantitatively similar behaviour.

4.3 Fluctuations in Hodgkin-Huxley variables

The threshold behaviour of a H-H axon depends on the membrane

I-V relations given by the H-H equations and the geometry of the axon.
The threshold for initiation of a membrane action potential in a
space-clamped axon is the line where the I-V-t relation has a negative
slope conductance intercept with the voltage axis; here J_i = 0 and so
below this point at a fixed time $|i_{Na}| < |i_K|$ and the membrane is
repolarized, and above this point $|i_{Na}| > |i_K|$ and so J_i is inward,
depolarizing the membrane. The voltage threshold for a propagated
action potential is higher than for a membrane action potential, as the
inward ionic current J_i at points where $|i_{Na}| > |i_K|$ supplies local circuit
current as well as discharging the membrane capacity. An approximate
analysis of this problem is given in Noble (1966): essentially, at
threshold, the surface integral of the inward current is equal to the
surface integral of the outward membrane current produced by electrotonic
spread. This condition is obtained more rigorously in Hastings (1975)
as a condition for the existence of a homoclinic orbit (a trajectory
in the state space beginning at, and tending, as t→∞, towards the same
equilibrium point, i.e. a propagating action potential) for a simplified
form of the H-H axon equations.

Thus fluctuations in the I-V relation can produce threshold
fluctuations, and if the internal and external ionic concentrations are
constant these fluctuations will be due to changes of the H-H variables,
n, m and h.

A simple probabilistic interpretation of the H-H K^+ channel
kinetics is given by Fitzhugh (1965). The proposed mechanism is that
each K^+ channel or pore is open (conducting) or closed (non-conducting),
the closing of a channel is by the occupation of a site by a charged
gating particle. If a channel has one or more sites occupied it is
non-conducting; particles bind to and leave the sites independently.
If all the channels are open

$$I_K = F(V)$$

and for a linear relation

$$I_K = \bar{g}_K (V - V_K)$$

If a fraction p of the total number of channels is open,

$$0 < \quad p < \quad 1$$

$$I_K \quad = \quad g_K \ (V - V_K) \tag{4.3.1}$$

$$g_K \quad = \quad g_K \ p$$

Let x_j denote the fraction of channels in which j of the sites are occupied. Then

$$x_0 \quad = \quad p$$

$$\sum_{j=0}^{\infty} x_j \quad = 1 \tag{4.3.2}$$

$$dx_0/dt \quad = \quad -\alpha x_0 \ + \ \beta x_1$$

where α and β are voltage dependent rate coefficients of particles binding and leaving the sites. Also,

$$dx_j/dt \quad = \quad \alpha x_{j-1} \ - \ (\alpha + \beta) x_j \ + \ (j + 1)\beta \ x_{j+1} \tag{4.3.3}$$

At a fixed voltage, α and β are constant, and (4.3.3) has the solution

$$x_j \quad = \quad (\alpha/\beta)^j \ \exp \ (-\alpha/\beta) \ / \ j! \ ; \quad j > 0 \tag{4.3.4}$$

which is a Poisson distribution with a mean number of occupied sites/ channel of (α/β). If the voltage is changing, x_j obeys a Poisson distribution with a time varying mean μ related to p by

$$p \equiv x_0 = \exp \ (-\mu)$$

with $\mu(t) = \mu_\infty - (\mu_\infty - \mu_0) \exp \ (-\beta t)$

Thus if there are 4 sites/channel, and $n \equiv \mu$ the H-H equations for I_K result.

Hill and Chen (1972) have derived expressions for the conductance spectral density G(f) for Fitzhugh's model for 2, an arbitrary number and an infinite number of sites/channel. The spectral density is the Fourier transform of the autocovariance function C(t) which is obtained in their appendix 1 as

$$C(t) = M \, n_\infty^4 \, \{ \, [n_\infty + (1 - n_\infty) \exp (-t/\tau) \,]^4 - n_\infty^4 \, \} \quad (4.3.5)$$

where $\tau = 1/(\alpha + \beta)$ and M is the total number of independent channels.

The spectral density is flat at low frequencies, tending to $1/f^2$ at frequencies above $2/\pi\tau$. Equation (4.3.5) is also derived in Stevens (1972).

Stevens also gives an alternative interpretation of the H-H equations, in which the state of a channel is completely specified by n, which is treated as a continuous variable subject to random fluctuations. If the time course of return to equilibrium is the same for deviations Δn, ΔV produced by external perturbations and by spontaneous fluctuations, the first order terms of the linearized H-H equations are

$$g_K \, (V,t) = \bar{g}_K \, n^4 \, (V,\infty) + 4 \, \bar{g}_K \, n^3 \, (V,\infty) \, \Delta n \quad (4.3.6)$$

$$\tau \, d(\Delta n)/dt + \Delta n = a \, \Delta V$$

where a is a constant. Use of a fluctuation - dissipation theorem (see Steven's appendix B) gives a spectral density G(f) with the same shape as obtained from (4.3.5), but the corner frequency between the flat spectral density and the $1/f^2$ asymptote is now $1/2\pi\tau$.

The spectral density of the voltage fluctuations V(f) is obtained from G(f) by

$$V(f) = G(f) \, (V - V_K) \, | \, Z(f) \, |^2 \quad (4.3.7)$$

where Z(f) is the membrane impedance.

Thus the spectral density of the voltage noise has the same shape
if fluctuations in g_K are modelled either by discrete gating changes or
continuous changes in n. The effect of fluctuations in g_K on the
latency and threshold fluctuations of the V-m reduced system of section
3.6 can be found by substituting into (3.6.15) a Langevin force with a
spectral density V(f) given by (4.3.7). Thus the effect of fluctuations
in g_K can be analyzed by obtaining the characteristics of the voltage
fluctuations they produce, and using this as an input to the V-m reduced
system for the threshold.

Changes in Na^+ conductance will have a larger effect on V than
similar changes in K^+ conductance, since the resting membrane potential
is farther from V_{Na} than V_K. Thus the opening of single Na^+ channels
may produce a detectable, depolarizing quantal change in menbrane potential. Such discrete, subthreshold depolarizations have been observed
(del Castillo and Suckling, 1957) and may account for the depolarizing
burst noise responsible for the low (< 1 Hz) frequency component of
the hyperpolarized membrane potential spectral density (Derksen, 1965,
Verveen et al 1967) even though these are TTX insensitive (Siebenga
and Verveen 1970).

If the Na^+ channels are identical and independent, a normalized
conductance s_i can be associated with each channel such that $s_i = 0$
for the closed and 1 for the open channel. If the normalized membrane
conductance is σ (see section 3.6)

$$< \sigma > = < s_i >$$

$$< \sigma^2 > = \frac{1}{N^2} \sum_{i=1}^{N} \sum_{j=1}^{N} < s_i s_j > \qquad (4.3.8)$$

with $< s_i s_j > = < s_i >$ for $i = j$

$$< s_i s_j > = < s_i >^2 \quad \text{for } i \neq j$$

since the channels are independent.

If $\tilde{\sigma}$ is the instantaneous fluctuation at σ from $< \sigma >$

$$< \tilde{\sigma}^2 > = < \sigma^2 > - < \sigma >^2$$

$$= < \sigma > (1 - < \sigma >)/N \qquad (4.3.9)$$

Lecar and Nossal (1971), on the assumption that $\tilde{\sigma}$ is small, obtained for the vicinity of the threshold point

$$d\tilde{\sigma}/dt \simeq - \lambda(V)\tilde{\sigma} + F_\sigma(t) \qquad (4.3.10)$$

where $\lambda(V)$ is the reciprocal of the sodium activation time constant as a function of voltage, defined by equation (3.6.9). The Langevin force representing the conductance fluctuations

$$< F_\sigma(t) F_\sigma(t+\tau) > \simeq \frac{2}{N} \lambda(V) \delta(\tau)\sigma_\infty(V) [1 - \sigma_\infty(V)]$$

was substituted into (3.6.16) and (3.6.17) to obtain the value of the relative spread R_σ of the probability of firing distribution (3.6.15). Numerical evaluation gave $R_\sigma \sim 0.04$ for a single node of Ranvier. This is an order of magnitude larger than the effect of thermal or flicker noise on the probability of firing distribution.

4.4 Models with time - varying threshold

In sections 4.1-4.2 I considered models in which the threshold $V_o(t)$ is the sum of a constant and a stationary random variable with a zero mean, and so the expected value of the threshold $< V_o(t) > = k$. In the reduced Hodgkin-Huxley system considered in sections 3.6 and 4.3, the threshold $V_o(t)$ is not given explicitly but was taken as the two separatrices leading to the unstable saddle point of the V-m plane (point B of Figure 3.3). Here I will treat the spontaneous activity of a class of neural models in which the threshold is an explicit function of time. This class of models has the characteristics:

(a) following an action potential, the threshold is a monotone decreasing function of time

(b)   the input is a deterministic function (here I will consider only a constant input, (E) added to a stationary random variable with a zero mean (N(t))

$$I (t) = E + N(t) \tag{4.4.1}$$

(c)   an action potential is generated whenever

$$I (t) \geqslant V_o(t) \tag{4.4.2}$$

Hagiwara (1954) proposed a model to account for amphibian muscle spindle discharge in which the threshold decayed after an action potential at time t = 0 with a time-course given by

$$V_o(t) = k \exp (\tau/t) \tag{4.4.3}$$

and so $V_o(t) \rightarrow k$ as $t \rightarrow \infty$.   A spike is initiated when

$$(E + N(t)) \geqslant k \exp (\tau/t) \tag{4.4.4}$$

Hagiwara took $N(t)$ to have a Gaussian distribution with standard deviation s and zero mean : however, Enright (1967) has argued that this model is insensitive to the shape of the distribution of N(t). If the distribution were asymetric, it would generate an interspike interval probability density which was virtually indistinguishable from the p.d.f. obtained from an appropriately chosen (E + N(t)) with N(t) Gaussian.   Numerical studies confirmed the insensitivity of the model to the shape of the distribution of N(t).

E and N(t) may be expressed in units of k, and so this model has three parameters E, τ and s.   If time is discrete, and so

$$t_n = n \, \Delta t, \qquad\qquad n = 0, 1, 2, \ldots$$

the probability density for the interspike intervals P (I,t) is given by

$$P (I, t_n) = \Phi(I,t_n) \exp \left(\int_0^\theta \Phi_n (I,\theta) \; d\theta \right) \tag{4.4.5}$$

where $\phi_n = \dfrac{1}{\sqrt{2\pi}} \displaystyle\int_{f_n(t_n,I)}^{\infty} \exp(-x^2/2)\ dx$

$$f_n(t_n,I) = \frac{1}{s}(k\ \exp(\tau/t_n) - I)$$

Numerical calculations generate unimodal, skewed interspike interval p.d.f.'s, with the coefficient of variation and skewness increasing with the mean interval.

ten Hoopen (1964) comments on a modification of this model. Since $N(t)$ has a symmetric distribution and equation (4.4.4) can be rewritten as

$$E \geqslant (k\ \exp(\tau/t) - N(t)) = V_o'\ (t) \tag{4.4.6}$$

where $V_o'(t)$ is a decaying, fluctuating threshold, the model may be investigated using an electronic neural analog with the noise added to the threshold. $N(t)$ is not only characterized by its amplitude distribution but also by its spectral density. The mean discharge rate increases as the upper limit of the noise band width increases. Another aspect of the model is the possibility of correlations between adjacent intervals, or of burst discharges. If $N(t)$ is slow and not arbitrarily reset after an impulse, there will be a tendency for burst discharges. The lack of correlation between adjacent intervals found by Hagiwara could mean either that the noise process is reset to a random value following an impulse, or that the noise process fluctuates more rapidly than the threshold decay time constant.

Buller, Nicholls and Ström (1953) have modelled the same preparation by a threshold which hyperbolically decreases with time

$$V_o(t) = A\tau/t \tag{4.4.7}$$

and additional parameters could be added to give a high, constant threshold during the absolute refractory period. This model may be obtained from the first two terms of the series expansion of Hagiwara's model. Buller (1965) has investigated this model using an electronic neural analog and found that the behaviour of the model was relatively insensitive to the bandwidth and r.m.s. value of $N(t)$. At low constant inputs E_o the average firing rate was greater than for an input $(E_o + N(t))$ - this was associated with an increase in the standard deviation due to an increase in the probability of longer intervals.

A simpler, linear decay of threshold is equivalent to a model proposed by Calvin and Stevens (1965). This model of the effect of synaptic noise on motorneurone discharge probability density had a deterministic depolarizing process represented by

$$V(t) = c_1 t + c_2$$

with c_1, c_2 constant. This ramp depolarization to a fixed threshold is equivalent to a ramp decay of threshold towards a constant resting potential

$$V_o(t) = k(1 - \tau t) \tag{4.4.8}$$

This model generates interspike interval density functions in which the coefficient of variability increases with the mean interval, but does not give an asymptotic approach to a Poisson density as the stimulus intensity decreases: rather, it tends to a regular discharge.

Geisler and Goldberg (1966) have investigated by numerical simulation the properties of a model in which, following an action potential there is an absolute refractory period ($V_o(t) = \infty; 0 < t < 0.7$ msec) followed by a decay of threshold given by

$$V_o(t) = k \left(\exp(-\tau t)/(1 - \exp(-\tau t)) \right) \tag{4.4.9}$$

$$t > 0.7 \text{ msec}$$

This threshold function was introduced by Fuortes and Mantegazzini (1962) to account for the repetitive discharge of eccentric cells of Limulus to maintained depolarizing currents. With appropriate values for the standard deviation and half-power frequency of the random component $N(t)$ of the membrane potential $I(t) = (E + N(t))$, the model generated unimodal interspike interval histograms, with the standard deviation increasing with mean interval. This is in close agreement with the experimental results obtained from cat auditory interneurones (Goldberg, Adrian and Smith, 1964). The model also generates an approximately linear relation between mean depolarization and mean discharge rate. A reduction of the half-power frequency of $N(t)$ reduces the discharge rate, and shifts the standard deviation -

mean interval curves slightly to the left: these changes could be compensated for by an increase in E, the steady input component of the membrane potential.

An increase in the standard deviation of the noise component moves the standard deviation - mean interval curves to the left, so that the discharge is more irregular. This is similar to the effect of increasing the rate constant τ, or decreasing the time constant of the model.

The discharge of the model does not show any correlation between adjacent intervals, and so is a renewal process. The model could be extended to generate a serial dependence between intervals by introducing a hyperpolarizing, decaying after potential into the deterministic component of the membrane potential.

Weiss (1964) has used a model with an exponentially decaying threshold

$$V_o(t) = k \exp (-\tau t) \tag{4.4.10}$$

to model the discharge patterns of auditory nerve fibres. The spontaneous activity, the responses to short pulses, sinusoidal stimuli and click trains were investigated by numerical simulation. The spontaneous interspike interval probability density function generated by a Gaussian input was

$$P(t = j\Delta t) = \prod_{i=1}^{j=1} \Phi(a_i/s) [1 - \Phi(a_j/s)] \tag{4.4.11}$$

where $\Phi(x) = \int_{-\infty}^{x} \frac{1}{\sqrt{2\pi}} \exp (-(y^2/2)) \, dy$

and $a_i = V_o(t = i\Delta t) = k \exp (-\tau i \Delta t)$ is the value of the threshold at the times of spike occurrence.

In the limit for large intervals this probability density approaches that of a Poisson distribution. The probability of a response to a a brief pulse whose duration is much smaller than τ is an integrated Gaussian function, and the response to a periodic train of identical stimulus pulses forms a set of Bernouilli trials with the probability of response - stimulus amplitude curve having the shape of an integrated Gaussian distribution.

Thus there are at least five distinct, explicit formulations for the decay of threshold following an action potential:

1) Hagiwara - equation (4.4.3)
2) Buller - equation (4.4.7)
3) Calvin and Stevens - equation (4.4.8)
4) Geisler and Goldberg - equation (4.4.9)
5) Weiss - equation (4.4.10).

Enright (1967) has compared these different three-parameter models, and has pointed out that although all the models show an increase in variability as mean interval increases, only those of Hagiwara, Buller, Geisler and Goldberg, and Weiss give discharge patterns which asymptotically approach that of a Poisson distribution as the stimulus intensity decreases. Of these models, only that of Weiss, with a simple exponential recovery of threshold, is adequate to account for the observed standard deviation-mean interval relation of many neurones, which shows, on a logarithmic plot, a positive curvature at short intervals, a central linear range and a negative curvature at long intervals.

4.5   References

Blair, E.A. & Erlanger, J.S.:   Responses of axons to brief shocks.
Proc. Roy. Soc. exp. Biol. (N.Y.) $\underline{29}$ 926 (1932).

Buller, A.J., Nicholls, J.G. & Ström, G.:   Spontaneous fluctuations of
excitability in the muscle spindle of the frog.   J. Physiol. $\underline{122}$
409-418 (1953).

Buller, A.J.:   A model illustrating some aspects of muscle spindle
physiology.   J. Physiol. $\underline{179}$ 402-16 (1965).

Calvin, W.H. & Stevens, C.F.:   A Markov process model for neuron
behaviour in the interspike interval.   Proc. 18th Ann. Conf. Engineering
in Medicine and Biology. $\underline{7}$ 118 (1965).

del Castillo, J. & Suckling, E.E.:   Possible quantal nature of subthreshold responses at single node of Ranvier.   Fed. Proc. $\underline{16}$ 29
(1957).

Derksen, H.E.:   Axon membrane voltage fluctuations.   Acta Physiol.
Pharmacol. Neerl. $\underline{13}$ 373 (1965).

Enright, J.T.:   The spontaneous neuron subject to tonic stimulation.
J. theoret. biol. $\underline{16}$ 54-77 (1967).

Fitzhugh, R.:   A kinetic model of the conductance changes in nerve
membrane.   J. cell. comp. Physiol. $\underline{66}$ 111-118 (1965).

Fuortes, M.G.F. & Mantegazzini, F.:   Interpretation of the repetitive
firing of nerve cells.   J. Gen. Physiol. $\underline{45}$ 1163-79 (1962).

Goldberg, J.M., Adrian, H.O. & Smith, F.D.:   Response of neurons of the
superior olivary complex of the cat to acoustic stimuli of long duration.
J. Neurophysiol. $\underline{27}$ 706-749 (1964).

Geisler, C.D. & Goldberg, J.M.:   A stochastic model of the repetitive
activity of neurons.   Biophys. J. $\underline{6}$ 53-69 (1966).

Gestri, G.:   Pulse frequency modulation in neural systems: a random
model.   Biophys. J. $\underline{11}$ 98-109.

Hagiwara, S.:   Analysis of interval fluctuations of the sensory nerve
impulse.   Japanese J. of Physiol. $\underline{4}$ 234-40 (1954).

Hastings, S.P.:   Some mathematical problems from neurobiology preprint
1975 to appear in Americal Math. Monthly.

Hill, T.L. & Chen, Y.:   On the theory of ion transport across the nerve
membrane.   IV.   Noise from the open-close kinetics of K^+ channels.
Biophys. J. $\underline{12}$ 948-959 (1972).

ten Hoopen, M.:   On an impulse interval generating mechanism.   Japanese
J. of Physiol. $\underline{14}$ 607-614 (1964).

ten Hoopen, M., den Hertog, A. & Reuver, H.A.:   Fluctuation in excitability
of nerve fibres - a model study.   Kybernetik $\underline{2}$ 1-8 (1963).

Householder, A.S. & Landahl, H.B.:   Mathematical Biophysics of the
Central Nervous System.   Prinapia Press Bloomington, Indiana (1945).

Landahl, H.D.:   Theory of the distribution of response times in nerve
fibres.   Bull. Math. Biophys. $\underline{3}$ 141 (1941).

Lecar, H. & Nossal, R.:   Theory of threshold fluctuations in nerves II
Analysis of various sources of membrane noise.   Biophys. J. $\underline{11}$ 1068-
1084 (1971).

Noble, D.:   Applications of Hodgkin-Huxley equations to excitable tissues.
Physiol. Rev. $\underline{46}$ 1-50 (1966).

Pecher, C.: La fluctuation d'excitabilite de la fibre nerveuse. Arch. Int. Physiol. biochem. $\underline{49}$ 129 (1939).

Rashevsky, N.: Mathematical Biophysics. Chicago Univ. Press. Chicago 1948. Reprinted and extended Dover, NY 1960.

Siebenga, E. & Verveen, A.A.: Noise voltage of axonal membrane. Pflugers Archiv. $\underline{318}$ 267 (1970).

Stevens, C.F.: Inferences about membrane properties from electrical noise measurements. Biophys. J. $\underline{12}$ 1028-1047 (1972).

Verveen, A.A.: On the fluctuation of threshold of the nerve fibre. In 'Structure and Function of the Cerebral Cortex', ed. D.B. Tower and J.P. Schade 282-288. Elsevier, Amsterdam (1960).

Verveen, A.A.: Fluctuation in excitability. Thesis, Univ. of Amsterdam (1961).

Verveen, A.A., Derksen, H.E. & Schick, K.L.: Voltage fluctuations of neural membrane. Nature $\underline{216}$ 588 (1967).

Weiss, R.F.: A model for firing patterns of auditory nerve fibres. Massachussetts Institute of Technology - Research Laboratory of Electronics Technical Report No. 418 March 1964.

5   STATISTICAL PROPERTIES OF A SPIKE TRAIN

The output process or axon of a large number of neurones is long and thin, and so may be considered as an infinite, cylindrical cable. The conducting core of the cable is the axoplasm, which is an ohmic fluid with a specific resistivity ρ, and the insulating sheath of the cable is the membrane (with the myelin sheath for a myelinated axon) with a capacitance and mechanisms through which an ionic current density J can pass. The pattern of voltage V with distance x and time t in an axon of diameter D is given by the partial differential equation

$$\frac{D}{4\rho} \frac{\partial^2 V}{\partial x^2} = c \frac{\partial V}{\partial t} + J$$

which, for a linear membrane, gives equation (3.0.1). In this case $J = V/R$ where R is the membrane resistance. The steady-state solution of this equation to a constant input is given by equation (3.0.4) as

$$V = V(0) \exp(-x/\lambda)$$

and so there is an exponential decay of voltage with distance, with voltage falling to $1/e$ of its value at $x = 0$ in the space-constant λ.

Hodgkin and Rushton (1946) showed that

$$\lambda = (G_a/G_m)^{\frac{1}{2}}$$

where G_a and G_m are the membrane and axoplasmic conductances/unit length of axon. Since the axoplasm may be considered as a weak salt solution, ρ, and hence G_a is small. The length constant increases as the square root of the fibre diameter, and so even for large diameter nonmyelinated fibres λ is of the order of a few mm. Thus any steady

voltage will decay rapidly with distance: transients will decay
more rapidly with distance because most of the current will be
shunted through the membrane capacity.   The transfer function L(s)
for a linear cable is given by Stein (1967) as

$$L(s) \propto \exp\ (-\sqrt{1 + s\tau}\ x/\lambda)\ /\ \sqrt{1 + s\tau}$$

where s is a complex variable and τ = RC. This has a sharp frequency
cutoff and so high frequency signals are rapidly attenuated.

Thus information may not be transmitted over long distances in
the nervous system by means of a continuously varying voltage, but only
by means of regenerative all-or-none events or action potentials.   This
argument only holds for long, thin processes: short processes may
transmit information by analog signals (see Bush and Roberts, 1968;
Werblin and Dowling, 1969;   Rall, 1970;   Shepperd, 1970, for examples).

Since the only way a neurone can transmit information about
rapidly varying signals over a long distance is by a series of all or
none events the shape of the action potential is irrelevant.   An action
potential may be considered either as a brief pulse, taken in the limit
as a Dirac delta function, or as a point event.   Thus, a spike train
may be considered either as a train of impulses

$$x(t)\ \ =\ \ \sum_i \delta(t - t_i)$$

or as a realization of a stochastic point process. The justification
for these representations is given in Stein (1970, 1972). If the spike
train is treated as a stochastic point process, that is, as a stochastic
process specified by events labelled by the random value of a continuous
parameter (time), it may be characterized by the cumulative number n(t)
of events which have occurred up to the time t.   n(t)   is an increasing,
staircase function. Bartlett (1963) defined a function J(f) analogous
to the periodogram of a stationary continuous process

$$J(f)\ \ =\ \ \frac{1}{\sqrt{\pi T}}\ \int_o^T\ \exp\ (jtf)\ dn(t)$$

which when multiplied by its complex conjugate gives the spectral
density of the point process.   Since

$$dn(t)\ \ =\ \ \sum_i\ \delta(t - t_i)\ dt$$

treating the spike train as a point process is equivalent to treating
it as a train of Dirac delta functions (French and Holden, 1971).

There are two, non-equivalent ways of treating a train of
impulses - in terms of t_i, the times of occurrence, or in terms of the
ordered sequence $\{\tau_i\}$ of inter-event intervals, where $\tau_i = t_{i+1} - t_i$

It is natural to examine the time - domain properties of an impulse
train in terms of intervals τ_i and the frequency - domain properties
in terms of the times t_i. A good review of methods of characterizing
impulse trains is that of Moore, Perkel and Segundo (1966).

5.1 Stationarity

For a stationary, stochastic point process, the probability
distributions which underly the observed sample of intervals do not
depend on the time of observation. This means that the time origin
(t = 0) is at an arbitrary point : it is usually sensible to take the
time origin at the occurrence time of an arbitrary event in order to
avoid transient effects : this is synchronous sampling (Lawrence, 1972).

Experimental data obtained from neurones will never be strictly
stationary; however, it may be modelled by a stationary process if
within a physiologically relevant segment of the spike train there is
no obviously non-stationary behaviour. Most parameters of the probability density functions of spike trains are relatively insensitive to
small deviations from stationarity, and large non-stationarities will
be obvious. The distinction between a stationary process with serial
dependence and a non-stationary process is arbitrary.

Nonstationarity may be modelled either by making the parameters
of a stationary model time varying (e.g. see section 6.8) or by combining a deterministic, time varying process with a stationary,
stochastic process to generate the model. This latter method has been
extensively used to model receptor neurones : variability is investigated
using constant inputs and the dynamics are investigated by using
periodic inputs and averaging. An alternative approach is to use a
broad-band stochastic input - the stochastic response may be analysed
in terms of the stochastic components due to intrinsic variability and
the stochastic components related to the input by means of the coherence

function. This approach is discussed in section 5.4 and French, Holden and Stein (1972).

Most neurones respond to a constant input by a discharge which adapts towards a steady-state level. Also, changes in neural activity appear to be biologically more relevant than steady-state discharges. This might suggest that models of stationary activity, although perhaps they might give insights into the mechanisms of the nervous system, may not contribute to understanding the functional aspects of neural activity. I do not accept this, as:

(a) Although non-stationary behaviour is more interesting and biologically relevant, models of stationary activity represent a starting point in approaching the more complex case of non-stationary behaviour in neural networks. Since biologically simple problems are often mathematically quite complex, it seems appropriate to start by using simple, perhaps biologically naive, models of stationary behaviour.

(b) A number of non-stationary problems may be approached by making them quasi-stationary by suitable choice of the time span considered. A slow non-stationarity may vanish in short time spans, and a rapid non-stationarity in long time spans.

(c) changes in neural activity may not be nonstationary in the sense of some kind of trend in system parameters, but rather represent abrupt changes between stationary, locally stable patterns of behaviour.

I have only discussed stationarity in rather general terms : a more detailed discussion of the properties of stationary point processes is given in Chapter 4 of Srinivasan (1974).

5.2   Interval Densities and Distributions

If the spike train

$$x(t) = \sum_{i}^{N} \delta(t - t_i) \tag{5.2.1}$$

is stationary, its statistical properties in the time domain are characterized by the properties of the ordered sequence of intervals

$\tau_i = t_{i+1} - t_i$. If $f(\tau)$ dτ is the probability that the interval is between τ and $\tau + $ dτ, where dτ is small, $f(\tau)$ is the probability density function of the interspike interval distribution. The parameters usually given are the mean μ which may be estimated from

$$\mu = (t_N - t_o)/N \qquad (5.2.2)$$

standard deviation σ from

$$\sigma^2 = \frac{1}{N-1} \sum_{i=1}^{N} (\tau_i - \mu)^2 \qquad (5.2.3)$$

and coefficient of variation σ/μ. The function $f(\tau)$ can be readily estimated from the interspike interval histogram.

The p.d.f. $f(\tau)$ has the properties

$$f(\tau) \rightarrow 0 \text{ as } \tau \rightarrow 0$$
$$f(\tau) \rightarrow 0 \text{ as } \tau \rightarrow \infty$$

$$\int_0^\infty f(\tau) \, d\tau = 1$$

This follows from the existence of the absolute refractory period, and the lack of very long intervals. Otherwise, $f(\tau)$ can, in principle, have any shape; however, some simple distributions often occur :

A. Poisson

$$f_A(\tau) = \mu \exp(-\mu \tau) \qquad (5.2.4)$$

B. Gaussian

$$f_B(\tau) = \frac{1}{\sigma\sqrt{2\pi}} \exp(-(\tau - \mu)^2/2\sigma^2) \qquad (5.2.5)$$

C. Gamma

$$f_C(\tau) = \frac{\kappa}{\mu}^\kappa \frac{\tau^{\kappa-1} \exp(-\kappa\tau/\mu)}{\Gamma(\kappa)} \qquad (5.2.6)$$

where $\Gamma (\kappa) = \int_0^\infty x^{\kappa-1} \exp (-x) dx, \kappa > 0$

The Poisson distribution arises when the probability of an impulse occurring in a short time Δt is constant : see section 6.6 for a derivation and section 8.1 for the occurrence of Poisson distributions resulting from superposition of independent events. When $\kappa = 1$, the Gamma distribution gives a Poisson distribution. When κ is large the Gamma distribution tends to a Gaussian distribution - see sections 2, 6 and 7 for models generating Gamma distributions.

There are a number of functions which may be obtained from $f(\tau)$: The cumulative distribution function

$$F(\tau) = \int_0^\tau f(x) dx \qquad (5.2.7)$$

Its complement, the survivor function $G(\tau)$

$$G(\tau) = 1 - F(\tau) \qquad (5.2.8)$$

and the conditional probability function

$$\phi(\tau) = f(\tau) / G(\tau) = f(\tau) / (1 - F(\tau)) \qquad (5.2.8)$$

which gives the probability that an impulse will occur between times τ and $\tau + d\tau$ given that no impulses have occurred since $\tau = 0$.

If adjacent intervals are independent the interval density function $f(\tau)$ or any of (5.2.7) - (5.2.9) completely characterize the process : the process is then a renewal process (see Cox, 1962).

If $h(t) dt$ is the probability that an impulse occurs between t and $t + dt$ given an impulse occurred at $t = 0$ (impulses may have occurred at times τ_1, $\tau_1 + \tau_2$ ... which are less than t), $h(t)$ is the renewal density.

$$h(t) = \lim_{\Delta t \to 0+} \frac{\text{Prob \{ one, or more events in } (t, t+\Delta t)\}}{\Delta t} \qquad (5.2.9)$$

For a Poisson process, $h(t) = 1/\mu$, for any renewal process

$$\lim_{t \to \infty} h(t) = \frac{1}{\mu} \qquad (5.2.10)$$

The renewal density is the derivative of the renewal function H(t)

$$H(t) = E(n(t)) \qquad (5.2.11)$$

where n(t) is the number of impulses in time t. For an ordinary renewal process (which in this context means that the time origin was at an arbitrary event) the renewal density and f(t) are related by

$$h(t) = f(t) + \int_0^t h(t - u)\ f(u)\ du \qquad (5.2.12)$$

or in terms of Laplace transforms (see section 5.4)

$$h(s) = f(s)/1 - f(s) \qquad (5.2.13)$$

where $f(s) = \int_0^\infty \exp(-sx)\ f(x)\ dx$

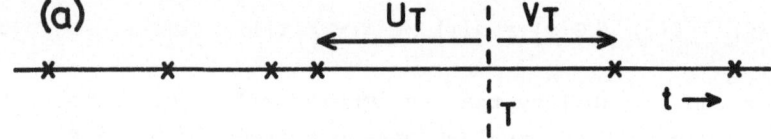

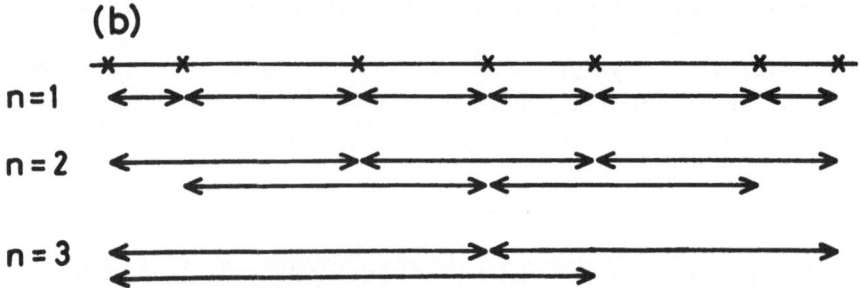

Figure 5.1. Properties of a point process. (a) Backward recurrence time U_T and forward recurrence time V_T (b) n^{th}-order intervals.

A useful result in renewal theory is the probability density function of the forward (or backward) recurrence time. If T is an arbitrary time remote from the origin t = 0 (see Figure (5.1)) the times to the first preceding and following impulses are the backward U_T and forward V_T recurrence times. The p.d.f. of the recurrence times U_T and V_T can be shown to be identical and given by

$$U(t) = \frac{1 - F(t)}{\mu} \qquad\qquad (5.2.14)$$

When adjacent intervals are not independent, i.e. a non-renewal process, the nature of any serial dependence between intervals may be characterized either by using higher-order intervals (the intervals between subsequent but non-adjacent events) or by the joint distributions of pairs or m-tuples of first order intervals.

The nth-order interval is the sum of n first-order intervals (see Figure (5.1.b)). For a renewal process, the probability density of the nth order interval $f_n(\tau)$ is given by the recursive convolution

$$f_{n+1}(\tau) = \int_0^\infty f_n(x)\ f(\tau - x)\ dx \qquad\qquad (5.2.15)$$

For any stationary point process the infinite sum over n of the nth order probability density functions gives the renewal density h(t) density defined in equation (5.2.9).

$$h(t) = \sum_{n=1}^\infty f_n(t) \qquad\qquad (5.2.16)$$

Thus if the spike train is taken as a train of Dirac delta functions h(t) is similar to an autocorrelation function, and is sometimes called the autocorrelation function (Gerstein and Kiang, 1960). However, to obtain a function with the same properties as an autocorrelation function the mean density should be subtracted from h(t) and the resulting function normalized.

The recursive convolution of equation (5.2.15) suggests an interesting class of renewal processes: those whose interval probability density functions are form-invariant (apart from a linear scale change) under self convolution. These are the stable distributions (Levy, 1940; Gnedenko and Kolmogorov, 1968) and they occur as solutions of the first passage time of diffusion models (see section 7), and may be associated

with the activity of a neurone in a pathway which is preserving
information (Holden, 1975; also see section 13).

Equation (5.2.15) does not hold for non-renewal processes, where
adjacent intervals are not independent. A different approach to serial
dependence in a spike train is by the serial correlation coefficient,
which summarizes the joint interval distribution of pairs of intervals.
Rodieck, Kiang and Gerstein (1962) introduced a scatter diagram
representation of the joint interval density for adjacent intervals, in
which the (n+1)th interval is plotted against the nth interval. If
adjacent intervals are independent, the frequency distribution along the
ordinate is the same for each abscissa value and vice versa. This may
be quantified by the first order serial correlation coefficient ρ_1.

$$\rho_1 = \sum_{i=1}^{n-1} (\tau_i - \mu)(\tau_{i+1} - \mu) / \sum_{i=1}^{n} (\tau_i - \mu)^2 \qquad (5.2.17)$$

which ranges from -1 to +1. If adjacent intervals are independent $\rho_1 = 0$.
The serial correlation coefficient may be estimated for any order j

$$\rho_j = \sum_{i=1}^{n-1} (\tau_i - \mu)(\tau_{i+j} - \mu) / \sum_{i=1}^{n} (\tau_i - \mu)^2 \qquad (5.2.18)$$

and so an ordered set of serial correlation coefficients $\{\rho_1, \rho_2, \rho_3, \cdots\}$
can be obtained. This is the serial correlogram. If $\rho_n = \rho_1^n$ the spike
train is a realization of a Markov process, in which the state of the
neurone (measured by the duration of the interspike interval) depends on
its previous state (measured by the duration of the preceding interval)
and not on the history of earlier intervals. Thus if ρ_1 is negative,
the serial correlogram of a Markov process will show a damped oscillation.

5.3 Spectral densities of point processes

If $x(t)$ is a stationary stochastic process which is varying
continuously with time the autocovariance function $\rho(\tau)$ is defined by
(Lee, 1960)

$$\rho(\tau) = \lim_{T \to \infty} \frac{1}{2T} \int_{-T}^{T} x(t)\, x(t + \tau)\, dt \qquad (5.3.1)$$

This is the mathematical expectation of the product of the values which
the stationary random process assumes at instants separated by τ, i.e.

if x(t) is ergodic

$$\rho(\tau) \;=\; E\{x(t)\, x(t+\tau)\}\qquad\qquad (5.3.2)$$

and is related to the autocorrelation function $\gamma(\tau)$ by the standard deviation σ

$$\gamma(\tau) \;=\; \sigma^2\, \rho(\tau)\qquad\qquad (5.3.3)$$

The autocovariance $\rho(\tau)$ and the autocorrelation $\gamma(\tau)$ are even functions of τ.

From the Wiener-Khinchine relation, the power spectrum is the Fourier transform of the autocovariance function

$$S(\omega) \;=\; \int_{-\infty}^{\infty} \rho(\tau)\, \exp(-j\omega\tau)\, d\tau\qquad\qquad (5.3.4)$$

which is real and even and defined for negative as well as positive frequencies. Since negative frequencies do not have any physical meaning the one sided power spectrum $X(\omega)$ defined only for positive frequencies may be obtained by folding $S(\omega)$ over the zero frequency axis to obtain

$$X(\omega) = \begin{cases} 2\,S(\omega) & 0 < \omega < \infty \\[6pt] 0 & f < 0 \end{cases}\qquad\qquad (5.3.5)$$

The problem now is to obtain expressions for $X(\omega)$ when $x(t)$ is a spike train treated as a stochastic point process or a series of Dirac delta functions. Lewis (1970) has reviewed aspects of the theory and application of the spectral analysis of series of events. If a discrete-time point process is considered, in which events occur/do not occur at integer time values $k = 1, 2, 3$, the sequence $\{\delta_k\}$ is a time series of binary valued ($\delta_k = 1$ means an event occurred at time k, $\delta_k = 0$ means no event at time k) random variables.

The counting process N_k

$$N_k \;=\; \sum_{m=1}^{k} \delta_m , \qquad (m = 1, 2, 3, \ldots)\qquad\qquad (5.3.6)$$

is associated with the process $\{\delta_k\}$ and the sequence $\{\tau_j\}$ of interevent intervals. The sequences $\{\delta_k\}$ and $\{\tau_j\}$ are related by

$$S_n = \sum_{j=1}^{n} \tau_j > k \quad \text{iff } N_k < n \, , \quad (n = 1, 2, 3 \, ...) \qquad (5.3.7)$$

or $\qquad \text{Prob}\{S_n > k\} = \text{Prob}\{N_k < n\}$

and so the properties of the sequences $\{\delta_k\}$ and $\{\tau_j\}$ are not equivalent: in fact stationarity of $\{\delta_k\}$ need not imply stationarity of $\{\tau_j\}$. Thus $\{\delta_k\}$ is the more basic process. This suggests that spectral analysis of a spike train should be in terms of $x(t) = \Sigma \, \delta(t - t_i)$ rather than the intervals (French and Holden, 1971; but see Shapley (1969, 1971) for an application of the spectral analysis of the reciprocal of the intervals or instantaneous rate). Brillinger (1972) discusses the spectral analysis of the interval functions $\{\tau_j\}$.

For the discrete-time process $\{\delta_k\}$ there is a correlation $\{\rho_k\}$ and spectral distribution $F(\omega)$

$$\rho_k = \frac{E(\delta_m \, , \, \delta_{m+k}) - E(\delta_m) \, E(\delta_{m+k})}{\text{var } (\delta_m)}$$

$$= \int_{-\pi}^{\pi} \cos (k\omega) \, dF(\omega) \qquad (5.3.8)$$

and so generalizing for a continuous-time point process via the counting process N_t, the autocovariance density of ΔN_t which is $(N_{t+\Delta t} - N_t)$ is given by

$$\gamma(\tau) = \lim_{\Delta t \to 0} \text{cov} \{\Delta N_t, \Delta N_{t+\tau}\}$$

$$= \mu\{h(\tau) - \mu\} + \mu\delta(\tau) \qquad (5.3.9)$$

where μ is the mean rate and $h(\tau)$ is the renewal density or intensity defined by equation (5.2.9). This is the analog of the correlation sequence $\{\rho_k\}$ of the discrete time process.

The spectrum is then

$$X(\omega) = \frac{1}{2\pi} \int_{-\infty}^{\infty} \mu \, \delta(\tau) \, \exp(-j\omega\tau) \, d\tau$$

$$+ \frac{1}{2\pi} \int_{-\infty}^{\infty} \gamma(\tau) \, \exp(-j\omega\tau) \, d\tau \qquad (5.3.10)$$

$$= \frac{\mu}{2\pi} + \frac{\mu}{2\pi} \int_{0}^{\infty} h(\tau) \, \exp(j\omega\tau) + \frac{\mu}{2\pi} \int_{0}^{\infty} h(\tau) \, \exp(-j\omega\tau) \, d\tau$$

Since $h(\tau) = \sum\limits_{n=1}^{\infty} f_n(\tau)$. (see 5.2.16)

$$X(\omega) = \mu \left\{ 1 + \sum_{n=1}^{\infty} F_n(j\omega) + F_n(-j\omega) \right\} \qquad (5.3.11)$$

where $F_n(\omega)$ is the Fourier transform of $f_n(\tau)$, the nth order interval density. Equation (5.3.11) is obtained in a series of papers by Beutler and Leneman (1966a, b; 1968), and does not require $x(t)$ to be renewal.

Coggshall (1973) has applied equation (5.3.11) to obtain the spectra of some impulse trains of interest:

1. The Periodic Spike Train

For a strictly periodic sequence of impulses

$$x(t) = \sum_{m=-\infty}^{\infty} \delta(t - mT)$$

with rate $\mu = 1/T$ the probability density is a delta function and

$$X(\omega) = \frac{2\pi}{T^2} \sum_{n=-\infty}^{\infty} \delta\left(\omega - \frac{2\pi n}{T}\right) \qquad (5.3.12)$$

i.e. a set of delta functions at frequencies which are multiples of μ.

2. The Randomly jittered spike train

If the intervals are represented by a constant T to which is added a small independent random jitter

$$t_n = nT + U_n$$

where U has a distribution $\phi(u)$ and a zero mean

$$f_n(\tau) = \delta(\tau - nT) * \phi(\tau) * \phi(\tau)$$

where * denotes convolution.

Then

$$X(\omega) = \frac{1}{T} \left\{ 1 + |\Phi(\omega)|^2 \sum_{\substack{n=-\infty \\ n \neq 0}}^{\infty} \exp(-j\omega nT) \right\} \qquad (5.3.13)$$

$$= \frac{1}{T} \left\{ 1 - |\Phi(\omega)|^2 + \frac{2\pi}{T} |\Phi(\omega)|^2 \sum_{n=-\infty}^{\infty} \delta\left(\omega - \frac{2\pi n}{T}\right) \right\}$$

where $\Phi(\omega)$ is the Fourier transform of $\phi(\tau)$. This expression has also been obtained by Nelsen (1964), and gives a mixed spectrum, the jitter giving a continuous spectrum and the periodicity a line spectrum.

3. The Poisson Spike Train

When the interspike interval density is given by equation (5.2.4) and adjacent intervals are independent

$$f_n(\tau) = \mu \frac{(\mu\tau)^{n-1}}{(n-1)!} \exp(-\mu\tau)$$

which gives the well known continuous spectrum

$$X(\omega) = \mu(1 + 2\pi \mu \delta(\omega)) \qquad (5.3.14)$$

4. The Gamma Density Spike Train

When the interspike interval density is given by equation (5.2.6) and adjacent intervals are independent

$$f_n(\tau) = \frac{\kappa/\mu}{(\kappa n-1)!} (\kappa\tau/\mu)^{\kappa n-1} \exp(-\kappa\tau/\mu)$$

and

$$X(\omega) = \mu \left\{ 1 + \frac{2\pi}{\mu} \delta(\omega) + \frac{1}{\mu} \sum_{r=1}^{\kappa-1} \epsilon^r \frac{2\kappa(1-\epsilon^r)/\mu}{\omega^2 + \kappa^2 (1-\epsilon^r)^2/\mu^2} \right\}$$

$$(5.3.15)$$

where $\varepsilon = \exp (2\pi j/\kappa)$.

Gestri (1972) has calculated the spectral density for $\kappa = 2,3,4$, and Koles (1970) has used Nelsen's (1964) method to compute the spectral density of spike trains with a Gamma density.

5. Sinusoidally modulated, regular spike train

Bayly (1968; 1969) has obtained an expression for the spectral density of a spike train generated by a perfect integrator model neurone (see section 3.2) in response to a sinusoidal input. If α is a random variable giving the interval between the arbitrary time origin and the first preceding spike (a backward recurrence time) the pulse train can be expressed as the infinite sum of its cosine components, the terms of the cosine expansion giving the terms of the line spectrum of the pulse train. The cosine expansion of the pulse train $P(t,\alpha)$ is

$$
\begin{aligned}
P(t,\alpha) = & \frac{I}{2\pi} \omega_o + \frac{m_i I}{r} \cos (\omega_m + \theta) \\
& + \frac{\omega_o}{\pi} I \sum_{k=1}^{\infty} \sum_{n=-\infty}^{\infty} J_n \left(\frac{km_i 2\pi}{r\, \omega_m} \right) \left[1 + \frac{n\, \omega_m}{k\, \omega_o} \right] . \\
& \cos \left\{ (k\, \omega_o + n\, \omega_m)t + k\, \omega_o \alpha + n\theta \right. \\
& \left. - \left[\frac{k\, m_i\, 2\pi}{r\, \omega_m} \right] \sin (\theta - \omega_m \alpha) \right\}
\end{aligned}
\qquad (5.3.16)
$$

where $m_i \cos (\omega_m t + \theta)$ is the modulating signal with amplitude m_i, frequency ω_m and phase θ, ω_o is the carrier frequency or discharge rate in the absence of the modulating signal, m_i/r is the modulation depth in impulses/sec and J_n is a Bessel function of the first kind of order n. I is the area of a single impulse. Thus the line spectrum has components at

a) A D.C. component measuring the mean rate of spikes and their area

b) a component at the modulating frequency ω_m, with an amplitude proportional to the amplitude of modulation

c) a component at a frequency equal to the mean rate of spikes, or carrier frequency ω_o.

d) components of all integral multiples of ω_o

e) a series of sidebands at integer multiples of the modulating frequency above and below each carrier band, i.e. (n ω_o ± m ω_m) with n and m positive integers.

6. Sinusoidally modulated random spike train

Knox (1969) has derived an expression equivalent to equation (5.3.14) for a random Poisson spike train and has considered the case when the mean rate μ is a function of time

$$\mu(t) \quad = \quad \mu_o + \mu_m \cos \omega_m t$$

For rectangular spikes of height I and duration T_o.

$$X(\omega) \quad = \quad \mu_o \; I^2 \; T^2 \; Sa^2 \; (\omega T_o/2) \; . \; \{1 + \mu_o \left[2\pi \; \delta(\omega) \right.$$

$$\left. + \; \pi \; \frac{\mu_m^2}{2 \; \mu_0^2} \quad \delta \; (\omega \pm \omega_m) \right] \; \} \qquad\qquad (5.3.17)$$

where

$$Sa(x) \quad = \quad (\sin x)/x$$

The spectra given above by equations (5.3.12-17) all extend to an arbitrarily high frequency. For spikes of shape g(t) the spectrum S(ω) of the spike train

$$s(t) \quad = \quad \Sigma \; g(t - t_i)$$

can be obtained from the spectrum X(ω) of the impulse train

$$x(t) \quad = \quad \Sigma \; \delta(t - t_i)$$

by $$S(\omega) \quad = \quad X(\omega) \; | \; G(\omega) \; |^2$$

where G(ω) is the Fourier transform of g(t).

ten Hoopen (1974) has reviewed the power spectra of spike trains treated as uni-variate stochastic point processes, and considers the conditions when frequency-domain analysis is more useful than an equivalent time-domain analysis.

5.4 Input-Output relations

I have considered the properties of a single spike train in the time and the frequency domain: another kind of property of interest is the relationship between two signals, one or both of which may be a spike train. If $x(t)$ is a continuous signal and is the input to a neural model, and $y(t)$ is the spike train output from the model, the relationship between $x(t)$ and $y(t)$ will characterize the signal transmission properties of the model.

If I take the input function $x(t)$ and the output $y(t)$ to be continuous, and consider a linear noise-free system, $y(t)$ is related to $x(t)$ by the convolution integral:

$$y(t) \;=\; \int_0^t x(\tau)\, g(t,\tau)\, d\tau \qquad\qquad (5.4.1)$$

and if the system is also time invariant a function $g(t)$ exists such that

$$y(t) \;=\; \int_0^t x(\tau)\, g(t - \tau)\, d\tau \qquad\qquad (5.4.2)$$

where $g(t)$ is the weighting function of the system

Treatment of equation (5.4.2) is greatly simplified if it is examined in the s domain, where s is a complex variable. This may be achieved by use of the Laplace Transform (Smith, 1966). The Laplace Transform is a linear transform and is defined by:

$$\text{L.T.}\,\{y(t)\} \;=\; Y(s)$$

$$= \int_0^\infty y \exp(-st)\, dt \qquad\qquad (5.4.3)$$

and its inverse is given by

$$\text{L.T,}^{-1}\,\{Y(s)\} \;=\; \frac{1}{2\pi} \int_{c-j\infty}^{c+j\infty} Y(s) \exp(ts)\, dt \qquad\qquad (5.4.4)$$

where c is a constant which must be defined to ensure convergence in this contour integration. Thus $y(t)$ and $Y(s)$ form a Laplace Transform pair

$$y(t) \leftrightarrow Y(s)$$

In practice it is generally not necessary to evaluate the forward and inverse transforms as they are available in tables for most commonly encountered functions.

The most useful properties of the Laplace Transform are:

    a)   Linearity:

$$k\ y(t) \leftrightarrow k\ Y(s)$$
$$y(t) + f(t) \leftrightarrow Y(s) + F(s) \tag{5.4.4a}$$

    b)   Time translation:

$$y(t - \tau) \leftrightarrow \exp(-\tau s)Y(s) \tag{5.4.4b}$$

    c)   Convolution:

$$\int_0^t y(t - \tau)\ f(\tau)\ d\tau \leftrightarrow Y(s)F(s) \tag{5.4.4c}$$

Application of the convolution property to the Laplace Transform of equation (5.4.2) yields

$$Y(s) = G(s)\ X(s) \tag{5.4.5}$$

When the input function $x(t)$ is a unit impulse or Dirac delta function $\delta(t)$ such that

$$\int_{-\infty}^{\infty} \delta(t) = 1 \tag{5.4.6}$$

$$\delta(t) = 0,\ t \neq 0$$

the Laplace Transform of $x(t)$ is unity and so equation (5.4.5) now gives

$$Y(s) = G(s)$$

Thus the Laplace Transform of the response to a unit impulse is the Laplace Transform of the weighting function $g(t)$ of the system. This is the transfer function, $G(s)$ of the system which is only defined for a linear, time invariant and noise-free system. The transfer function completely characterizes the input-output relations of the system.

An alternative representation of the transfer function is the frequency response function $G(\omega)$ given by

$$G(\omega) \quad = \quad \int_0^\infty g(t) \quad \exp(-j\omega t) dt \qquad (5.4.7)$$

where in place of the Laplace Transform I am using the one-sided Fourier Transform.

The frequency response function is complex and in engineering contexts its magnitude is generally plotted as the dimensionless ratio gain in dB. However, although this convention is often used in neurophysiology (e.g. Poppele and Bowman, 1970) it is inappropriate when the input and output signals do not have the same dimensions.

When the input $x(t)$ to the system is a realization of a stationary stochastic process one can define the autocovariance function $\rho_{xx}(\tau)$ of $x(t)$ as (Jenkins and Watts, 1968; Lee, 1960)

$$\rho_{xx}(\tau) \quad = \quad \lim_{T \to \infty} \quad \frac{1}{2T} \quad \int_{-T}^{T} \quad x(t) \; x(t + \tau) \; dt \qquad (5.4.8)$$

and the autocovariance function $\rho_{yy}(\tau)$ of the response $y(t)$ as

$$\rho_{yy}(\tau) \quad = \quad \lim_{T \to \infty} \quad \frac{1}{2T} \quad \int_{-T}^{T} \quad y(t) \; y(t + \tau) \; dt$$

and the forward cross-covariance $\rho_{xy}(\tau)$ from the input to the output

$$\rho_{xy}(\tau) \quad = \quad \lim_{T \to \infty} \quad \frac{1}{2T} \quad \int_{-T}^{T} \quad x(t) \; y(t + \tau) \; dt \qquad (5.4.9)$$

These covariance functions are the mathematical expectation of the product of the values which the stationary random processes assume at instants separated by τ. i.e. for the autocovariance function of $x(t)$ where $x(t)$ is a realization of an ergodic process.

$$\rho_{xx}(\tau) \quad = \quad E \; \{ x(t) \; x(t + \tau) \; \}$$

They are related to the corresponding correlation functions $\gamma_{xx}(\tau)$, $\gamma_{yy}(\tau)$, and $\gamma_{xy}(\tau)$ by

$$\gamma_{xx}(\tau) \quad = \quad \sigma^2_x \rho_{xx}(\tau)$$

$$\gamma_{yy}(\tau) \quad = \quad \sigma^2_y \rho_{yy}(\tau)$$

$$(5.4.10)$$

$$\gamma_{xy}(\tau) \quad = \quad \sigma_x \sigma_y \rho_{xy}(\tau)$$

where σ is the standard deviation.

The autocovariances and autocorrelation functions are even functions of τ, the cross-covariance and cross-correlation are generally odd functions of τ.

From the Wiener-Khinchine relation (Lee, 1960) one can obtain the power spectra of the input and the output and the cross-power spectrum from the Fourier Transforms of the corresponding input, output and cross-covariance functions.

The Fourier Transform $F(\omega)$ of a function $f(t)$ is defined by:

$$F(\omega) \quad = \quad \int_{-\infty}^{\infty} f(t) \; \exp \; (-j\omega t) \; dt$$

$$(5.4.11)$$

and the inverse transform by

$$f(t) \quad = \quad \frac{1}{2\pi} \int_{-\infty}^{\infty} F(\omega) \; \exp \; (j\omega t) \; dt$$

$$(5.4.12)$$

where $j = \sqrt{-1}$

Thus the input power spectrum $S_{xx}(\omega)$, the output power spectrum $S_{yy}(\omega)$ and the cross-power spectrum $S_{xy}(\omega)$ are given by:

$$S_{xx}(\omega) \quad = \quad \int_{-\infty}^{\infty} \rho_{xx}(\tau) \; \exp \; (-j\omega t) \; d\tau$$

$$S_{yy}(\omega) \quad = \quad \int_{-\infty}^{\infty} \rho_{yy}(\tau) \; \exp \; (-j\omega t) \; d\tau$$

$$(5.4.13)$$

$$S_{xy}(\omega) \quad = \quad \int_{\infty}^{\infty} \rho_{xy}(\tau) \; \exp \; (-j\omega t) \; d\tau$$

Note that $S_{xx}(\omega)$ and $S_{yy}(\omega)$ are real, even functions, while $S_{xy}(\omega)$ is complex and odd. $S_{xy}(\omega)$ is often considered in terms of its real part, the real-valued even co-spectrum $C_{xy}(\omega)$ and its imaginary part, the real-valued odd quad-spectrum $Q_{xy}(\omega)$ where

$$S_{xy}(\omega) \quad = \quad C_{xy}(\omega) \quad - \quad jQ_{xy}(\omega)$$

$$(5.4.14)$$

or alternatively

$$C_{xy}(\omega) \quad = \quad S_{xy}(\omega) \quad + \quad S_{yx}(\omega)$$

$$Q_{xy}(\omega) \quad = \quad j\,S_{xy}(\omega) \quad - \quad S_{yx}(\omega)$$

where $S_{yx}(\omega)$ is the backward cross-spectrum.

These spectra are two-sided, that is they are defined for negative as well as positive frequencies. Although two-sided spectra simplify the analysis, in most applications negative frequencies do not have a physical interpretation, and so from the two-sided spectra one can obtain one-sided spectra defined only for positive frequencies by folding the two-sided spectrum over the zero axis so that the one-sided spectrum is given by

$$R_{xx}(\omega) \quad = \quad \begin{cases} 2S_{xx}(\omega) & 0 < \omega < \infty \\ 0 & \omega < 0 \end{cases} \qquad (5.4.15)$$

Since the cross-spectrum is odd, to obtain its one-sided spectrum the two-sided spectrum has to be folded over both axes.

From either the one or two-sided spectra one can obtain the frequency response function from the following relations:

$$G(\omega) \quad = \quad \frac{S_{xy}(\omega)}{S_{xx}(\omega)} \qquad (5.4.16)$$

or

$$G(\omega) \quad = \quad \frac{S_{yy}(\omega)}{S_{yx}(\omega)} \qquad (5.4.17)$$

and

$$S_{yy}(\omega) \quad = \quad |\,G(\omega)\,|^2 \; S_{xx}(\omega) \qquad (5.4.18)$$

Note that the frequency response function $G(\omega)$ can be interpreted as a complex linear regression coefficient for obtaining $y(t)$ from $x(t)$ (Benignus, 1969a).

If the response y(t) of the system is contaminated by additive, intrinsic noise then equation (5.4.2) does not hold and a transfer function may not be defined. The response will be given by

$$y(t) = \int_0^t x(\tau) \, h(t - \tau) \, d\tau + n(t) \tag{5.4.19}$$

where n(t) is that part of the output due to the intrinsic noise and h(t) is the weighting function of the linear system.

In this case a describing function may be defined whose frequency response function H(ω) is given by

$$H(\omega) = \int_0^\infty h(t) \, \exp(-j\omega t) \, dt \tag{5.4.20}$$

Note that the describing function does not completely characterize the input-output relations of the system, but it provides the best (in the sense of minimum mean squared deviations) linear model for the system.

If the input signal to the system is a realization of a broad-band stochastic process one can define the coefficient of coherence (Tick, 1963; Enochson and Goodman, 1965; Bendat and Piersol, 1966; Benignus, 1969b) by

$$\gamma^2(\omega) = \begin{cases} \dfrac{|\, S_{xy}(\omega)\,|^2}{S_{xx}(\omega) \cdot S_{yy}(\omega)}, & S_{xx}(\omega) \cdot S_{yy}(\omega) > 0 \\[2ex] 0, & S_{xx}(\omega) \cdot S_{yy}(\omega) = 0 \end{cases} \tag{5.4.21}$$

Alternative definitions of coherence have been proposed. Wiener (1930) defined a coefficient of coherency as

$$\gamma(\omega) = \begin{cases} \dfrac{S_{xy}(\omega)}{\{ S_{xx}(\omega) \cdot S_{yy}(\omega) \}^{1/2}} \\[2ex] 0, & S_{xx}(\omega) \cdot S_{yy}(\omega) = 0 \end{cases}$$

and this is a complex function which is not invariant when the x(t) and y(t) are subjected to linear transformations. Foster and Guinzy (1967) and Hinich and Clay (1968) define the coefficient of coherence as the positive square root of the right hand side of equation (5.4.21), which although it is a real number and is invariant when the x(t) and y(t) are

subjected to linear transformations has been described (Tukey, 1967) as
'an unlikely quantity'.

The definition of equation (5.4.21) has an analogy with the
coefficient of determination (the square of the correlation coefficient)
between two random variables. The coherence function is a normalized
measure of the linearity of the relation between $x(t)$ and $y(t)$ and
since the spectral matrix

$$S(\omega) = \begin{bmatrix} S_{xx}(\omega) & S_{yx}(\omega) \\ S_{xy}(\omega) & S_{yy}(\omega) \end{bmatrix}$$

is non-negative definite (Koopmans, 1964)

$$|S_{xy}(\omega)|^2 \leqslant S_{xx}(\omega) \cdot S_{yy}(\omega)$$

and so

$$0 \leqslant \gamma^2(\omega) \leqslant 1 \tag{5.4.22}$$

$\gamma^2(\omega)$ represents that proportion of $S_{yy}(\omega)$ which may be accounted for by
linear regression on to $S_{xx}(\omega)$, and $(1 - \gamma^2(\omega))$ represents that proportion
which cannot be accounted for by linear regression. In this case this
is the proportion of $S_{yy}(\omega)$ due to $S_{nn}(\omega)$, where $S_{nn}(\omega)$ is the power
spectrum of the noise $n(t)$.

From equation (5.4.19) it can be shown that (Amos and Koopmans,
1963)

$$\gamma^2(\omega) = \frac{1}{1 + \dfrac{S_{nn}(\omega)}{S_{xx}(\omega) \, |H(\omega)|^2}} \tag{5.4.23}$$

From this expression Stein and French (1970) have obtained a relationship between the coherence function and the information transmission
rate of the system, and this has been applied to neural models (Stein,
French and Holden, 1972 ; see also section 3.2). Shannon (1948) derives
the channel capacity in bits/second of a noisy channel for continuous
signals as

$$C = \int_0^W \log_2 \left[1 + \frac{S_{ss}(\omega)}{S_{nn}(\omega)} \right] df \qquad (5.4.24)$$

where $S_{ss}(\omega)$ is the signal spectrum and $S_{nn}(\omega)$ the noise spectrum, both of which are band-limited to W. From equation (5.4.23) and (5.4.24) the rate of information transmission is approximately I bits/second where

$$I \sim -\int_0^W \log_2 (1 - \gamma^2(\omega)) \, d\omega \qquad (5.4.25)$$

An important property of the coherence funtion is its invariance under linear operations (Koopmans, 1964).

For a zero memory non-linearity, that is, one whose output at a given time depends only on the input at the same time, Booton (1952) has shown that for a given input function x(t) there is an equivalent quasi-linear element $K_{eq}(f)$ such that the mean squared difference between the response of the nonlinear element and the response of the quasi-linear element is minimized. Thus analogous to equation (5.4.19) the response of the nonlinear system is related to the input by

$$y(t) = \int_0^\infty x(t) \, k(t - \tau) \, d\tau + N(t) \qquad (5.4.26)$$

where N(t) is due to the nonlinear terms in the response and to any intrinsic noise.

In this case the describing function $K(\omega)$ may be defined by equations similar to equations (5.4.19 - 20) and the coherence function provides a measure of that part of $S_{yy}(\omega)$ which is linearly dependent on $S_{xx}(\omega)$ and $(1 - \gamma(\omega))$ provides a measure of that part of $S_{yy}(\omega)$ which is due either to nonlinear transformation of $S_{xx}(\omega)$ or to intrinsic noise.

Thus, to characterize the relation between two signals there are:

1) the cross-correlation function, which is real and odd
2) the cross spectrum, which is complex and odd
3) the frequency response function, when one of the signals is an input and the other signal an output

4)   the coherence function.

These functions have been derived in terms of the continuous response
of a linear system to a continuous input, but may be applied to spike
trains (French, Holden and Stein, 1972).

When both signals are spike trains, if they are independent, then
a spike in train x(t) occurs at a random instant with respect to spikes
in train y(t). If the spike trains are realizations of renewal processes,
and independent, then the distribution of times from a spike in x(t) to
the first preceding and following spikes in y(t) are the first order
backward and forward recurrence times for y(t), with a probability
density function given by equation (5.2.14). If the spike trains are
independent, the forward and backward recurrence times have the same
distribution.

The cross-correlation between two spike trains gives the
probability per unit time of observing a spike in y(t) as a function of
time before or after a spike in x(t)

$$\gamma_{xy}(\tau) \; d\tau \;\; = \;\; \text{Prob} \; \{ \; \text{event in } y(t) \text{ in } (\tau_0 + \tau, \; \tau_0 + \tau + d\tau) |$$

$$\text{event in } x(t) \text{ at time } \tau_0 \; \}$$

If the spike trains are independent, $\gamma_{xy}(\tau) \; d\tau$ will be flat. If there
is a dependence it may be due to some kind of functional interaction,
which may be characterized by the frequency response function and coherence
function, or due to a common input.

Different types of dependence between spike trains result in
different shapes or signatures in the cross-correlation - these have
been extensively investigated by numerical methods (Gerstein, 1970;
Perkel, 1966, 1970; Moore et al, 1970). Three types of dependence were
considered - dependence resulting from synaptic inhibition, synaptic
excitation and a common input.

When there is an excitatory synaptic connection between the
neurones generating trains x(t) and y(t) the cross-correlation function
$\gamma_{xy}(\tau)$ will show a peak on the positive side of the time origin, at a
time given by the conduction delay and synaptic delay. At long positive
and negative times the cross-correlation will be flat. Moore et al
(1970) termed these the primary effects. The shape of the EPSP will
influence the shape of the peak: Moore et al suggested that the shape
of the primary peak in the cross-correlation function might be obtained

by a linear transformation of the shape of the EPSP. However, Bryant et al (1973) did not find a simple linear relation between the primary peak and the PSP. Knox (1974) has analyzed this problem in detail using the perfect integrator model (see section 3.2) and has shown

1)   for positive τ near $\tau = 0$, the cross correlation function is related to the probability density function $f(V, t-t_i)$ of the membrane potential $V(t-t_i)$ by

$$\gamma_{xy} (\tau) = f(V_o - h(\tau), t-t_i + \tau)\{h'(\tau) + E(V'(t-t_i + \tau))\}$$

(5.4.27)

where $h(t)$ is the PSP and $E(V'(t))$ is the mean value of the derivative of $V(t)$, which is zero when $V(t-t_i)$ is stationary

2)   there may be local minima in $\gamma_{xy}(\tau)$ even though the connection is purely excitatory

3)   at large τ, $\gamma_{xy}(\tau)$ is approximately the convolution of the input autocorrelation $\gamma_{xx}(\tau)$ with $\gamma_{xy}(0+)$

4) $\gamma_{xy}(\tau)$ is a biased estimator of $h(t)$, tending to overestimate the time to peak and rise time.

As well as the primary peak there are secondary effects, reflecting rhythmicities in the input train or output train. Secondary effects due to the input autocorrelation will appear symmetrically in the crosscorrelation function; secondary effects due to an intrinsic rhythmicity in the output train will only appear in $\gamma_{xy}(\tau)$ for positive τ provided there is no feedback pathway from $y(t)$ on to $x(t)$.

When there is an inhibitory synaptic connection between the neurones generating $x(t)$ and $y(t)$ the cross correlation will show a primary trough at positive τ near $\tau = 0$. As in the case of an excitatory connection, $\gamma_{xy}(\tau)$ will reflect any periodicities in $x(t)$ and $y(t)$ as secondary troughs.

When there is a common excitatory input to the neurones generating $x(t)$ and $y(t)$ there will be a primary peak at $\tau = 0$: if there are differences in conduction times the peak will be shifted. However, this will be a symmetric peak, rather than the asymmetric peak found when $x(t)$ drives $y(t)$. The autocorrelations of the common input, $x(t)$ and $y(t)$ will all contribute to secondary effects in the cross correlation function γ_{xy} (τ).

When there is a common inhibitory input to the neurones generating $x(t)$ and $y(t)$ the periods of silence of $x(t)$ and $y(t)$ will tend to be synchronized and so there will be a weak correlation between the spikes. This primary effect will appear as a weak, diffuse peak in γ_{xy} (τ). Any periodicities in the common input, $x(t)$ or $y(t)$ will contribute secondary effects to the cross-correlation functions, the secondary peaks resulting from convolution with the appropriate autocorrelations.

Since the relations between $x(t)$ and $y(t)$ are expressed in terms of convolutions the algebra may be simplified by examining the relations in the frequency domain. This approach has been pursued by Perkel (1970), Holden and French (1971) and ten Hoopen (1974). For the case of an excitatory link between $x(t)$ and the neurone generating $y(t)$, the frequency response function $G(\omega)$ defined by equation (5.4.16) is given by

$$G(\omega) = P(\omega)(1 + k S_{xx}(\omega))$$
(5.4.28)

where $P(\omega)$ is the Fourier transform of the primary response and k a constant. Thus the convolution of the autocorrelation with the primary response is represented by the product of their Fourier Transforms. Even though the algebraic representation is simplified, quite complex formulations are necessary to describe the relations in simple feedforward neural networks.

ten Hoopen (1974) has computed the auto spectra, cross spectra and coherence functions produced by spike trains derived from three independent renewal point processes $f_o(t)$, $f_x(t)$ and $f_y(t)$ with exponential or Gaussian interval distributions. Combined pulse trains were obtained by superposition to give

$$x(t) = f_o(t) + f_x(t)$$

$$y(t) = f_o(t+d) + f_y(t)$$

where d is a delay time. The basic processes $f_o(t)$, $f_x(t)$ and $f_y(t)$

were subject to random deletions. Changes in the random deletion parameter alters the relative contribution to the synthesized spike trains
s(t) and y(t), and hence alters the correlation between x(t) and y(t).
The computations suggested that

  a) for fairly regular trains, the crosscorrelation and
cross spectrum were equally sensitive to dependence between x(t)
and y(t)

  b) for irregular trains, the cross correlation function was
more sensitive than the cross spectrum

  c) the autospectrum was more sensitive than the autocorrelation
in detecting composite periodicities in a spike train

  d) when the common source was fairly regular and dominant the
coherence function was a sensitive index of the dependence between x(t)
and y(t)

  Thus, in terms of characterizing the relations between spike
trains choice between a time-domain or frequency-domain representation
is largely a matter of convenience, both representations being equally
informative. There are some cases where a frequency domain representation has dinstinct advantages - for example, the first passage time distribution of the Ornstein-Uhlenbeck process (see section 7) is not
available in closed form, but its Laplace transform and spectrum may be
obtained. Also, the coherence function provides a link with information
theory (see Stein, French and Holden, 1972 for an application) and
gives an index of the adequacy of a linear characterization of the
relation between two signals (French, Holden and Stein 1972).

5.5   References

Amos, D.E. & Koopmans, L.H.: Tables of the distribution of the coefficient of coherence for stationary bivariate Gaussian processes.
Sandia Corp. Monographs SCR - 483 (1963)

Bartlett, M.S.: The spectral analysis of point processes. J. Roy.
Stat. Soc. B 25 264-280 (1963)

Bayly, E.J.: Spectral analysis of pulse frequency modulation in the
nervous system. I.E.E.E. Trans. Bio-med Engng 15 257-265 (1968)

Bayly, E.J.: Spectral analysis of pulse frequency modulation. In
'Systems Analysis in Neurophysiology' ed. C.A. Terzuolo. Univ. of
Minnesota, (1969)

Bendat, J.S. & Persol, A.G.: Measurement and analysis of random data.
Wiley, New York (1966)

Benignus, V.A.: Computation of coherence and regression spectra using
the F.F.T. COMMON, Proc. Houston meeting, Dec. 9-11, 1968 (1969a)

Benignus, V.A. Estimation of the coherence spectrum and its confidence
interval using the Fast Fourier Transform. I.E.E.E. Trans. Audio.
Electro. Acous. AU - 17 145 (1969b)

Beutler, F.J. & Leneman, O.A.Z.: The theory of stationary point processes. Acta Math 116 159-97 (1966a)

Beutler, F.J. & Leneman, O.A.Z.: Random sampling of random processes:
Stationary point processes. Information and control 9 325-346 (1966b)

Beutler, F.J. & Leneman, O.A.Z.: The spectral analysis of impulse
processes. Information and control 12 236-58 (1968)

Booton, R.C. Jr. Non-linear control systems with statistical inputs.
M.I.T. Dynamic Analysis and Control Laboratory Report no. 61 (1952)

Brillinger, D.R.: The spectral analysis of stationary interval functions. Proc. 6th Berkeley Symposium on Mathematical Probability and
Statistics, 483-513 (1972)

Bryant, H.L., Marcos, A.R. & Segundo, J.R.: Correlations of neuronal
spike discharges produced by mono-synaptic connections and by common
inputs. J. Neurophysiol. 36 205-225 (1973)

Bush, B.M.H. & Roberts, A.: Resistance reflexes from a crab muscle
receptor without impulses. Nature 218 1171-73 (1968)

Cox, D.R.: Renewal Theory. Methuen, London (1962). reprinted Science Paperbacks, London SP 58 (1970)

Enochson, L.D. & Goodman, N.R.: Gaussian approximations to the distribution of smaple coherence. Air Force Flight Dynamics Laboratories,
Wright-Patterson Air Force Base Technical Report AFFDL-TR-65-57 (1965)

Foster, M.R. & Guinzy: The coefficient of coherence: its estimation
and use in geophysical data processing. Geophys. 32 602 (1967)

French, A.S. & Holden, A.V.: Alias-free sampling of Neuronal spike trains. Kybernetik 8 165-171 (1971)

French, A.S., Holden, A.V. & Stein, R.B.: The estimation of the frequency response function of a mechanoreceptor. Kybernetik, 11 15-23 (1972)

Gerstein, G.L.: Functional association of neurons: detection and interpretation. In the Neurosciences-Second Study program ed. F.O. Schmitt. Rockeller Univ. Press (1970)

Gerstein, G.L. & Kiang, N.Y.-S.: An approach to the quantitative analysis of electrophysiological data from single neurones. Biophys. J. 1 15-28 (1960)

Gestri, G.: Autocorrelation and spectrum of modulated retinal discharge. A comparison of theoretical and experimental results. Kybernetik 11 77-85 (1972)

Gnedenko, B.V. & Kolmogorov, A.N.: Limit distributions for sums of independent random variables. Trans. K.L. Chung. Addison Wesley Publ. Co. Reading, Mass. (1968)

Hinich, M.J. & Clay, C.S.: The application of the discrete Fourier Transform in the estimation of the power-spectra, coherence and bi-spectra of geophysical data. Rev. Geophys. 6 347 (1968)

Hodgkin, A.L. & Rushton, W.A.H.: The electrical constants of a crus-tacean nerve fibre. Proc. Roy. Soc. B 133 444-479 (1946)

Holden, A.V.: A note on convolution and stable distributions in the nervous system. Biological Cybernetics 20 171-174 (1975).

Holden, A.V. & French, A.S.: Spectral analysis of neuronal inter-actions. Proc. I.U.P.S. 9 256 (1971)

ten Hoopen, M.: Examples of power spectra of uni-variate point pro-cess. Kybernetik 18 145-154 (1974)

ten Hoopen M.: Frequency domain analysis of pulse trains. Institute of Medical Physics TNO Progress report 4 156-164 (1974)

Jenkins, G.M. & Watts, D.G.: Spectral analysis and its applications. Holden-Day, San Francisco. (1968)

Knox, C.K.: The power spectral density of random spike trains. In 'Systems Analysis in Neurophysiology' ed. C.A. Terzuolo , Univ. of Minnesota. (1969)

Knox, C.K.: Signal transmission in random spike trains with application to the statocyst neurons of the lobster. Kybernetik 7 267-274 (1970)

Knox, C.K.: Cross-correlation functions for a neuronal model. Bio-phys. J. 14 567-582 (1974)

Koles, Z.J.: A study of the sensory dynamics of a muscle spindle. Thesis, Univ. of Alberta, (1970)

Koopmans, L.H.: On the coefficient of coherence for weaky stationary stochastic processes. Ann. Math. Stat. 35 532 (1964)

Lawrence, A.J.: Some models for stationary series of univariate events. In 'Stochastic Point Processes: statistical analysis, theory and applications.' ed. P.A.W. Lewis. Weley N.Y. (1972)

Lee, Y.W.: Statistical Theory of Communication. Wiley, N.Y. (1960)

Levy, P.: Sur certain processes stochastiques homogenes. Comp. Math. 7 283 (1940)

Lewis, P.A.W.: Remarks on the theory, computation and application of the spectral analysis of series of events. J. Sound. Vib. 12 353-75 (1970)

Moore, G.P., Perkel, D.H. & Segundo, J.P.: Statistical analysis and functional interpretation of neuronal spike data. Ann. Rev. Physiol. 28 413-522 (1966)

Moore, G.P., Segundo, J.P., Levitan, H. & Perkel, D.H.: Statistical signs of synpatic interaction in neurons. Biophys. J. 10 876-900 (1970)

Nelsen, D.E.: Calculation of power density spectra for a class of randomly jittered waveforms. M.I.T. Research Lab of Electronics Q.P.R. 74 168-179 (1964)

Perkel, D.H.: Statistical techniques for detecting and classifying neuronal interactions. Rand Corporation Memorandum RM-4939-PR (1966)

Perkel, D.H.: Spike trains as carriers of information. In 'The Neurosciences - second study program' ed F.O. Schmitt. Rockefeller Univ Press (1970)

Poppele, R.E. & Bownman, R.J.: Quantitative description of linear behaviour of mammalian muscle spindles. J. Neurophysiol. 33 59 (1970)

Rall, W.: Dendritic neuron theory and dendrodendritic synapses in a simple cortical system. In "The Neurosciences - Second Study Program" ed F.O. Schmitt. Rockefeller Univ. Press, New York (1970)

Rodieck, R.W., Kiang, N.Y.-S. & Gerstein, G.L.: Some quantitative methods for the study of spontaneous activity of single neurons. Biophys. J. 2 351-68 (1962)

Shannon, C.E.: A mathematical theory of communication. Bell system tech J. 27 379 (1948) Reprinted Univ. of Illinois Press, Urbana (1962)

Shapley, R.: Fluctuations in the response to light of visual neurones in Limulus. Nature 221 439 (1969)

Shapley, R.: Fluctuations of the impulse rate in Limulus eccentric cells. J. Gen. Physiol 57 538-556 (1971)

Shephard, G.M.: The olfactory bulbs as a simple cortical system: experimental analysis and functional implications. In 'The Neurosciences - Second Study Program' ed. F.O. Schmitt. Rockefeller Univ. Press, New York (1970)

Smith, M.G.: Laplace Transform Theory. van Nostrand co. Toronto (1966)

Srinivasan, S.K.: Stochastic Point Processes and their applications. Griffin, London (1974)

Stein, R.B.: Some models of neuronal variability Biophys. J. $\underline{7}$ 37-68 (1967)

Stein, R.B.: The role of spike trains in transmitting and distorting sensory signals; in 'The Neurosciences - Second Study Program' ed. F.O. Schmitt. Rockefeller Univ Press (1970)

Stein, R.B.: The stochastic properties of spike trains recorded from nerve cells. In 'Stochastic Point Processes - statistical analysis, theory and applications' ed. P.A.W. Lewis. John Wiley, N.Y. (1972)

Stein, R.B. & French, A.S.: Models for the transmission of information by nerve cells. In 'Excitatory Synaptic Mechanisms' ed P. Anderson and J.K.S. Jansen. Univ. of Oslo Press, Oslo (1970)

Stein, R.B., French, A.S. & Holden, A.V.: The frequency response, coherence and information capacity of two neuronal models. Biophys. J. $\underline{12}$ 295-322 (1972)

Tick, L.J.: conditional spectra, linear systems and coherency. In 'Proc of the symposium on Time Series Analysis' ed M. Rosenblatt. John Wiley, New York (1963)

Tukey, J.W.: An introduction to the calculations of numerical spectrum analysis in 'Spectral Analysis of Time Series' ed B. Harris. Wiley, New York (1967)

Werblin, F.S. & Dowling, J.E.: Organisation of the retina of the mud puppy, Necturus Maculosus . II. Intracellular recording. J. Neurophysiol $\underline{32}$ 339-355 (1969)

Wiener, N.: Generalized Harmonic Analysis. Acta Math. $\underline{55}$ 117-258 (1930) Reprinted in Generalized Harmonic Analysis and Tauberian Theorems M.I.T. Press, Cambridge, Mass. (1964)

6. RANDOM WALK MODELS

Gerstein and Mandlebrot (1964) introduced a model for spike train
generation in which the membrane potential is reset instantaneously to
its resting value $V = V_r$ after an action potential at $t = t_o$, and then
follows the sample path of a random walk until a constant threshold
voltage V_o is reached. At this time t_i an action potential is generated and the voltage reset to V_r. The problem is to obtain, under
certain assumptions, an expression for the probability density function
of the interspike intervals $(t_{i + 1} - t_i)$ or equivalently the probability
density of the random variable t_i. This is the p.d.f. of the first
passage time of a one-dimensional restricted random walk to an absorbing
barrier.

The rationale for the proposal of a random walk model was the
experimental observations of Rodieck et al (1962) which showed for some
auditory neurones   (a)   interspike interval histograms which were
asymmetric about their mode and also that   (b)   histograms of the
intervals which were the sums of 2^m adjacent intervals ("scaled
intervals") with $\underline{m}$ = 0, 1, 2 $\cdots$, had the same shape. If adjacent
intervals are independent the scaled interval p.d.f.s are obtained
by successive convolution of the interval p.d.f. with itself
(Papoulis 1965, section 7-1). Since the scaled interval histograms
had the same shape as the interval histogram, the density function
was invariant under convolution, or the distribution was stable.

Only three distributions which are stable are available in closed
form. These are the Gaussian density with mean μ and standard
deviation σ

$$g(x) = 1/(\sigma \sqrt{2\pi}) \exp\{- (x - \mu)^2 / 2\sigma^2\}$$

and the Cauchy density which is the density of the ratio of two
random variables each of which has a Gaussian density with standard
deviation σ_1, and σ_2

$$c(z) = (\sigma_1/ \sigma_2\pi)/ (z^2 + \sigma_1^2/ \sigma_2^2)$$

and the stable distribution of order one half with probability distribution function

$$F(t) = 2(1 - G(t^{-\frac{1}{2}}))$$

where $G(\cdot)$ is the Gaussian distribution function (Levy 1940; Gnedenko and Kolmogorov 1954). Thus the only stable distribution available in closed form which has values for only positive times and has an asymmetric density is the positive stable distribution of order one half. This is well known as the P.D.F. of the first passage time of a one-dimensional random walk to an absorbing barrier (Feller 1968 - section III-7). Thus the invariant shape (for small $\underline{m}$) of the scaled histograms naturally suggested a random walk model.

6.1 The Random Walk Model

The state of the neurone is specified by a single number, the membrane potential V_m. Thus changes in V_m can be considered as motion on a straight line. When the membrane potential reaches a constant threshold, V_o at a time t_1, an impulse is instantaneously generated and the membrane potential is instantaneously reset to its resting value V_r. I will, without loss of generality, set $V_r = 0$ and $V_o = +1$.

There are two types of quantal input to the neural model, both of which instantaneously produce a change in membrane potential of constant magnitude. These are depolarising (or excitatory) events, which produce a change in V_m of $+1/k_e$, and repolarising (or inhibitory) events, which produce a change in V_m of $-1/k_i$. The excitatory events arrive at a rate P/sec and the inhibitory events at a rate Q/sec. Thus in any period of time Δt there is a probability $p = P\Delta t$ and $q = Q\Delta t$ that an excitatory or inhibitory event arrives.

6.2 The Probability of first passage

I will assume $|1/k_e| = |-1/k_i| = 1/k$. Following Gnedenko (1968 - section 16), I will assume that Δt is small and so there can only be one event (excitatory or inhibitory) in any period Δt. Further, I will assume that in any period Δt

$$p + q = 1 \qquad\qquad (6.2.1)$$

Then, in the unrestricted random walk on the infinite line (i.e. when there is no threshold or absorbing barrier) the probability that at time $n\Delta t$ the membrane potential is m/k is given by Bernouilli's formula

$$P(V = m/k) = \begin{cases} \dbinom{n}{(m+n)/2} p^{(m+n)/2} q^{n-(m+n)/2} & \text{for } -n \leq m \leq n \\ 0 \text{ for } |m| > n \end{cases}$$

$$(6.2.2)$$

When m = k this gives the probability that, in the unrestricted case, V = 1 at a time $n\Delta t$. This is $\underline{not}$ the probability for the first passage as it includes all possible sample paths which have passed through the point V = 1 to values of V > 1 and then returned. Introducing the absorbing barrier at V = 1 means that these paths are illegal and so to obtain the probability that V reaches the threshold at time $n\Delta t$ these illegal paths must be subtracted from the total number of paths

These illegal paths are all those which lead to the point V = k + 1/k (or by symmetry, V = (2k - m/k)) and then move in the repolarising direction. The probability for the first passage is given by

$$P_1(V = 1) = P(V = k/k) - P(V = (2k - m)/k)$$

$$= \left\{ \dbinom{n}{(m+n)/2} - \dbinom{n}{(n+2k-m/k)/2} \right\} p^{(n+m)/2} q^{(n-m)/2}$$

$$(6.2.3)$$

when (n+m) is even.

If excitatory and inhibitory events have an equal probability and so $p = q = \frac{1}{2}$ this simplifies to

$$P_1(V = 1) = \left\{ \dbinom{n}{(m+n)/2} - \dbinom{n}{(n+2k-m/k)/2} \right\} (\tfrac{1}{2})^n$$

for which the local DeMoivre-Laplace theorem (Gnedenko 1968 - section 12) gives the approximation

$$P_1(V = 1) \simeq \frac{2}{\sqrt{2\pi n \Delta t}} \left\{ \exp\{-(m/k)^2/2n\Delta t\} - \exp\{-(2 - m/k)^2/2n\Delta t\} \right\}$$

$$(6.2.4)$$

when n is large.

6.3   The interval probability density

For an impulse to be generated at time t_i an excitatory event must have arrived (with probability p) at a time $t_i - \Delta t$ when V was equal to $1 - 1/k$. Thus the first passage time density, or the interspike interval density f (nΔt) is given by

$$f(n\Delta t) = P_1 \; (V = 1 - 1/k).p \qquad\qquad (6.3.1)$$

Using equation 6.2.3 gives

$$f(n\Delta t) = \frac{k}{n} \begin{pmatrix} n \\ (n+k) \end{pmatrix} p^{(n+k)/2} q^{(n-k)/2} \qquad\qquad (6.3.2)$$

when (n+k) is even
= 0 when (n+k) is odd.

When k is large (and hence n is also large) a continuous approximation for (6.3.2) when $p \simeq \frac{1}{2}$

$$f(t) = \frac{k}{\sqrt{8\pi pq}} \; t^{-3/2} \; \exp\{-(k-(2p-1)\ t)^2/8pqt\} \qquad (6.3.3)$$

This is the same form as the expression for the normal diffusion current at an absorbing barrier given by Chandraskhar (1943). Note that the p.d.f of the first passage time is a function of k, the ratio of the magnitude of the threshold to the step magnitude.

The P.D.F. for the first passage is given by the sum

$$F(k) \;\; = \;\; \sum_{n=0}^{\infty} f(n\Delta T) \qquad\qquad (6.3.4)$$

since $F(k) = F(1)^k$ and $F(1) = P/q$ then if $p > \frac{1}{2}$ all paths will reach threshold with a probability of one. However, if $p < \frac{1}{2}$ only $(p/q)^k$ of possible paths reach threshold. Thus F(k) for $p < \frac{1}{2}$ is an unusual P.D.F. in that it does not tend to one as to $t \to \infty$.

The integral of the continuous approximation for f(t) is

$$F = \int_0^\infty f(t)\ dt = 1 \qquad\qquad p \geqslant \tfrac{1}{2}$$
$$= \exp\{(p-q)\ k/2pq\} \qquad p < \tfrac{1}{2} \qquad (6.3.5)$$

and the mean interval

$$<t> = \int_0^\infty t\ f(t)\ dt = \frac{k}{2p-1} \qquad\qquad p \geqslant \tfrac{1}{2}$$
$$= \infty \qquad\qquad p < \tfrac{1}{2} \qquad (6.3.6)$$

Thus since the mean is unbounded the sample mean will increase with sample length.

6.4 A diffusion approximation

I have already shown that a continuous approximation for the interval density (equation 6.3.3) is similar in form to the normal diffusion current at an absorbing barrier. Here I will obtain the diffusion approximation for a random walk.

For the unrestricted random walk

$$P_{(n+1)\Delta t}\ (V = m/k) = P_{n\Delta t}(V = (m-1)/k).p$$
$$+ P_{n\Delta t}(V = (m+1)/k).q \qquad (6.4.1)$$

let $\emptyset(V, n\Delta t) \equiv P_{n\Delta t}(V)$

$\emptyset(V, (n+1)\Delta t) = p.\emptyset(V - 1/k, n\Delta t) + q.\emptyset(V + 1/k, n\Delta t)$

subtract $\emptyset(V, n\Delta t) = p.\emptyset(V, n\Delta t) + q.\emptyset(V, n\Delta t)$

and using the second order Taylor's series expansion

$$\Delta t\ \frac{\partial\emptyset}{\partial t} + \tfrac{1}{2}(\Delta t)^2\frac{\partial^2\emptyset}{\partial t^2} = \qquad\qquad (6.4.2)$$

$$p\left\{\frac{-1}{k}\frac{\partial\emptyset}{\partial v} + \tfrac{1}{2}\left(\frac{1}{k}\right)^2\frac{\partial^2\emptyset}{\partial v^2}\right\} + q\left\{\frac{1}{k}\frac{\partial\emptyset}{\partial v} + \frac{1}{2}\left(\frac{1}{k}\right)^2\frac{\partial^2\emptyset}{\partial v^2}\right\}$$

which gives

$$\frac{\partial \emptyset}{\partial t} + \frac{1\Delta t}{2} \frac{\partial^2 \emptyset}{\partial t^2} = - \frac{(p-q)}{k\Delta t} \frac{\partial \emptyset}{\partial V} + \frac{1}{k^2 2\Delta t} \frac{\partial^2 \emptyset}{\partial V^2} \qquad (6.4.3)$$

Let $1/k \to 0$, $\Delta t \to 0$, $p = q = 1/2$

and let $a = \lim(\ (p-q)\ /\ k\Delta t)$

$\qquad b = \lim\ (1/2\Delta t k^2)$

and so

$$\frac{\partial \emptyset}{\partial t} = -a \frac{\partial \emptyset}{\partial V} + b \frac{\partial^2 \emptyset}{\partial V^2} \qquad (6.4.4)$$

which is the well known equation for one-dimensional diffusion with drift. This equation is discussed in greater detail in section 7, here, note that the boundary conditions are

$$\emptyset(V, 0) = \delta(V)$$

$$\emptyset(1, t) = 0$$

The solution for $-\infty < V \leqslant 1$ is given by

$$\emptyset(V, t) = \frac{1}{\sqrt{4\pi bt}} \ \exp\ \{\ (aV/2b)\ -\ (a^2 t/4b)\ \} \left\{ \exp\ \{-V^2/4bt\}\ - \right.$$
$$\left. \exp\ \{-\ (V-2)^2/4bt\} \right\} \qquad (6.4.5)$$

(see section 7). The first passage time density is given by

$$f(t) = -b(\partial \emptyset/\partial V)_{V=1} \qquad (6.4.6)$$

$$= \frac{1}{\sqrt{4\pi b}} \ t^{-3/2} \ \exp\ \{-\ (1-at)^2/4bt\}$$

which is the same as equation (6.3.3) when $\underline{a} = 0$, i.e. there is no drift, or $p = q$.

6.5   Introduction of a reflecting barrier

A problem with the simple random walk to an absorbing barrier is
the number of possible sample paths which go in the hyperpolarizing
direction to arbitrarily large negative potentials. The possibility
of V reaching large negative values can be overcome by the introduction
of a reflecting barrier at some hyperpolarized voltage, say V = -b.
Then, if a sample path arrives at V = 1/k - b, it moves to V = 2/k - b
with probability p, and remains at V = -b with probability q (i.e. it
moves to V = -b with probability q and is reflected back to V = 1/k - b
with probability 1). If |-b| >> 1 , the threshold, the first passage
time density is not appreciably affected.

The reflecting barrier is an artificial device when V is the
membrane potential of a neuronal model; however, it is useful in
Monte Carlo simulations. The problem of sample paths reaching large
negative values is better dealt with by having p > q, or by further
modifications of the model such as decay of potential towards the
resting potential (i.e. leaky integrator models).

6.6   Generalization of the random walk model as a birth and death process

So far I have only considered the case in which there is an event,
either excitatory or inhibitory, in every period Δt i.e. p + q = 1.
The more general case of a random walk to an absorbing barrier when
p + q < 1 is treated in Cox and Miller 1968 (chapter 2). Here I will
treat the case of p + q < 1 in terms of a homogenous birth and death
process.

Initially, consider only excitatory events arriving randomly in
time at a rate r/sec. The probability that an event arrives in a
period Δt is $(r\Delta t + o(\Delta t))$, and the probability that there is no
event is $(1 - r\Delta t - o(\Delta t))$. Then, if

$$P(n/k, \tau) \equiv P(V = n/k, t = \tau)$$

$$P(n/k, t + \Delta t) = P((n - 1)/k, t) \, r.\Delta t + P(n/k, t)(1 - r.\Delta t)$$

$$(6.6.1)$$

and so $\dfrac{dP(n/k, t)}{dt} = \lim_{\Delta t \to 0} \dfrac{P(n/k, t + \Delta t) - P(n/k, t)}{\Delta t}$ $(6.6.2)$

$$= r(P(n - 1, t) - P(n, t)); \quad n > 0$$

and $d\,P(0,\,t)/dt = -rP(0,\,t)$ (6.6.3)

since V is reset to 0 at $t = 0$, $P(0,\,0) = 1$

Thus, from 6.6.2 to 3

$$P(n/k,\,t) = \exp(-rt)\,(rt)^{n}/n!$$ (6.6.4)

which is the Poisson distribution with parameter rt. The distribution of times between excitatory events is

$$f_e(\tau) = r\,\exp(-r\tau),\ 0 \leqslant \tau < \infty$$ (6.6.5)

similarly, for inhibitory events,

$$f_i(\tau) = s\,\exp(-s\tau),\ 0 \leqslant \tau < \infty$$ (6.6.6)

where $\underline{s}$ is the rate of inhibitory events. Now, with both excitatory and inhibitory events

$$\frac{d\,P(n/k,\,t)}{dt} = r(n-1)\,P((n-1)/k,\,t)$$

$$- (r+s)\,n\,P(n/k,\,t) + s(n+1)\,P(n+1,\,t)$$ (6.6.7)

for $n \geqslant 1$, and

$$d\,P(0,\,t)/dt = s\,P(1/k,\,t) \qquad \text{for } n = 0$$

If V is identified with population size, excitatory and inhibitory events with births and deaths, and $k = 1$ this is the equation for a simple, homogenous birth and death process (see Bailey 1964, chapter 8).

Note that a birth and death process ceases when $V = 0$ and a sample path may not cross the point $V = 0$. Thus in this case the point $V = 0$ behaves identically as an absorbing barrier.

Thus now let $V = 0$ be the threshold

$V = 1$ be the reset potential

s = rate of <u>inhibitory</u> events, which move

V in the +ve direction

r = rate of <u>excitatory</u> events, which move

V in the -ve direction

The solution of (6.6.7) is given by Bailey (1964 - section 8.6) using the moment generating function technique as

$$P(n/k,\ t) = \sum_{j=0}^{\min(k,n)} \binom{k}{j} \binom{k+n-j-1}{k-1} \alpha^{k-j} \beta^{n-j} (1 - \alpha - \beta)^j$$

$$\hspace{6cm} (6.6.8)$$

$$P(0,\ t) = \alpha^k$$

where $\alpha = r\ (\exp\{(s-r)t\} - 1)/(s\ \exp\{(s-r)t\} - r)$

$$\beta = s\ \exp\{(s-r)t\} - 1)/(s\ \exp\{(s-r)t\} - r)$$

The probability density of first passage time, or time to extinction if V is taken as a population, is then $P(1/k,\ t)r\Delta t$ which is

$$f(t) = \left\{ \sum_{j=0}^{1} \binom{k}{j}\binom{k-j}{k-1} \alpha^{k-j} \beta^{n-j} (1 - \alpha - \beta)^j \right\} r\Delta t \hspace{2cm} (6.6.9)$$

which has a mean

$$E(f(t)) = \frac{k}{r-s} + \frac{s}{(r-s)^2} \frac{s^k - 1}{r},\ r = s \hspace{2cm} (6.6.10)$$

$$E(f(t)) = \frac{k}{2r};\ r \neq s$$

Logan and Shepp (1974) have described a different situation where a birth and death process models neural spike train generation. This is the model of Schroeder and Hall (1975) for auditory neurones, in which

a)   at each time t = 1, 2, ··· a vesicle appears (a birth) independently with probability b

b)   at each time t an action potential is initiated, with small
probability d_j =   j/N where j = number of vesicles (the population size)
and N is large. On firing j is reduced by one (a death). Thus the
state (population size) j = s(t) forms a finite - state Markov chain
with a stationary distribution. If births and deaths occur independently for each j and if b is independent of j, Logan and Shepp showed that
the interspike intervals τ in the stationary case are independent with
a geometric distribution

$$P \; (\tau = t) \;\; = \;\; b \; (1 - b)^{t-1}$$

Such interspike interval distributions have been observed.

6.7   Birth and Death Process Model of Leaky Integrator

In the simple models discussed above no account has been taken of
the decay of membrane potential V towards V_r. Goel, Richter-Dyn and
Clay (1972) have investigated the moments of the first passage time
PDF of a leaky integrator model subject to Poisson distributed excitatory
and inhibitory inputs, using the framework of the transition probabilities of a birth and death process. This model is also discussed in Goel
and Richter-Dyn (1974). Their two types of model are specified by:

a)   the magnitude of the response ΔV to excitatory and inhibitory
inputs is the same, and so V(t) changes in integer steps and so can only
assume values - ∞,   ···, -2, -1, 0, 1, 2, ··· K where K is a constant,
time-independent threshold. Normalizing so that V_0 = 1 complicates the
notation and so will not be used in this section

b)   after firing, V is instantaneously reset to its resting
value 0

c)   the subthreshold membrane potential V ≤ (K - 1) decays
spontaneously towards V = 0, with a time constant τ, in discrete steps

d)   the effects of inputs summate linearly

e)   in model A, the stochastic input pulses are Poisson distributed with mean rates r_e for excitatory and r_i for inhibitory inputs.

In model B, the probability that V depolarizes due to the arrival of an excitatory input is proportional to V.

The transition probabilities λ_n and μ_n for model A, which are the probabilities/unit time that V changes from $n \to n + 1$ and $n \to n - 1$ are

$$\lambda_n = \begin{cases} r_e & ; \quad 0 \leqslant n \leqslant K - 1 \\ 0 & ; \quad n \geqslant K \\ r_e - n/\tau & ; \quad n < 0 \end{cases} \qquad (6.7.1)$$

$$\mu_n = \begin{cases} r_i + n/\tau & ; \quad 0 \leqslant n \leqslant K - 1 \\ 0 & ; \quad n \geqslant K \\ r_i & ; \quad n < 0 \end{cases} \qquad (6.7.2)$$

The average of the first passage time t' is given by (see section 8.1 of Goel and Richter-Dyn)

$$< t' > = \sum_{i=0}^{K-1} \{ \frac{1}{\lambda_i} + \frac{\mu_i}{\lambda_i \lambda_{i-1}} + \frac{\mu_i \mu_{i-1}}{\lambda_i \lambda_{i-1} \lambda_{i-2}} + \cdots \} \qquad (6.7.3)$$

$$i \leqslant K - 1$$

which can be manipulated into an expression containing the standard confluent hypergeometric function. The variance of the first passage time is

$$\sigma^2 = \sum_{i=0}^{K-1} \left\{ 2 \sum_{n=0}^{\infty} \Pi_{i-n,i} (M_{i-n,i-n-1})^2 + (M_{i+1,i})^2 \right\} \qquad (6.7.4)$$

where $\Pi_{i,j} = \dfrac{\mu_i \, \mu_{i+1} \cdots \mu_j}{\lambda_i \, \lambda_{i+1} \cdots \lambda_j} \qquad ; \quad i \leqslant j$

$$\Pi_{i,i-1} = 1$$

$$M_{i+1,i} \;=\; \frac{1}{\lambda_i} \;+\; \frac{\mu_i}{\lambda_i \, \lambda_{i-1}} \;+\; \frac{\mu_i \, \mu_{i-1}}{\lambda_i \, \lambda_{i-1} \, \lambda_{i-2}} \;+\; \cdots$$

$$i \;\leqslant\; K - 1$$

These formidable expressions of Goel et al were numerically evaluated and show

    a)   the model does not fire until a critical input rate is reached

    b)   above this critical input rate the output rate is proportional to the input rate

    c)   the critical input rate increases with increasing threshold

    d)   inhibition has little effect on the coefficient of variation $\sigma / < t' >$ at low input rates, but increases the coefficient of variation at high input rates.

If $t \to \infty$, the expressions simplify to

$$< t' > \;=\; K / (r_e - r_i) \tag{6.7.5}$$

$$\sigma^2 \;=\; K \, (r_e + r_i) \,/\, (r_e - r_i)^3 \tag{6.7.6}$$

and the probability density of t' is

$$P(t') \;=\; K \, \left\{ \frac{r_e}{r_i} \right\}^{K/2} \frac{1}{t} \exp \left(- (r_e + r_i) \, t \right) . \, I_K \left(2 \, (r_e r_i)^{\frac{1}{2}} \, t \right) \tag{6.7.7}$$

where $I_k(x)$ is a modified Bessel function

$$I_K(x) \;=\; \sum_{r=0}^{\infty} (x/2)^{K+2r} / \; r! \;\; (K+r)! \tag{6.7.8}$$

Note equation (6.7.5) is the same as equation (6.3.6).

    The behaviour of model B was obtained, and was similar to that of model A for small K. However, for large K, the potential in model B cascades towards the threshold and so there is no phenomenon equivalent to the critical input rate.

6.8 More complex Random Walks

In the models discussed above p and q are time-invariant. Gerstein and Mandelbrot (1964) extended their discussion of random walk
models to consider some examples when p and q were time variant by
making b, the drift term in the one-dimensional diffusion equation
(6.4.4), a function of time. This was done to simulate the interspike
histograms of Gerstein and Kiang (1960) obtained from auditory interneurones in the cat cochlear nucleus during repetitive presentation of
click stimuli. Using periodic changes in b, to mimic a transient
increase in p followed, after a delay, by a longer but smaller increase
in q, Monte Carlo methods gave histograms which were in good agreement
with the experimental histograms. Further generalization of this
approach does not seem promising as the diffusion equations with time
varying parameters are analytically intractable.

The random walk model may be generalized to make the state variable V of the neurone a point in a multidimensional space rather than a
point on a line. The absorbing barrier now becomes an absorbing
surface which bounds the region within which the state variable undergoes a multidimensional random walk. If the sample paths of the state
variable undergo step movements in orthogonal directions, and if the
absorbing surface is defined by an array of orthogonal planes then
the model behaves as independent, simultaneous random walks in orthogonal
directions, with action potential generation whenever one of the one-
dimensional random walks reaches the absorbing plane orthonormal to
its path. Gerstein and Mandelbrot investigated the case of a two-dimensional random walk in Cartesian coordinates with resting position
(x_o, y_o) and absorbing planes $(x = 0, y = 0)$. In this case the first
passage time density is

$$f(t) \; = \; \frac{1}{\sqrt{\pi}} \cdot x_o \cdot \frac{1}{t^{3/2}} \cdot \exp\{-x_o^2/4t\}\}. \; \frac{2}{\sqrt{\pi}} \int_{o}^{x_o/2\sqrt{\tau}} \exp\{-u^2\}du$$

$$(6.8.1)$$

When the absorbing surface is not an orthogonal array there is interaction between the passage times of the component one-dimensional random walks. Two example cases treated using Monte Carlo methods by

Gerstein and Mandlebrot were a two dimensional walk with a circular
absorbing surface a a three-dimensional walk with a spherical absorbing
surface. Although these cases are analytically tractable the solutions
are cumbersome and are in the form of infinite series e.g. for the
three dimensional case with no drift

$$f(t) \; = \; \frac{r_1}{r_2} \; \frac{1}{4\pi t^3}^{\frac{1}{2}} \cdot \; \sum_{n=-\infty}^{\infty} \; (\; (2n + 1)r_1 - r_2) \; \exp\{ \; \frac{-((2n+1)r_1 r_2)^2}{4t} \; \}$$

$$(6.8.2)$$

where r_1 and r_2 are the radii of the absorbing surface and the resting
state point.

Stein (1967) has treated the case when the step size has an
arbitrary density $a(1/k)$. This includes both excitatory (k is +ve)
and inhibitory (k is -ve) inputs. If the distribution of times to
the nth input event (which may be excitatory or inhibitory) is $F_n(t)$
the first passage density is

$$f(t) \; = \; \sum_{n-1}^{\infty} \; F_n(t) \left\{ P_{n-1} \; (V \leqslant 1) - P_n \; (V \leqslant 1) \right\}$$

$$(6.8.3)$$

$$P_o \; (V \leqslant 1) \; = \; 1, \qquad 0 \leqslant t \leqslant \infty$$

When the input process is Poisson with parameter p, $F_n(t)$ is the density
of a gamma distribution with first two moments about the origin of n/p
and $(n + n^2) / p^2$. Thus

$$\mu \; = \; \int_0^\infty tf(t) \; dt \; = \; \frac{1}{p} \; \sum_{n=0}^{\infty} \; P_n \; (V \leqslant 1)$$

$$(6.8.4)$$

and $\quad \sigma^2 = \int_0^\infty t^2 f(t) \; dt - \mu^2$

$$= \; \frac{2}{p^2} \; \sum_{n=0}^{\infty} \; (n + 1)P_n(V \leqslant 1) - \mu^2$$

$$(6.8.5)$$

A further modification would be to make a(1/k) a function of V:
this would model the fact that the size of the voltage change produced
by a conductance change depends on the difference between V and the
reversal potential for the conductance mechanism, which is negative
for inhibitory mechanisms and positive for excitatory mechanisms.

6.9 Some comments on the random walk models

A number of objections can be raised to the random walk models
discussed above - see Stevens (1964). One class of objection is that
the specific models are physiologically naive, in that no account is
taken of the decay of the membrane potential towards its resting value
when there are no inputs, and that the reset mechanism destroys the
memory of previous inputs and so there can be no correlation between
adjacent interspike intervals. These objections would apply to all
models of stochastic neuronal activity which are based on the perfect
integrator with reset model, and appropriate modifications could be
made to satisfy them. However, such modifications would destroy the
simplicity of the model.

A more fundamental objection is that the random walk model is
only one of a number of possible mechanisms which lead to first passage
time densities generated by the diffusion equations. Thus it seems
appropriate to consider the diffusion equations as the basis of a
neuronal model, and to investigate appropriate modifications of the
diffusion approximation model. This is done in the following section.

6.10 References

Abramowitz, M. & Stegun, I.A. (eds). Handbook of Mathematical
Functions. Nat. Bureau of Standards, Washington D.C. (1964).

Bailey, N.T.J.: The Elements of Stochastic Processes, with applications
to the life sciences. 249 pp. John Wiley, N.Y. (1964).

Chandraskhar, S. Stochastic problems in physics and astronomy. Rev.
Modern Physics 15 1-89 (1943). Reprinted in: Noise and Stochastic
Processes (N. Wax. editor). Dover, N.Y. (1954).

Cox, D.R. & Miller, H.D.: The Theory of Stochastic Processes. 398 pp.
Methuen, London (1968).

Feller, W.: An Introduction to Probability Theory and its Applications.
Vol. I 509 pp. John Wiley, N.Y. (1968).

Gerstein, G.L. & Mandlebrot, B.: Random walk models for the spike
activity of a single neuron. Biophys. J. 4 41-68 (1964).

Goel, N.S. & Richter-Dyn, N.: Stochastic Models in Biology. 269 pp.
Academic Press, N.Y. (1974).

Goel, N.S., Richter-Dyn, N. & Clay, J.R.: Discrete stochastic models
for the firing of a neuron. J. theoret. biol. 34 155-184 (1972).

Gnedenko, B.V.: The Theory of Probability. (translator B.D. Secker)
529 pp. Chelsea Publ. Cp., N.Y. (1968).

Gnedenko, B.V. & Kolmogorov, A.N.: Limit Distributions for Sums of
Independent Random Variables. (K.L. Chung, translator and editor).
Addison-Wesley, Cambridge, U.K. (1954).

Levy, P.: Sur certain processus stochastiques homogenes. Compositio
Math. 7 283 -   (1940).

Logan, B.F. & Shepp, L.A.: A birth and death model of neurone firing.
J. Appl. Probability 11 369-373 (1974).

Papoulis, A.: Probability, Random Variables and Stochastic Processes.
583 pp. McGraw-Hill, N.Y. (1965).

Rodieck, R.W., Kiang, N.Y.-s. & Gerstein, G.L.: Some quantitative
methods for the study of spontaneous activity of single neurons.
Biophys. J. 2 351 -.   (1962).

Stein, R.B.: Some models of neuronal variability. Biophys. J. 7
37-68 (1967).

Stevens, C.F.: Letter to the editor. Biophys. J. 4 417-419 (1964).

7.   DIFFUSION MODELS.

In the models of section 6 the state of the neurone is
represented by a single number, the membrane potential V(.). The
membrane potential could change only at integral multiples of a time
period ΔT, and so is represented by the sequence {V(mΔT)}, m = 0,1,2,...
When V(mΔT) reaches a threshold value, an action potential is generated
and the membrane potential is reset instantaneously to its resting
value.

For the case of the perfect integrator model (see section 3.2)
the subthreshold membrane potential can only change when there is an
input pulse, an excitatory pulse moving the potential in the positive
direction

$$V(m\Delta T) \rightarrow V((m+1)\Delta T) = V(m\Delta T) + a$$

and an inhibitory pulse moving the potential in the negative direction

$$V(m\Delta T) \rightarrow V((m+1)\Delta T) = V(m\Delta T) - b$$

where   a   and b are positive numbers representing the magnitudes of the
excitatory and inhibitory effects. In section 6 these were taken to have
the same magnitude: this does not lead to a loss of generality as the
rates of the excitatory and inhibitory pulses can be adjusted independently.   Thus the subthreshold membrane potential is not a continuous
function of time, but undergoes a random walk. As the step size of a
random walk decreases the sample path might be thought of as approximating a smooth function representing a diffusion.

In diffusion models of neural activity the membrane potential is
a continuous stochastic process which may be obtained from a random walk
by taking the limit as the time interval ΔT→0, and   a   and b decrease as
the rates of the excitatory and inhibitory processes increase.   The
resulting diffusion process will be described by the general diffusion
equations:

$$\frac{\partial\ p(y,t|x)}{\partial t} = \frac{1}{2}\ \frac{\partial^2\{\alpha(y)p(y,t|x)\}}{\partial y^2}\ -\ \frac{\partial\{\beta(y)p(y,t|x)\}}{\partial y}$$

$$\frac{\partial\ p(y,t|x)}{\partial t} = \frac{1}{2}\ \alpha(x)\ \frac{\partial^2 p(y,t|x)}{\partial x^2}\ +\ \beta(x)\ \frac{\partial p(y,t|x)}{\partial x}$$

where $p(y,t|x)$ is the probability of the membrane potential having a
value y at a time t given that it had a value x at a time $t=0$, $\alpha(x)$
is the infinitessimal variance and $\beta(x)$ the rate of growth of the
mean. These equations will be derived and discussed in section 7.4.
The use of diffusion equations for models of biological processes is
discussed in the monograph by Goel and Richter-Dyn (1974).

To formulate a diffusion model of the neurone the procedure is
to:

1) define a suitable stochastic input - section 7.1

2) obtain the properties of the stochastic process $\{V(t)\}$
representing the membrane potential in terms of diffusion equations,
with expressions for $\alpha(x),\beta(x)$, initial and boundary conditions - see
sections 7.2. 7.3 and 7.4.

3) investigate the behaviour of the first passage time of $\{V(t)\}$
to the threshold level - section 7.5.
This methodology is applied to the perfect integrator model in section
7.6 and to the leaky integrator model in section 7.7.

7.1 The Input Process.

The input process i(t) may be thought to represent the current
resulting from the activity of a large number of excitatory and
inhibitory synapses. Thus I am assuming that the resting membrane
potential is far from the reversal potentials of the synaptic
conductance mechanisms. If the number of synapses is large and the
activities of the synapses are mutually independent, superposition
theory (see section 8) gives the result that the intervals between
input pulses will have a Poisson distribution. Thus the input process
to the neural model can be represented by the superposition of an
excitatory Poisson process with parameter r_e and an inhibitory Poisson
process with parameter r_i. In any short period of time $(t,t+\delta t)$
the probability of an excitatory input event is $r_e\delta t + o(\delta t)$, the
probability of an inhibitory input event is $r_i\delta t + o(\delta t)$ and the

probability of two or more input events is o(δt). These two Poisson
processes are sequences of zero-width pulses, or Dirac delta functions,
the effect of which on V(t) is associated with a sign and magnitude
(+a for an excitatory pulse, -b for an inhibitory pulse).

The membrane potential, which is given by an integral of the
input process i(t), is to be a continuous stochastic process. Thus it
seems reasonable to let the rate of the input processes increase as the
magnitude of a and b decreases. Capocelli and Ricciardi (1971) point
out that there are difficulties in assuming values for a and b; if
a = b = 0, the effect of the random input is lost, and if a and b
are small but finite the membrane potential V(t) will not be continuous.
These problems can be overcome by letting

$$a \to 0, \quad b \to 0, \quad r_e \to \infty, \quad r_i \to \infty$$

such that

$$r_e a + r_i b \to m \qquad\qquad (7.1.1)$$

where m = $<i(t)>$ and $-\infty < m < +\infty$
and i(t) is assumed to be stationary.

Thus the correlation function $\rho(t,\tau)$ of the input is a delta
function. The input i(t) can be replaced by a signal having the same
statistical properties

$$i(t) = m(t) + s(t)w(t) \qquad\qquad (7.1.2)$$

where w(t) has a zero mean and unit incremental variance, and the
incremental variance $s^2(t)$ of i(t) is

$$s^2(t) = \int_{-\infty}^{\infty} d\tau \; \rho(t,\tau) \qquad\qquad (7.1.3)$$

The process w(t) is a white noise which has a Gaussian amplitude
distribution and a flat spectral density, and

$$<w(t)> = 0 \qquad\qquad (7.1.4)$$

$$<w(t)w(t+\tau)> = \delta(\tau)$$

The membrane potential, V(t), depends only on its present
value and the input i(t) and satisfies

$$dV/dt = \phi_1(V,t) + \phi_2(V,t)w(t)$$

or dV = $\phi_1(V,t)dt$ + $\phi_2(V,t)dW(t)$ (7.1.5)

where $\phi_1(V,t)$ = $f(V)$ + $g(V)m(t)$

$\phi_2(V,t)$ = $g(V)s(t)$ (7.1.6)

and $f(V)$ and $g(V)$ are continuous functions. Equation 7.1.5 is a fluctuation or Langevin equation, and Stratonovich (1963) shows that the characteristics of the Langevin equation, $\phi_1(V,t)$ and $\phi_2(V,t)$, are related to the incremental momemts of the diffusion equation $\alpha(V,t)$ and $\beta(V,t)$, by

$$\phi_1(V,t) = \alpha(V,t) + \frac{1}{4}\frac{\partial}{\partial V}\{\beta(V,t)\}^2 \quad (7.1.7)$$

$$\phi_2(V,t) = \{\beta(V,t)\}^2 \quad (7.1.8)$$

7.2 The Wiener Process.

Consider the motion of a particle undergoing in discrete time an unrestricted random walk along the real line $-\infty < x < +\infty$, starting at time t=0 at the origin x=0. In every time interval ΔT the particle will take a step Z of magnitude ΔX in either the positive or negative direction, with

Prob $\{Z = \Delta X\}$ = p

Prob $\{Z = -\Delta X\}$ = q = 1 - p (7.2.1)

and all steps are mutually independent. The distance $X(n\Delta T)$ of the particle from the origin is a discrete-valued stochastic process which is the sum of n = $t/\Delta T$ independent random variables Z_i, i = 1,2,...n This process $X(n\Delta T)$ will have a mean μ and variance σ^2 given by:

$$\mu = (t/\Delta T)\Delta X (p - q)$$

$$\sigma^2 = (t/\Delta T) (\Delta X)^2 4pq \quad (7.2.2)$$

As the step magnitude ΔX and time interval ΔT both tend to zero such that

$$(p - q)\Delta X/\Delta T \rightarrow \mu$$
$$4pq(\Delta X)^2/\Delta T \rightarrow \sigma^2 \quad (7.2.3)$$
$$p \rightarrow q \rightarrow \tfrac{1}{2}$$

the process X(nΔT) gives a stochastic process W(t) with the properties:

    (a) $W(0) = 0$

    (b) $(W(t) - W(\tau))$ has a Gaussian distribution with a zero mean and variance $\sigma^2(t - \tau)$ for $\tau \leqslant t$

    (c) $W(t_2) - W(t_1),\ldots,W(t_m) - W(t_{m-1})$ are independent for $t_1 < t_2 < \ldots < t_m$ i.e. the process has independent increments.

This process W(t) is a Wiener process with parameter σ^2 and occurs as a simple model for Brownian motion. When $\mu > 0$ the process obtained from X(nΔT) on taking the limits (7.2.3) is a Wiener process with drift μ. The Wiener process may be obtained from the random walk either by using the de Moivre-Laplace approximation to obtain the probability distribution of W(t) by taking the distribution of X(nΔT) to the limits (7.2.3), or by approximating the difference equation of the process X(nΔT):

$$f(t + \Delta T, x) = p.f(t, x - \Delta X) + q.f(t, x + \Delta X) \qquad (7.2.4)$$

where $f(t,x) = \text{Prob}\{X(t=n\Delta T) = x\}$. Since the function f(.) has continuous derivatives, this difference equation can be expanded by Taylor's theorem to give the approximation

$$\frac{\partial f(t,x)}{\partial t} = (q - p)\delta x \frac{\partial f(t,x)}{\partial x} + \frac{1}{2}\delta x \partial^2 \frac{\partial^2 f(t,x)}{\partial x^2} + \ldots \qquad (7.2.5)$$

which, neglecting third order terms, gives on taking the limits (7.2.3)

$$\frac{\partial f(t,x)}{\partial t} = -\mu \frac{\partial f(t,x)}{\partial x} + \frac{1}{2}\sigma^2 \partial^2 \frac{\partial^2 f(t,x)}{\partial x^2} \qquad (7.2.6)$$

which is the Fokker-Planck diffusion equation.

    Since the Wiener process is obtained from the random walk process, by letting $\Delta T \to 0$, in any short interval there will be a large number of steps : thus the Wiener process is not differentiable. However, it is possible to give meaning to a process w(t) which, when integrated, gives the Wiener process W(t). If $\phi(t)$ is a continuously differentiable function on the interval from a to b, the integral

$$\int_a^b \phi(t)\,dW(t)$$

does not exist, but can be interpreted as

$$\lim_{\varepsilon \to 0} \int_a^b \phi(t) \; \{ \; \frac{W(t+\varepsilon) \; - \; W(t)}{\varepsilon} \; \} \; dt$$

$$= \; \int_a^b \phi(t) \; \frac{d}{dt} \; \{ \frac{1}{\varepsilon} \int_t^{t+\varepsilon} W(s) \, ds \; \} \; dt$$

$$= \; \phi(b) W(b) \; - \; \phi(a) W(a) \; - \; \int_a^b \phi'(t) W(t) \, dt \qquad (7.2.7)$$

Thus $w(t) = dW(t)/dt$ can be defined, and is white noise.

In section 7.1 I showed that a suitable input signal to a model neurone is a white noise $w(t)$; for the perfect integrator model the subthreshold potential is the integral of the input signal, and so the subthreshold membrane potential of the perfect integrator model is a Wiener process when the input is a white noise.

7.3 The Ornstein-Uhlenbeck Process.

The Wiener process is a poor model of Brownian motion, since for small values of time the change in displacement $\Delta W(t)$ in an interval Δt is of the order of magnitude $\sqrt{(\Delta t)}$, and so the velocity of the particle becomes infinite as $\Delta t \to 0$. Uhlenbeck and Ornstein (1930) and Wang and Uhlenbeck (1945) have given an alternative model for Brownian motion of a particle, which holds for all values of time. If $U(t)$ is the velocity of a particle undergoing Brownian motion, in any time interval the velocity may change due to:

(a) frictional resistance to movement, the frictional resistance being proportional to $U(t)$

(b) the effects of random impacts which are equivalent to the increments $dW(t)$ of a Wiener process $W(t)$.

Thus there is an elastic force restraining the motion and this gives a drift towards the origin

$$\frac{dU(t)}{dt} \; = \; -\beta U(t) \; + \; W(t) \qquad (7.3.1)$$

If the particle has a unit mass β is a frictional coefficient, and $U(t)$ has no derivative. Doob (1942) writes (7.3.1) as

$$dU(t) \quad = \quad -\beta U(t) \quad + \quad dW(t) \qquad\qquad (7.3.2)$$

and interprets $dW(t)$ as a white noise. The stochastic differential equation (7.3.2) has a solution defining the Ornstein-Uhlenbeck process $U(t)$

$$U(s + t) \quad = \quad U(s)\exp(-\beta t) \qquad\qquad (7.3.3)$$

$$+ \exp \; (-\beta \, (s + t) \,) \; \int_s^{s+t} \exp(\beta\tau)dW(\tau)$$

This can be integrated to give the displacement $X(t)$:

$$X(t) \quad = \quad X(0) \quad + \quad \frac{1}{\beta}\{1 - \exp(-\beta t) \; \} \; U(0)$$

$$\qquad\qquad (7.3.4)$$

$$= \quad \frac{1}{\beta} \int_0^t \{1 \; - \; \exp(-\beta \, (t - \tau) \,) \;) \; \} \; dW(\tau)$$

which has a mean $X(0)$ and a variance which is proportional to t for large t and is proportional to t^2 for small t. Thus, unlike the Wiener process model for Brownian motion, the velocity is finite as $\Delta t \to 0$.

When the input signal to a leaky integrator model neurone (see section 3.3) is a white noise as defined in section 7.1, the subthreshold membrane potential is a realization of an Ornstein-Uhlenbeck process $U(t)$ as defined by equation 7.3.3 . The time constant of the leaky integrator corresponds to the reciprocal of the frictional coefficient β.

7.4 The Diffusion Equations.

Let $\{ V(t) \}$, $t \geqslant 0$, be a stationary, continuous stochastic process defined on the real line, with a transition probability distribution $F(y,t_2|x,t_1)$, $t_2 > t_1$, $t_2 - t_1 = t$

$$F(y,t_2|x,t_1) \quad = \quad \text{Prob} \, \{ V(t_2) < y \; |V(t_1) = x\}$$

$$= \quad \text{Prob} \, \{ V(t_2 - t_1) < y \; |V(0) = x\} \quad (7.4.1)$$

$$= \quad F(y,t|x)$$

which is a continuous function of t for any fixed value of x. Thus $F(y,t|x)$ is a conditional distribution function satisfying

$$\lim_{y \to -\infty} F(y,t|x) = 0$$

$$\lim_{y \to +\infty} F(y,t|x) = 1$$

and $\quad f(y,t|x) = \dfrac{\partial F(y,t|x)}{\partial y}$

defines the transition probability density function.

If $\{V(t)\}$ is a Markov process the Markov property can be written as

$$f(y,t|x) = \int_{-\infty}^{\infty} f(y,\tau|z)f(z,t-\tau|y)dz \; ; \tag{7.4.2}$$

$$0 \leqslant \tau \leqslant t$$

which is the Smoluchowski or Chapman-Kolmogorov equation. This equation is an expression of the transition of probability from instant to instant, and follows from the definition of a Markov process as a process whose future developmemt is independent of the past, and depends only on its present state.

Using the Smoluchowski equation:

$$F(y,t_2|x,t_1 - \Delta t) = \int_{-\infty}^{\infty} F(y,t_2|z,t_1)F(z,t_1|x,t_1 - \Delta t)dz$$

$$F(y,t_2|x,t_1) = \int_{-\infty}^{\infty} F(y,t_2|x,t_1)F(z,t_1|x,t_1 - \Delta t)dz$$

and so

$$\frac{F(y,t_2|x,t_1 - \Delta t) - F(y,t_2|x,t_1)}{\Delta t} =$$

$$\frac{1}{\Delta t} \int_{|z-x|>\delta} \{F(y,t_2|z,t_1) - F(y,t_2|x,t_1)\}F(z,t_1|x,t_1 - \Delta t)dz$$

$$+ \; \frac{1}{\Delta t} \int_{|z-x|<\delta} \{F(y,t_2|z,t_1) - F(y,t_2|x,t_1)\}F(z,t_1|x,t_1 - \Delta t)dz$$

$$\tag{7.4.3}$$

In an infinitesimal period of time Δt, the probability that the magnitude of the change in $V(t)$ is $\geqslant \delta$, where $\delta > 0$, is much less than the magnitude of Δt, and so

$$\lim_{\Delta t \to 0} \frac{1}{\Delta t} \int_{|z-x|>\delta} F(z,t_1|x,t_1 - \Delta t)dz = 0$$

Thus the first integral of the RHS of equation 7.4.3 will approach zero as $\Delta t \to 0$. If the second derivative of $F(y,t|x)$ exists, the second integral of the RHS of equation 7.4.3 can be expressed using Taylor's formula, and so:

$$\frac{F(y,t_2|x,t_1 - \Delta t) - F(y,t_2|x,t_1)}{\Delta t} =$$

$$\frac{1}{\Delta t} \int_{|z-x|<\delta} \{F(y,t_2|z,t_1) - F(y,t_2|x,t_1)\}F(z,t_1|x,t_1 - \Delta t)dz$$

$$+ \frac{1}{\Delta t} \frac{\partial F(y,t_2|x,t_1)}{\partial x} \int_{|z-x|<\delta} (z - x)F(z,t_1|x,t_1 - \Delta t)dz$$

$$+ \frac{1}{\Delta t} \frac{\partial^2 F(y,t_2|x,t_1)}{\partial x^2} \int_{|z-x|<\delta} \{(z-x)^2 + o(z-x)^2\}.$$

$$\cdot F(z,t_1|x,t_1 - \Delta t)dz \qquad (7.4.4)$$

On taking the limit as $\Delta t \to 0$, equation 7.4.4 gives:

$$\frac{-\partial F(y,t_2|x,t_1)}{\partial t} = \frac{1}{2} \alpha(t_1,x) \frac{\partial^2 F(y,t_2|x,t_1)}{\partial x^2}$$

$$+ \beta(t_1,x) \frac{\partial F(y,t_2|x,t_1)}{\partial x} \qquad (7.4.5)$$

where

$$\alpha(t_1,x) = \lim_{\Delta t \to 0} \frac{1}{\Delta t} \int_{|z-x|<\delta} (z-x)^2 F(z,t_1+\Delta t|x,t_1)dz$$

$$\geqslant 0 \qquad\qquad (7.4.6)$$

is the infinitesimal variance and

$$\beta(t_1,x) = \lim_{\Delta t \to 0} \frac{1}{\Delta t} \int_{|z-x|<\delta} (z-x) F(z,t_1+\Delta t|x,t_1)dz$$

$$(7.4.7)$$

is the infinitesimal mean. Equation 7.4.5 is the backward Kolmogorov equation, and the derivation given above follows that given in Bharucha-Reid (1960). Moran (1968) gives a different derivation of this equation.

If $\{V(t)\}$ is a stationary stochastic process, the backward Kolmogorov equation for the transition probability distribution function can be written as:

$$\frac{-\partial F(y,t|x)}{\partial t} = \frac{1}{2} \alpha(x) \frac{\partial^2 F(y,t|x)}{\partial x^2} + \beta(x) \frac{\partial F(y,t|x)}{\partial x}$$

$$(7.4.8)$$

The backward Kolmogorov equations 7.4.5 and 7.4.8 are also satisfied by the transition probability density function $f(y,t|x) = \frac{\partial F(y,t|x)}{\partial y}$

$$\frac{-\partial f(y,t|x)}{\partial t} = \frac{1}{2} \alpha(x) \frac{\partial^2 f(y,t|x)}{\partial x^2} + \beta(x) \frac{\partial f(y,t|x)}{\partial x}$$

$$(7.4.9)$$

Any stochastic process $\{V(t)\}$ whose transition distribution $F(y,t|x)$ or transition density $f(y,t|x)$ satisfy the backward Kolmogorov equation also has a density function which satisfies the forward Kolmogorov equation:

$$(7.4.10)$$

$$\frac{\partial f(y,t|x)}{\partial t} = \frac{1}{2} \frac{\partial^2 \{\alpha(y)f(y,t|x)\}}{\partial y^2} - \frac{\partial\{\beta(y)f(y,t|x)\}}{\partial y}$$

which is also called the Fokker-Planck equation. This forward
equation can be derived from the backward equation - see Moran (1968)
for details of this derivation.

The Kolmogorov equations define a diffusion process, and the
type of diffusion process is specified by the infinitesimal moments
$\alpha(x)$ and $\beta(x)$. When $\alpha(x) = 1$ and $\beta(x) = 0$, a Wiener process is
obtained; when $\alpha(x) = \alpha$ and $\beta(x) = -\beta x$ an Ornstein-Uhlenbeck process
is obtained.

Given the Kolmogorov equations with appropriate infinitesimal
moments $\alpha(x)$ and $\beta(x)$ the problem is to solve, for appropriate initial
and boundary conditions, the equations for $f(y,t|x)$. The Kolmogorov
partial differential equations for the transition probability density
function are parabolic partial differential equations and can be
solved by a variety of methods. Bharucha-Reid discusses two approaches:
the method of separating the variables and the Laplace transform
method. Such partial differential equations also occur in the analytic
theory of heat conduction (Carslaw and Jaeger, 1959; Luikov, 1968;
Widder, 1975), the theory of diffusion (Crank, 1974) and cable theory
(Jack, Noble and Tsien, 1975), and so solutions of this type of
equation, for a wide range of initial and boundary conditions, are
well known. Solutions for $f(y,t|x)$ for a number of diffusion processes
with a variety of initial and boundary conditions are tabulated in
Goel and Richter-Dyn (1974).

7.5   The First Passage Time Distribution.

Given a stochastic process $\{V(t)\}$, $t \geqslant 0$, one class of
problem is concerned with the zero- and level-crossing behaviour of
$\{V(t)\}$: a recent review of this field is given by Blake and Lindsey
(1973); see also the chapter by Helmstrom in Balakrishnan (1968). In
this section I am concerned with one problem : if $\{V(t)\}$ is a continuous
Markov stochastic process, how can the distribution of times of the
first passage to a level k be obtained, given that $V(0) = x_o$. The
context is that V(t) represents the membrane potential, and for models
in which V(t) is reset instantaneously to a resting value $V_r = x_o$ after
an action potential, the distribution of first passage times from x_o
to a threshold level k $(k > x_o)$ will be the interspike interval
distribution. A discussion of the first passage time distribution
problem is given by Siegert (1951) and Darling and Siegert (1953)

If $\{V(t)\}$ is stationary and has a transition probability
distribution $F(y,t|x)$ which satisfies the Smoluchowski equation 7.4.2

and the backward Kolmogorov equation 7.4.5, and has a transition

probability density $f(y,t\,|x)$ which satisfies both the forward and
backward Kolmogorov equations (7.4.9 and 7.4.10), the probability
density function $p(k,t\,|x_o)$ of the first passage time $T(k\,|x_o)$ satisfies:

$$f(y,t\,|x_o) \quad = \quad \int_0^t p(k,t-\tau\,|\,x_o)f(y,\tau\,|k)\,d\tau \qquad (7.5.1)$$

$$x_o < k < y \qquad or \qquad y < k < x_o$$

This follows from the continuity of the sample paths of $\{V(t)\}$: if $V(t)$
is to reach a value y at a time t given that it was at x_o at $t = 0$, it
must have reached some intermeadiate value k for the first time at a
time $(t-\tau)$, and then in the interval τ changed from k to y. Thus the
first passage time density function is inside a convolution: this
can be simplified by taking the Laplace transforms of the densities
(see section 5.4) and using the convolution theorem to give

$$f^*(y,s\,|x_o) \quad = \quad p^*(k,s\,|x_o)\,f^*(y,s\,|k) \qquad (7.5.2)$$

where $f^*(y,s\,|x)$ and $p^*(y,s\,|x_o)$ are the Laplace transforms of $f(y,t\,|x)$
and $p(k,t\,|x_o)$. Taking the Laplace transforms is not just a notational
convenience, as $p(k,t\,|x_o)$ is a solution of a partial differential
equation which can be solved by the Laplace transform method, and, for
an Ornstein-Uhlenbeck process, $p(k,t\,|x_o)$ is not available but $p^*(k,s\,|x_o)$
can be obtained (see section 7.7)

　　　　If equation 7.5.1 is substituted into the backward Kolmogorov
equation 7.4.9:

$$-p(k,0\,|x_o)f(y,t\,|x_o) \quad = \quad \int_0^t \{ \frac{\partial}{\partial t} \quad - \beta(x)\frac{\partial}{\partial x} \quad - \tfrac{1}{2}\alpha(x)\frac{\partial^2}{\partial x^2} \} \cdot$$

$$p(k,t-\tau\,|x_o)f(y,\tau\,|k)\,d\tau \qquad (7.5.3)$$

By definition, $p(k,0\,|x_o) = 0$ for $x_o \neq k$, and so

$$\frac{\partial p(k,t\,|x_o)}{\partial t} \quad = \quad \beta(x)\frac{\partial p(k,t\,|x_o)}{\partial x_o} \quad + \tfrac{1}{2}\alpha(x)\frac{\partial^2 p(k,t\,|x_o)}{\partial x_o^2}$$

$$(7.5.4)$$

which is the backward Kolmogorov equation with initial condition:

$$p(k,0|x_o) = \delta(k - x_o) \qquad (7.5.5)$$

and boundary conditions:

$$p(k,t|k) = \delta(t)$$

$$\lim_{x \to -\infty} p(x,t|x_o) = 0 \qquad (7.5.6)$$

For a Wiener process, $\alpha(x) = \sigma^2$, the variance, and $\beta(x) = \mu$, the mean, and so taking the Laplace transform of (7.5.4)

$$sp^*(y,s|x_o) = \frac{1}{2} \sigma^2 \frac{d^2 p^*(y,s|x_o)}{dx_o^2} + \mu \frac{dp^*(y,s|x_o)}{dx_o}$$

$$(7.5.7)$$

which is a second order linear differential equation which has a general solution:

$$p^*(y,s|x_o) = A \exp\{x_o \theta_1(s)\} + B \exp\{x_o \theta_2(s)\} \qquad (7.5.8)$$

where A and B are constants and

$$\theta_1(s), \theta_2(s) = \frac{-\mu \pm \sqrt{(\mu^2 + 2s\sigma^2)}}{\sigma^2}$$

and the particular solution for the boundary conditions (7.5.6)

$$p^*(k,s|x_o) = \exp\{(x_o - k)\theta_2(s)\} \qquad (7.5.9)$$

When $x_o = 0$, the inverse Laplace transform of equation 7.5.9 gives the well known first passage time density of a Wiener process as

$$p(k,t|0) = \frac{k}{\sigma\sqrt{(2\pi t^3)}} \exp\{-(k - \mu t)^2 / 2\sigma^2 t\} \qquad (7.5.10)$$

which has been given above as equation 6.4.6. This result is obtained by different methods in example 5.1 of Cox and Miller (1965) and in Blake and Lindsey (1973).

When $\{V(t)\}$ is a Wiener process with zero drift

$$p(k,t|0) \quad = \quad \frac{k}{\sigma \sqrt{(2\pi t^3)}} \quad \exp \{ k^2/2\sigma^2 t \} \qquad (7.5.11)$$

which has no finite moments.

For the Ornstein-Uhlenbeck process the renewal equation 7.5.1 has no solution.

7.6   The SIPIT Model.

The properties of the Stochastic Input to a Perfect Integrator with constant Threshold followed by an instantaneous reset mechanism, or SIPIT model (Johannesma, 1968) have been extensively investigated: Gerstein and Mandlebrot, 1964; Johannesma, 1969; Cowan, 1972; Stein, French and Holden,1972; Knox, 1974; Rubio and Holden, 1975. In this model there are two types of input, excitatory input pulses arriving with a rate r_e and causing the transition

$$V \quad \to \quad V \quad + \quad a$$

and inhibitory input pulses arriving with a rate r_i and causing the transition

$$V \quad \to \quad V \quad - \quad b.$$

Thus the incremental moments of $V(t)$ are the infinitesimal mean   m:

$$m \quad = \quad r_e a \quad - \quad r_i b \qquad (7.6.1)$$

and the infinitesimal variance s^2:

$$s^2 \quad = \quad r_e a^2 \quad + \quad r_i b^2 \qquad (7.6.2)$$

The Langevin equation of the diffusion approximation for the membrane potential is given by equation 7.1.5 with $f(V) = 0$ and $g(V) = 1$, and so

$$dV/dt \quad = \quad m \quad + \quad sw(t) \qquad (7.6.3)$$

and $w(t)$ is a Gaussian, delta-correlated noise or white noise as defined in section 7.1.

The subthreshold membrane potential $\{V(t)\}$ is a Wiener process; when $V(t)$ reaches a constant threshold k an impulse is generated and $V(t)$ is instantaneously reset to its resting value x_o. $V(t)$ is characterized by the transition probability density function $f(y,t|x)$ which satisfies the forward and backward Kolmogorov equations with $\alpha(x) = s^2$ and $\beta(x) = m$.

$$f(y,t|x) = \text{Prob } \{V(t) = y|V(0) = x \text{ and no action potentials}$$

$$\text{have been generated in the interval } [0,t]\}$$

$$= G\{(y - x)/s\sqrt{t}\} - G\{(y - x - 2k)/s\sqrt{t}\} .$$

$$\exp\{(my - m^2t)/s^2\} \qquad (7.6.4)$$

where
$$G\{(z - \mu)/\sigma\} = \frac{1}{\sigma\sqrt{2\pi}} \exp\{-\tfrac{1}{2}(\frac{z - \mu}{\sigma})^2\}$$

is the Gaussian distribution.

The first passage time density of $\{V(t)\}$, which is the inter-spike interval probability density function, is given by equation 7.5.10 as

$$p(k,t|x_o) = \frac{k - x_o}{s\sqrt{(2\pi t^3)}} \exp\{-(k - x_o - mt)^2/2s^2t\}$$

$$(7.6.5)$$

the integral of which gives the first passage time time distribution $P(\theta)$.

$$P(\theta) = \int_0^\theta p(k,t|x_o)dt \qquad (7.6.6)$$

The transition density function $f(y,t|x)$ and the first passage time density function $p(k,t|x_o)$ of the Wiener process are well known, and they describe the stationary characteristics of the SIPIT model. I now want to use these density functions to describe the dynamics of the model in terms of its stochastic frequency response function or its impulse response.

Johannesma (1969) gives an introductory account of the problem of the stochastic dynamics of neural models in his chapter 4. Since spike generating models are grossly nonlinear, any linear approximations of the dynamics by a frequency response function or an impulse response will depend on the form and amplitude of the input signal.

The case of a Poisson distributed, purely excitatory input pulse train has been discussed in section 3.2 and Holden (1976). Here I will treat two approaches to the dynamics of the SIP IT model when the input is a Gaussian, delta correlated white noise with spectral density $S_x(\omega)$ given by:

$$S_x(\omega) \quad = \quad A^2/2\pi \qquad\qquad (7.6.7)$$

The first approach to the dynamics of the SIPIT model is in the frequency domain and follows Stein, French and Holden (1972). Without loss of generality let

$$x_o \quad = \quad 0$$

$$k \quad = \quad 1$$

and so for a white noise input with mean m, variance A^2 and spectral density given by (7.6.7) the interspike interval density will be given by equation 7.6.5. This interspike interval distribution has a Laplace transform P*(s) given by Sugiyama et al (1970) as:

$$P*(s) \quad = \quad \exp(m/A^2)\exp\{-(m^2 + 2sA^2)^{\frac{1}{2}} /A^2\} \qquad (7.6.8)$$

Note that s is now the complex variable of the Laplace transform. The mean interval is $\mu = 1/m$ and the variance of the interval distribution is $\sigma^2 = A^2/m^3$.

For a frequency ω, P(jω) will be of the form:

$$P(j\omega) \quad = \quad \exp(-\alpha)\exp(-j\beta) \qquad\qquad (7.6.9)$$

where $\quad \alpha \quad = \quad \dfrac{m}{A^2} \quad (C \cos\theta - 1)$

$$\beta \quad = \quad \dfrac{m}{A^2} \quad (C \sin\theta)$$

$$C \quad = \quad \{1 + (2\omega)^2 (A/m)^4\}^{1/4}$$

$$\theta \quad = \quad \tfrac{1}{2}\tan^{-1}\{2\omega\,(A/m)^2\}$$

and so substituting (7.6.9) into equation 3.2.8 gives the output spectral density $S_y(\omega)$ as:

$$S_y(\omega) \quad = \quad \frac{m}{2\pi}\ \frac{\exp(2\alpha)\ -\ 1}{\exp(2\alpha)\ -\ 2\exp(\alpha)\cos(\beta)\ +\ 1} \qquad (7.6.10)$$

which for small ω

$$S_y(\omega) \quad = \quad A^2/2\pi$$

and for large ω

$$S_y(\omega) \quad = \quad m/2\pi$$

Limits for the cross-spectral density between the input white noise and the output spike train can be obtained from the cross-correlation function $\rho_{xy}(u)$

$$\rho_{xy}(u) \quad = \quad \lim_{T\to\infty}\ \frac{1}{T}\ \int_0^T\ x(t)\,y(t+u)\,dt$$

$$= \quad \lim_{T\to\infty}\ \frac{1}{T}\ \sum_{i=1}^{n}\ x(t_i - u)$$

$$= \quad \frac{1}{\mu}\ E\,\{x(t-u)\,\} \qquad (7.6.11)$$

$E\,\{x(t-u)\}$ is the expected value of the input $x(t)$ at a time u before an output pulse. For small negative times $\tau = t - u$

$$E\,\{x(\tau)\,\} = \quad (-\pi A^2/8\,\tau) \qquad (7.6.12)$$

and for large negative times τ

$$E\,\{x(\tau)\} \quad = \quad m \qquad (7.6.13)$$

Thus for high ω the cross-spectral density $S_{xy}(\omega)$ will be

$$S_{xy}(\omega) \quad = Am/4\sqrt{(2j\pi)} \tag{7.6.14}$$

and for low ω

$$S_{xy}(\omega) \quad = \quad S_{xx}(\omega) \quad = \quad A^2/2\pi \tag{7.6.15}$$

The frequency response function is proportional to the cross-spectral density $S_{xy}(\omega)$, and the coherence function $\gamma^2(\omega)$ (see section 5.4) is found from $S_x(\omega)$, $S_y(\omega)$ and $S_{xy}(\omega)$ from equations 7.6.7, 7.6.10, 7.6.14 and 7.6.15:

$$\gamma^2(\omega) \quad = \quad 1 \quad \text{for small } \omega$$
$$\tag{7.6.16}$$
$$= \quad \pi^2 m/8\omega \quad \text{for large } \omega$$

These spectral densities and coherence functions have been computed and are illustrated in Figure 7.1.

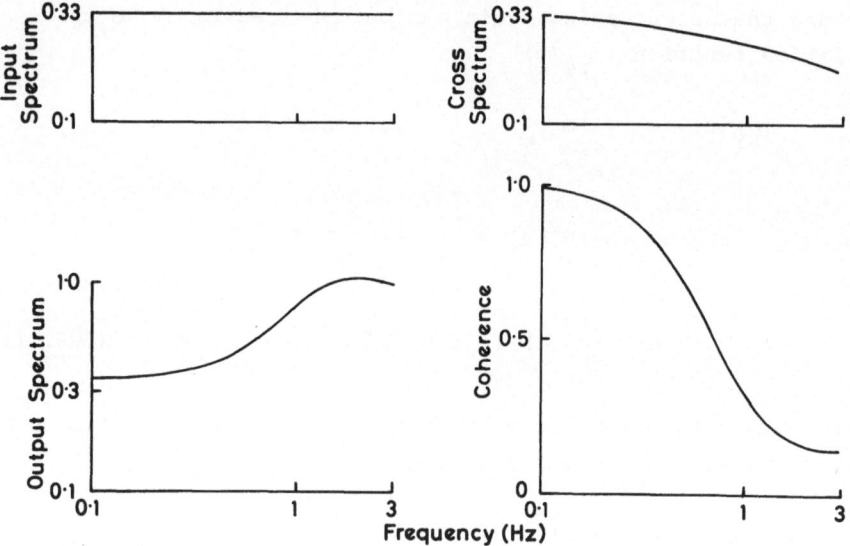

Figure 7.1. The input, output and cross-spectra and coherence function of a perfect integrator model subject to a white noise input with $A^2 = 1/3$. Input and output spectra obtained from 7.6.7 and 7.6.10; cross-spectrum from 7.6.11-15 and the coherence function from these spectra using 5.4.21.

The second approach to the dynamics of the SIPIT model is in the time domain and follows Rubio and Holden (1975). Once again, the input is a white noise w(.) with spectral density given by equation 7.6.7; V(.) is the membrane potential and y(.) is the output spike train. These are all real valued stochastic processes defined on the real line. The small-signal impulse response H(t) of the model is given by the convolution integral:

$$H(t) \quad = \quad \frac{1}{A^2} \lim_{T \to \infty} \frac{1}{2T} \int_{-T}^{T} y(\xi) \, w(\xi - t) \, d\xi \qquad (7.6.17)$$

The response to a sample of the process w(.) will be a sequence of spikes at times $\{t_i\}$, $i = \ldots, -2, -1, 0, 1, 2, \ldots$ with $t_{i+1} > t_i$. If $m_+(T)$ is the number of elements of this sequence in the interval $[0,T]$ and $m_-(T)$ the number of elements in $[-T,0)$, then

$$\lim_{T \to \infty} \frac{m_+(T) + m_-(T)}{2T} \quad = \quad r \qquad (7.6.18)$$

where r is the average rate of output pulses, which is proportional to m, the mean of the input white noise. For the case of $x_o = 0$ and $k = 1$, $r = m$. The output pulses have a height h and a duration a.

For $\quad t \geqslant 0$

$$A^2 H(t) \quad = \quad \lim_{T \to \infty} \frac{1}{2T} \sum_{i=m_-(T)}^{i=m_+(T)} \int_{t_i}^{t_i+a} h \, w(\xi - t) \, d\xi$$

$$= \quad \lim_{T \to \infty} \frac{m_+(T) + m_-(T)}{2T} \, (ha) \, \frac{1}{m_+(T) + m_-(T)}$$

$$\cdot \; \sum_{i=m_-(T)}^{i=m_+(T)} \frac{1}{a} \int_{t_i}^{t_i+a} w(\xi - t) \, d\xi$$

$$= \quad (r \, h \, a) \lim_{m_+,m_- \to \infty} \frac{1}{m_+ + m_-} \sum_{i=m_-}^{i=m_+} z(t_i - t)$$

$$(7.6.19)$$

where the stochastic process z(.) is defined by:

$$z(t) = \frac{1}{a} \int_{t}^{t+a} w(\xi) d\xi \qquad (7.6.20)$$

It can be shown (see the appendix of Rubio and Holden, 1975)
that:

$$\lim_{m_+, m_- \to \infty} \frac{1}{m_+ + m_-} \sum_{i=m_-}^{i=m_+} z(t_i - t)$$

$$= E\{z(t' - t)\} \qquad (7.6.21)$$

where t' is a value of the time when V(t) reaches the threshold k, and
so E{z(t' - t)} is the expected value of z(.) a time t before an
action potential.

If there is no generation of an action potential in the
interval (t, t+a]

$$z(t) = \frac{1}{a}\{y(t + a) - y(t)\} \qquad (7.6.22)$$

Now let the impulses have negligible width but constant area i.e. let
a → 0 as ah → d, a constant.

Then,

$$E\{z(t' - t)\} = \lim_{a \to 0} \frac{1}{a}\left\{E\{y(t' + a - t)\} - E\{y(t' - t)\}\right\}$$

$$= -d\phi(t)/dt \qquad (7.6.23)$$

where $\phi(t) = E\{y(t' - t)\}$ \qquad (7.6.24)

Thus the impulse response H(t) is given by:

$$H(t) = \frac{-rd}{A^2} \frac{d\phi(t)}{dt} \qquad (7.6.25)$$

and the problem now is to find an expression for the function $\phi(t)$.
Let t' - t = 0 and so the evaluation of the function $\phi(t)$ is

the evaluation of the probability density function of V(0) given that
the neurone generated an action potential at a time t $\geqslant$ 0: this is
the probability density function f(V(0) = x|V(t) = k).

To simplify the notation, let τ be a random variable so that
$\tau \in I_t$ means that an impulse occurs in the interval

$$I_t = [t - \Delta t/2, \ t + \Delta t/2].$$

Similarly define the closed interval

$$I_x = [x - \Delta x/2, \ x + \Delta x/2].$$

The conditional probability P { V(0) $\in$ I_x | $\tau \in I_t$} is given by

$$P\{V(0) \in I_x | \tau \in I_t\} = \frac{P\{\tau \in I_t | V(0) \in I_x\} P\{V(0) \quad I_x\}}{P\{\tau \in I_t\}}$$

and so

$$\frac{1}{\Delta x} P\{V(0) \in I_x | \tau \in I_t\} = \frac{\frac{1}{\Delta t} P\{\tau \in I_t | V(0) \in I_x\} \frac{1}{\Delta t} P\{V(0) \quad I_x\}}{\frac{1}{\Delta t} P\{\tau \in I_t\}}$$

Taking the limits as Δx, $\Delta t \to 0$

$$f(V(0)=x|V(t)=k) = \frac{\left[\begin{array}{c} \lim \\ \Delta t \to 0 \\ \Delta x \to 0 \end{array} \dfrac{P\{\tau \in I_t | V(0) \in I_x\}}{\Delta t} \right] f(V(0) = x)}{f(\tau = t)}$$

$$(7.6.26)$$

The limit in the square brackets on the RHS of equation 7.6.26 is a
value of a function n(x,t), which is the density for the event of an
action potential initiation at a time t after the occurence of a value
x of the membrane potential. Johannesma (1969) describes this function
n(x,t) as the probability density for the initiation of an action

potential under repetitive activity. In the interval $[0, t]$ there can be an arbitrary number $N \geqslant 0$ of impulses, and so $n(x,t)$ is given by the sum over N of the convolution of the first passage time density from a membrane potential x to threshold, $p(k,t|x)$, with the N-fold convolution of the interspike interval density $p(k,t|x_o)$. Thus, in terms of Laplace transforms,

$$n^*(x,s) = \Sigma_{N=0}^{\infty} p^*(k,s|x) \{ p^*(k,s|x_o) \}^N$$

$$= p^*(k,s|x) / \{ 1 - p^*(k,s|x_o) \} \qquad (7.6.27)$$

Since the LHS of equation 7.6.26 is a probability density function, the integral from $-\infty$ to k of the RHS of (7.6.26) must equal unity: hence

$$f(\tau = t) = \int_{-\infty}^{k} n(x,t) \; f(V(0) = x) \qquad (7.6.28)$$

and so equation 7.6.25 becomes

$$f(V(0) = x|V(t) = k) = \frac{n(x,t) \; f(V(0) = x)}{\int_{-\infty}^{k} n(x,t) \; f(V(0) = x) dx} \qquad (7.6.29)$$

If θ is a random variable which equals the time since the last action potential, the unconditional probability density of the membrane potential is given by:

$$f(V(0) = x) = \int_{0}^{\infty} f(V(0) = x|\theta = \theta) \; f(\theta = \theta) d\theta \qquad (7.6.30)$$

where $f(V(0) = x|\theta = \theta)$ is the conditional probability density function of the membrane potential at a time θ seconds after an action potential, and $f(\theta = \theta)$ is the probability density of the time since the last action potential. From the recurrence time relation for renewal processes (equation 5.2.14)

$$f(\theta = \theta) = r \{ 1 - G(\theta) \} \qquad (7.6.31)$$

where $G(\theta)$ is the first passage time distribution function given by equation 7.6.6. Further, $f(V(0) = x| \theta = \theta)$ is the transition probability density function $f(x,\theta|x_o)$ defined by equation 7.6.4 , which is the probability density of the membrane potential having a value x at a time θ after an action potential.

The above results give the final expression for $\phi(t)$ as:

$$\phi(t) \quad = \quad E\{V(t' - t)\}$$

$$= \quad \int_{-\infty}^{k} x\, f(V(0) = x | V(t) = k)\, dx$$

$$= \quad \frac{\int_{-\infty}^{k} x n(x,t)\left\{\int_{0}^{\infty} f(x,\theta|x_o)\{1 - G(\theta)\}d\theta\right\} dx}{\int_{-\infty}^{k} n(x,t)\left\{\int_{0}^{\infty} f(x,\theta|x_o)\{1 - G(\theta)\}d\theta\right\} dx}$$

$$(7.6.32)$$

where $n(x,t)$ can be obtained from (7.6.27), and $f(x,\theta|x_o)$ and $G(\theta)$ are given by equations 7.6.4 and 7.6.6. $\phi(t)$ can be evaluated numerically using equation 7.6.32 and the on-response, given by equation 7.6.25, computed. Graphs of the on-response for two values of the mean m and variance s^2 of the input are plotted in Figure 7.2 : the input with the higher mean gives a steeper, more oscillatory on-response.

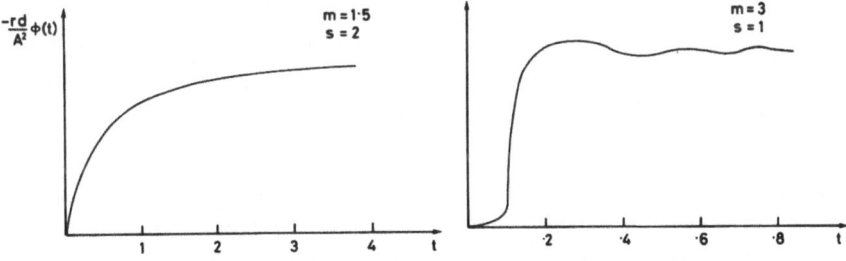

Figure 7.2. Graphs of the step on-response of the perfect integrator obtained from 7.6.25 using 7.6.4, 7.6.6, 7.6.27 and 7.6.32 for two values of the mean m and variance s^2 of the input white noise. The threshold $k = 1$ and the reset potential $x_o = 0$.

7.7. The SILIT Model.

The properties of the Stochastic Input to a Leaky Integrator with constant Threshold, followed by an instantaneous reset mechanism, or SILIT model (Johannesma, 1968), have been extensively investigated, but only limited results have been obtained : Fetz and Gerstein, 1963; Stein, 1965; Gastwirth, 1967; Gluss, 1967; ten Hoopen, 1966; Siebert, 1968; Johannesma, 1969; Roy and Smith, 1969; Siebert, 1969; Fienberg, 1970; Stein and French, 1970; Sugiyama, Moore and Perkel, 1970; Ricciardi and Ventriglia, 1970; Afanas'eva and Petunin, 1971; Capocelli and Ricciardi, 1971; Ricciardi, 1971; Cowan, 1972; Stein, French and Holden, 1972; Clay and Goel, 1973; Capocelli and Ricciardi, 1973; Knox, 1974; Matsuyama, Shirai and Akizuki, 1974; Poggio and Torre, 1975; Tuckwell, 1975. The major problem in the analysis of the properties of the SILIT model is that when the input $w(t)$ is a white noise, $\{V(t)\}$, the stochastic process representing the membrane potential, is an Ornstein-Uhlenbeck process (see section 7.3), and the first passage time density is not available. However, expressions for the moments and the Laplace transform of the first passage time density can be obtained.

The SILIT model can be considered either as a white noise input to the leaky integrator model of section 3.3, or as an input of Poisson distributed excitatory and inhibitory pulses to the leaky integrator model. On taking appropriate limits the sequence of Poisson distributed input pulses becomes a white noise.

In the absence of an input, the membrane potential decays with an exponential time course to the resting value x_o, with a time constant τ. It is tempting to identify this time constant with the membrane time constant, given by the product of the membrane resistance and capacitance : however Knight (1973) has derived the leaky integrator model as a limiting form of the Hodgkin-Huxley equations, with the time constant determined by the sodium activation kinetics. As in section 7.6, the excitatory and inhibitory input pulses arrive with rates r_e and r_i, and cause the transitions

$$V \rightarrow V + a$$
$$V \rightarrow V - b$$

and when $V(t)$ reaches a threshold level k an action potential is generated and $V(t)$ is instantaneously reset to x_o. To simplify the notation, let

$$|a| = |b| = \varepsilon$$

Note that this does not produce a loss of generality in the diffusion approximation.

As before $f(x,t|x_o)$ is the probability density function for the membrane potential to be at a level x at a time t given that $V(0) = x_o$, and $F(x,t|x_o)$ denotes the corresponding distribution function. Now let $f(x,t) \equiv f(x,t|x_o)$, and so

$$F(x,t+\delta t) = F(\{x - x_o\} \exp(\delta t/\tau) + x_o, t) \{1 - (r_e + r_i)\varepsilon\delta t\}$$

$$+ F(\{x - x_o - \varepsilon\} \exp(\delta t/\tau) + x_o, t) r_e \delta t$$

$$+ F(\{x - x_o + \varepsilon\} \exp(\delta t/\tau) + x_o, t) r_i t + o(\delta t)$$

$$(7.7.1)$$

This gives the probability distribution of $V(.)$ being at a level less than, or equal to, x at a time $(t + \delta t)$ as the sum of the probabilities that there have been no input pulses, an excitatory pulse, an inhibitory pulse or two or more pulses in the preceeding interval $[t,t + \delta t]$. The Taylor's series expansion of equation 7.7.1 gives:

$$F(x,t + \delta t) = \{F(x,t) + \frac{\delta t}{\tau}(x - x_o) \frac{\partial F(x,t)}{\partial x}\}\{1 - (r_e + r_i)\varepsilon\delta t\}$$

$$+ \{F(\{x - \varepsilon\},t) + \frac{\delta t}{\tau}(x - x_o - \varepsilon)\frac{\partial F(x - \varepsilon,t)}{\partial x}\} r_e \delta t$$

$$+ \{F(\{x + \varepsilon\},t) + \frac{\delta t}{\tau}(x - x_o + \varepsilon)\frac{\partial F(x - \varepsilon,t)}{\partial x}\} r_i \delta t$$

$$+ o(\delta t) \qquad\qquad (7.7.2)$$

where all quantities of order $o(\delta t^2)$ have been absorbed into $o(\delta t)$.

For small ε:

$$F(\{x - \varepsilon\},t) = F(x,t) + \sum_{n=1}^{\infty} \frac{(-1)^n \varepsilon^n}{n!} \frac{\partial^n F(x,t)}{\partial x^n}$$

$$(7.7.3)$$

and

$$F(\{x + \varepsilon\}, t) = F(x,t) + \sum_{n=1}^{\infty} \frac{\varepsilon^n}{n!} \frac{\partial^n F(x,t)}{\partial x^n} \qquad (7.7.4)$$

Since ε is assumed to be small, all terms of order ε^3 and above are negligible, and so from (7.7.2-7.7.4):

$$\frac{F(x, t + \delta t) - F(x,t)}{\delta t} = \frac{\varepsilon^2}{2} (r_e + r_i) \frac{\partial^2 F(x,t)}{\partial x^2}$$

$$- \{ \varepsilon (r_e + r_i) - \frac{(x - x_o)}{\tau} \} \frac{\partial F(x,t)}{\partial x} + o(\delta t)$$

$$(7.7.5)$$

which, taking the limit as $\delta t \to 0$ and differentiating wrt x gives the forward Kolmogorov equation :

$$\frac{\partial f(x,t)}{\partial t} = \frac{\varepsilon^2}{2} (r_e + r_i) \frac{\partial^2 f(x,t)}{\partial x^2}$$

$$(7.7.6)$$

$$- \frac{\partial}{\partial x} \{ \{ \varepsilon (r_e - r_i) - \frac{(x - x_o)}{\tau} \} \quad f(x,t) \}$$

This derivation of the diffusion equation follows that given by Fienberg (1970); Johannesma (1969) gives an alternative derivation of the forward Kolmogorov equation starting from the Smoluchowski equation. When $r_e = r_i$ equation 7.7.6 is the forward equation of the Ornstein-Uhlenbeck process - see section 7.3.

Note that equation 7.7.6 is an approximation, based on the plausible assumption that $\varepsilon << k$, and that the exact equation is the differential-difference equation:

$$\frac{\partial f(x,t)}{\partial t} = \frac{\partial}{\partial x} \{ \frac{(x - x_o) f(x,t)}{\tau} \} + r_e \quad f(x - \varepsilon, t) - f(x,t)$$

$$+ r_i \quad f(x + \varepsilon, t) - f(x,t) \qquad (7.7.7)$$

The problems in approximating this equation by a partial differential equation are discussed in Capocelli and Ricciardi (1971). An integrated version of this differential-difference equation is given in Stein (1965), who transforms (7.7.7) to obtain the characteristic function of $F(x,t)$, and by differentiating the logarithm of the characteristic function, finds the mean μ and variance σ^2 of the distribution of the subthreshold membrane potential as:

$$\mu(t) = \varepsilon\tau(r_e - r_i)\,\{1 - \exp(-(t - t_o)/\tau)\} \qquad (7.7.8)$$

$$\sigma^2(t) = (\varepsilon^2\tau/2)(r_e + r_i)\,\{1 - \exp(-(t - t_o/\tau)\} \qquad (7.7.9)$$

The probability density $f(x,t)$ of the subthreshold membrane potential of the diffusion approximation can be found by solving equation 7.7.6. Sugiyama, Moore and Perkel (1970) have obtained $f(x,t)$ for a model in which the resting potential is more positive than the reset potential : here I will use their method to solve equation 7.7.6.

$$\text{Let} \quad T = \exp(2t/\tau) \qquad (7.7.10)$$

$$X = (x - x_o + \varepsilon(r_e - r_i)/\tau)\exp(t/\tau) \qquad (7.7.11)$$

$$s^2 = \varepsilon^2(r_e + r_i) \qquad (7.7.12)$$

and so

$$f(x,t) = g(X,T)\frac{\partial X}{\partial x}$$

$$= \sqrt{T}g(X,T) \qquad (7.7.13)$$

and $g(X,T)$ satisfies

$$\frac{s^2\tau}{4}\frac{\partial^2 g(X,T)}{\partial x^2} = \frac{\partial g(X,T)}{\partial T^2} \qquad (7.7.14)$$

which is the heat conduction equation (Carslaw and Jaeger, 1959; Widder, 1975), which, with the initial conditions

$$t = 0 \quad \rightarrow \quad T = T_o = 1$$

$$\qquad (7.7.15)$$

$$x = x_o \quad \rightarrow \quad X = X_o = \{\varepsilon(r_e - r_i)/\tau\}\,\exp(t/\tau)$$

has the solution

$$g(X,T) \;=\; \sqrt{\{\pi s^2 (T - T_o)\}} \exp\left\{\frac{(X - X_o)^2}{s^2 \tau (T - T_o)}\right\} \qquad (7.7.16)$$

which is a time-varying Gaussian. Substituting back for x and t by using equations 7.710-13 and 7.7.15 gives:

$$f(x,t) \;=\; \sqrt{\{\pi s^2 \tau (1 - \exp(-2t/\tau))\}} \;.$$

$$\exp\left\{-\frac{((x - x_o + \varepsilon(r_e - r_i)/\tau) - \{\varepsilon(r_e - r_i)/\tau\}\exp(-t/\tau))^2}{s^2 \tau (1 - \exp(-2t/\tau))}\right\}$$

$$(7.7.17)$$

which has a mean

$$\mu(t) \;=\; x_o + \varepsilon\tau(r_e - r_i)\{1 - \exp(-t/\tau)\} \qquad (7.7.18)$$

and a variance

$$\sigma^2(t) \;=\; (\varepsilon^2\tau/2)(r_e - r_i)\{1 - \exp(-t/\tau)\} \qquad (7.7.19)$$

which are identical to those obtained by Stein (1965) - equations 7.7.8 and 7.7.9 - for the exact differential-difference equation.

Having found the transition probability density function $f(x,t|x_o) = f(x,t)$ given by equation 7.7.17, the problem now is to obtain the Laplace transform of the first passage time density $p^*(k,s|x_o)$

$$p^*(k,s|x_o) \;=\; \text{L.T.}\{p(k,t|x_o)\}$$

and $p(k,t|x_o)$ is the first passage time density defined in section 7.5. This first passage time density is only obtainable for the special case of $k = 0$ (Wang and Uhlenbeck,1945), which is of no interest in the neurophysiological context, although expressions for an analogous case have been obtained by Siebert (1969) and Sugiyama et al. (1970). In these two papers the reset potential was more negative than the resting potential x_o, and the threshold was equal to the asymptotic steady-state potential $(x_o + \varepsilon(r_e - r_i)\tau)$.

Siegert (1951) and Darling and Siegert (1953) give a general method for approaching the problem of finding $p*(k,s|x_o)$ for any Markov process. Starting from equation 7.5.2, rearranging gives

$$p*(k,s|x_o) \quad = \quad f*(y,s|x_o) \; / \; f*(y,s|k) \qquad (7.7.20)$$

$$\text{for } x_o < k < y$$

These Laplace transforms exist only if $\{V(t)\}$ is continuous. The transform of the transition density, $f*(y,s|x_o)$, can be expressed as the product of two functions:

$$f*(y,s|x_o) \quad = \quad \begin{cases} u(x_o) \; u_1(y) & x_o < y \\ v(x_o) \; v_1(y) & x_o > y \end{cases} \qquad (7.7.21)$$

and so

$$p*(k,s|x_o) \quad = \quad \begin{cases} u(x_o) \; / \; u(k) & x_o < k \\ v(x_o) \; / \; v(k) & x_o > k \end{cases} \qquad (7.7.22)$$

Setting $x = x_o$ in the backward Kolmogorov equation 7.4.9 and taking the Laplace transforms:

$$sf*(y,s|x_o) \quad = \quad \tfrac{1}{2}\alpha(x_o)\frac{d^2 f*(y,s|x_o)}{dx_o{}^2} + \beta(x_o)\frac{df*(y,s|x_o)}{dx_o}$$

$$(7.7.23)$$

with $\qquad f*(x_o,s|x_o) \quad = \quad 1 \qquad$ and $\qquad \displaystyle\lim_{y\to\infty} f*(y,s|k) = 0$

and so $-f*(y,s|x_o)$ is the Green's function of equation 7.7.23 over the interval $(-\infty < x_o < +\infty)$. The Green's function can be expressed as (see section 3.25 of Sokolnokoff and Redheffer, 1966) :

$$f*(y,s|x_o) \quad = \quad \begin{cases} v(x_o) \; u(y) & x_o \geqslant y \\ v(y) \; u(x_o) & x_o \leqslant y \end{cases} \qquad (7.7.24)$$

where u(y) and v(y) are any two linear, linearly independent solutions
of:

$$\tfrac{1}{2}\,\alpha\,(y)\,\frac{d^2\phi\,(y)}{dy^2} \;+\; \beta\,(y)\,\frac{d\phi\,(y)}{dy} \;-\; s\phi\,(y) \;=\; 0 \qquad (7.7.25)$$

which satisfy u(∞) = v($-\infty$) = 0.

For the normalized Ornstein-Uhlenbeck process, $\alpha\,(y)$ = 2 and
$\beta\,(y)$ = -y in equation 7.7.25, and Darling and Siegert give u(y) and
v(y) as:

$$u(y) \;=\; \exp(y^2/4)\,D_{-s}(y)$$

$$v(y) \;=\; \exp(y^2/4)\,D_{-s}(-y) \qquad (7.7.26)$$

where $D_{-z}(y)$ is the parabolic cylinder function or Weber function
(Abramowitz and Stegun, 1964) defined by:

$$D_{-z}(y) \;=\; \frac{\exp(-y^2/4)}{\Gamma(z)}\;\int_0^{\infty}\exp\{\frac{(-yx\,-x^2)}{2}\}\;x^{z-1}\;dx \qquad (7.7.27)$$

and $\Gamma(z)$ is the Gamma function defined in equation 5.2.6.

Thus from equations 7.7.22 and 7.7.26 Darling and Siegert
obtained the Laplace transform of the first passage time density of the
Ornstein-Uhlenbeck process as:

$$p^*_{0-U}(k,s|x_0) \;=\; \begin{cases} \exp\{\,(x_0^2 - k^2)/4\}\;\;D_{-s}(x_0)/D_{-s}(k) & ;\;x_0 < k \\[2ex] \exp\{\,(x_0^2 - k^2)/4\}\;\;D_{-s}(-x_0)/D_{-s}(-k) & ;\;x_0 > k \end{cases}$$

and noted that it appears very difficult to invert this transform.
Fienberg (1970) points out that Cowan (unpublished) has shown that the
inverse of this transform does not exist, except for the special case
of k = 0 considered above.

The paper by Darling and Siegert appears to have been missed by
neural modellers until it was referred to by Capocelli and Ricciardi
(1971), and so there have been several independent derivations of
$p^*(k,s|x_0)$ of the SILIT model : Gluss, 1967; Roy and Smith,1969;
Sugiyama et al. , (1970; Capocelli and Ricciardi, 1971, 1973;

Cowan, 1972; Goel, Richter-Dyn and Clay, 1972; Clay and Goel, 1973; Goel and
Richter-Dyn, 1974 .   The expression derived by Capocelli and Ricciardi
(1971) for $p*(k,s|x_o)$ can be written as:

$$p*(k,s|x_o) \quad = \quad exp\{ \frac{x_o^2 - k^2}{2 \, \sigma^2 \tau} \} \; \{ \; \frac{D_{-s\tau}(-x_o\sqrt{\{2/\sigma^2\tau\}})}{D_{-s\tau}(-k\sqrt{\{2/\sigma^2 \, \tau\}})} \; \}$$

$$(7.7.29)$$

which can be maipulated into the expression given by Roy and Smith
(1969) and Sugiyama et al. (1970).   For the case of $\tau = \frac{1}{2}\sigma^2 = 1$,
equation 7.7.29 coincides with (7.7.28).

Numerical estimates of $p(k,t|x_o)$ could be obtained by numerical
inversion of equation 7.7.29 by the methods of Bellman, Kalaba and
Lockett (1966);   however Sugiyama et al point out that it is
computationally simpler to solve a finite-difference approximation of:

$$p(k,t|x_o) \quad = \quad \frac{\partial}{\partial t} \; \{ \; 1 \; - \; \int_{-\infty}^{x_o} f(y,t|x_o) \; dy \; \}$$

$$= \quad \frac{\partial}{\partial t} \int_{-\infty}^{x_o} f(y,t|x_o) \, dy \qquad (7.7.30)$$

and they present a plot of $p(k,t|x_o)$ computed using this method.
Matsuyama, Shirai and Akizuki (1974) have presented a family of
numerically evaluated membrane potential transition densities and first
passage time densities.   From such numerically evaluated densities
one could then estimate the small signal impulse response of the SILIT
model by methods similar to those of Rubio and Holden (1975) for the
SIPIT model.   An alternative line of approach has been taken by Stein,
French and Holden (1972) who estimated experimentally the frequency
response function and coherence function of an electronic analog of the
SILIT model. Plots of the frequency response function for two rms
amplitudes of the band-limited white noise input are shown in Figure 7.3.
These experimental estimates show:

(a) a decrease in the peaks of the sensitivity at integer
multiples of the mean discharge rate (or carrier rate) as the rms
amplitude of the input noise increases.

(b) a decrease in the phase changes at integer multiples of
the carrier rate as the input noise amplitude increases.

(c) the large increase in sensitivity found at high frequencies by using cyclic inputs, which produce phase-locking of the response (see section 3.3), is not apparent.

These three effects all tend to reduce the distortions of periodic signals which are produced by phase-locking.

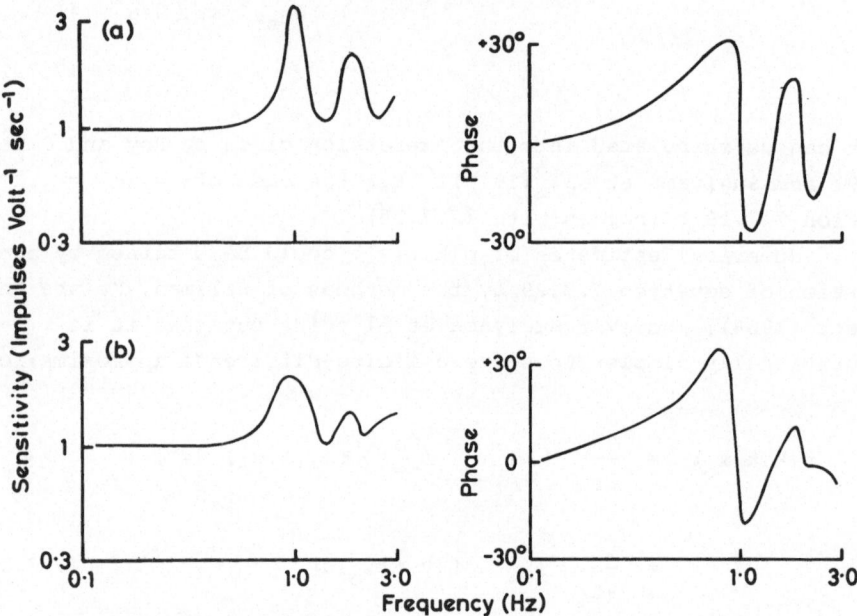

Figure 7.3. Frequency response function estimates for an electronic leaky integrator model subject to band-limited white noise with rms amplitude (a) 0.19 V (b) 0.28 V. Threshold = 1 V, time constant = 1 sec, carrier rate = 1 imp/sec, reset potential = 0.

7.8. The Inverse Approach to Diffusion Models.

The Wiener process and Ornstein-Uhlenbeck processes considered above are both diffusions, and the major problem with the SILIT model is that the first passage time density of the Ornstein-Uhlenbeck process is not available in closed form. Numerical estimates of the interspike interval density of the SILIT model (e.g. Sugiyama et al., 1970; Matsuyama et al., 1974) obtained by Monte Carlo methods or a finite-difference representation of equation 7.7.30 show asymetric, unimodal histograms with slowly decaying tails. In principle, one could obtain better and better numerical estimates of the first passage time density of the Ornstein-Uhlenbeck process: however, this might be a misdirected effort. A more useful approach might be to determine whether a given function could be the first passage time density of some diffusion process, and then to determine the diffusion process, rather than to specify the diffusion model and then attempt to find its first passage time density. This fits in the neurophysiological context of fitting an experimentally obtained interspike interval histogram by some arbitrary function, and then attempting to infer the properties of {V(t)} or the properties of the synaptic inputs.

This inverse approach to the first passage time problem has been explored by Capocelli and Ricciardi (1972). If the arbitrary function f(t) which is to be identified with a first passage time density $f(k,t|x_o)$ with the properties:

$$f(k,t|x_o) \;>\; 0$$

$$0 \;<\; \int_0^\infty f(k,t|x_o)\,dt \;<\; 1 \qquad\qquad (7.8.1)$$

$$\lim_{x_o \to k} f(k,t|x_o) \;=\; \delta t$$

the problem is to find a continuous Markov process {V(t)} for which $f(k,t|x_o)$ is the first passage time density. For $f(k,t|x_o)$ to be the first passage time density of a process {V(t)} the transition density $f(x,t|x_o)$ must satisfy the backward Kolmogorov equation in an open interval (a,b)

$$\frac{\partial f(x,t|x_o)}{\partial t} \;=\; \alpha(x_o)\,\frac{\partial^2 f(x,t|x_o)}{\partial x_o^{\,2}} \;+\; \beta(x_o)\,\frac{\partial f(x,t|x_o)}{\partial x_o}$$

$$(7.8.2)$$

where α and 2β are the infinitesimal mean and variance of {V(t)}. For equation 7.8.2 to be satisfied by only one process, the boundaries a and b must be inaccessible i.e. at $\pm\,\infty$: the initial conditions then determine the transition density $f(x,t|x_o)$.

Capocelli and Ricciardi derive a necessary condition for $f^*(k,s|x_o)$ to be the Laplace transform of the first passage time density of a diffusion process {V(t)}as:

$$\text{for} \qquad \alpha \;=\; \psi/\omega$$

$$(7.8.3)$$

$$\beta \;=\; \phi \;-\; \chi\,\psi/\,\omega$$

where ϕ and ψ are the real and imaginary parts of

$$
\left\{ \frac{sf^*(k,s|x_o)}{\frac{d^2}{dx_o^2} \; f^*(k,s|x_o)} \right\}
$$

and χ and ω are the real and imaginary parts of

$$
\frac{\frac{d}{dx_o} \{ \; f^*(k,s|x_o) \}}{\frac{d^2}{dx_o^2} \; \{ f^*(k,s|x_o) \}}
$$

If both boundaries are inaccessible, then if the above conditions hold, $f^*(k,s|x_o)$ is the Laplace transform of the first passage time density of a diffusion with infinitesimal moments α and 2β . However, if one of the boundaries r is inaccessible, $f^*(k,s|x_o)$ is the Laplace transform of the first passage time density of a diffusion with these infinitesimal moments if $f^*(k,s|r) = 0$.

Capocelli and Ricciardi used this theorem to show that the Wiener and Ornstein-Uhlenbeck processes could be obtained from the Laplace transforms of their first passage time densities (7.6.8) and (7.7.29). Further, they illustrated the use of the theorem to show that the function

$$
f(k,t|x_o) \;\; = \;\; \frac{|\ln(k/x_o)|}{2\sqrt{\pi}} \;\; t^{-3/2} \;\; \exp\{-(\ln^2(k/x_o))/4t\}
$$

could be interpreted as the first passage time density of a diffusion with drift x_o and infinitesimal variance $2x_o^2$, which has the lognormal distribution as its transition density. This example illustrated the method of obtaining the diffusion from the putative first passage time density : the method requires the functions α and β to be functions of x_o only. However, experimentally determined first passage time densities will be obtained as $f(t)$ and not as $f(k,t|x_o)$, as the threshold k and the reset potential x_o are unobservable with extracellular recording and uncontrollable with intracellular recording.

7.9 References.

Abramowitz, M. and Stegun, I.A. (eds.) : Handbook of Mathematical Functions. National Bureau of Standards, Washington D.C. (1964). reprinted Dover, New York (1974).

Afanas'eva, A.G. and Petunin, Yu. I. : A theoretical analysis of pulse information processing by a neurone. Kibernetica 3 74-81 (1971).

Balakrishnan, A.V. (ed.) : Communication Theory. McGraw-Hill, N.Y. (1968)

Bellman, R., Kalaba, R.E. and Lockett, J.A.: Numerical Inversion of the Laplace Transform. American Elsevier, New York (1966).

Bharucha-Reid, A.T. : Elements of the Theory of Markov Processes. McGraw-Hill, New York. (1960).

Blake, I.F. and Lindsey, W.C. : Level-crossing problems for random processes. I.E.E.E. Trans. Information Theory IT-19 295-315 (1973).

Calvin, W.H. and Stevens, C.F. : A Markov process model for neuron behaviour in the interspike interval. Proc. 18th Annual Conference on Engineering in Medicine and Biology, 118. (1965).

Capocelli, R.M. and Ricciardi, L.M. : Diffusion approximation and first passage time problem for a model neurone. Kybernetik 8 214-223. (1971).

Capocelli, R.M. and Ricciardi, L.M. : On the inverse of the first passage time probability problem. J. Appl. Prob. 9 270-287 (1972).

Capocelli, R.M. and Ricciardi, L.M. : A continuous Markovian model for neuronal activity. J. Theoretical Biology 40 369-387 (1973).

Carslaw, H.S. and Jaeger, J.C. : Conduction of Heat in Solids. Clarendon Press, Oxford. (1959).

Clay, J.R. and Goel, N.S. : Diffusion models for the firing of a neuron with varying threshold. J. Theoretical Biology 39 633-644 (1973).

Cowan, J.D. : Stochastic models of neuroelectric activity. in: Proc. 6th I.U.P.A.P. Conf. on Statistical Mechanics. ed. S.A. Rice, K.F. Freed and J.C. Light. University of Chicago Press, Chicago (1972).

Cox, D.R. and Miller, H.D.: The Theory of Stochastic Processes. Methuen, London. (1965).

Crank, J.: The Mathematics of Diffusion. Clarendon Press, Oxford (1974).

Darling, D.A. and Siegert, A.J.F.: The first passage problem for a continuous Markov Process. Ann. Math. Statist. $\underline{24}$ 624-639 (1953).

Doob, J.L.: The Brownian movement and stochastic equations. Ann. Math. $\underline{43}$ 351-69. Reprinted in N.Wax: Selected papers on noise and stochastic processes. Dover, New York (1954).

Fetz, E.E. and Gerstein, G.L.: An RC model for spontaneous activity of single neurones. M.I.T.-Q.P.R.-R.L.E. $\underline{71}$ 249-257 (1963).
Fienberg, S.E.: A note on the diffusion approximation for single neurone firing problems. Kybernetik $\underline{6}$ 227-229 (1970).

Gastwirth, J.L.: A renewal theoretic approach to a first passage time problem occurring in a neuron firing problem. John Hopkins University Dept. of Statistics Technical Report 77 (1967).

Geisler, C.D. and Goldberg, J.M.: A stochastic model of the repetitive activity of neurons. Biophysical J. $\underline{6}$ 53-69 (1966).

Gerstein, G.L. and Mandlebrot, B.: Random walk models for the spike activity of a single neuron. Biophysical J. $\underline{4}$ 41-68 (1964).

Goel, N.S. and Richter-Dyn,N.: Stochastic Models in Biology. Academic Press, New York. (1974).

Goel, N.S., Richter-Dyn,N. and Clay, J.R.: Discrete stochastic models for firing of a neuron. J. Theoret. Biol. $\underline{34}$ 155-184 (1972).

Gluss, B.: A model for neuron firing with exponential decay of potential resulting in diffusion equations for probability density. Bulletin Math. Biophys. $\underline{29}$ 233-243 (1967).

Helmstrom, C.W.: Markov processes and their application. In: Communication Theory. Ed. A.V.Balakrishnan. McGraw-Hill, New York. (1968).

Holden, A.V.: Information transfer in a chain of model neurones. In: Proc. 3rd European Meeting on Cybernetics and System Research, held at Vienna, April 1976. Transcripta Press, London (1976).

ten Hoopen, M.: Probabilistic firing of neurones considered as a first passage problem. Biophysical J. $\underline{6}$ 435-451 (1966).

Jack, J.J.B., Noble, D. and Tsien, R.W.: Electric Current Flow in Excitable Cells. Clarendon Press, Oxford (1975).

Johannesma, P.I.M. : Diffusion models for the stochastic activity of neurones. in: Neural Networks - Proc. of the School on Neural Networks June 1967 at Ravello. ed. E.R. Caianiello. Springer-Verlag, N.Y. (1968).

Johannesma, P.I.M. : Stochastic Neural Activity - a Theoretical Investigation. Thesis : Nijmegen (1969).

Knight, B.W. : Frequency response for sampling integrator and for voltage to frequency convertor. in: Systems Analysis Approach to Neurophysiological Problems ed. C.A. Terzuolo. Minnesota (1969).

Knight, B.W. : Some questions concerning the encoding dynamics of neuron populations. Proc. 4th International Biophysics Congress, Puschino, U.S.S.R. p.422-434 (1973).

Knox, C.K. : Cross-correlations for a neuron model. Biophys J. $\underline{14}$ 567-582 (1974).

Luikov, A.V. : Analytical Heat Diffusion Theory. Academic Press, (1968)

Matsuyama, Y., Shirai, K. and Akizuki, K. : On some properties of stochastic information processes in neurons and neuron populations. Kybernetik $\underline{15}$ 127-145 (1974).

Moran, P.A.P. : An Introduction to Probability Theory. Clarendon Press, Oxford. (1968).

Poggio, T. and Torre, V. : A nonlinear transfer function for some neuron models. Proc. of the Conference on Identification of Nonlinear Syatems. Caltech, March (1975).

Ricciardi, L.M. : Formalized neurons and neural networks. Proc. 1st. European Biophysics Congress Vol V Vienna 297-315 (1971).

Ricciardi, L.M. and Ventriglia, F. : Probabilistic models for determining the input-output relationships in formalized neurons.I. A theoretical approach. Kybernetik $\underline{7}$ 175-183 (1970).

Roy, B.K. and Smith, D.R. : Analysis of the exponential decay model of the neuron showing frequency threshold effects. Bull. of Math. Biophys. $\underline{31}$ 341-357 (1969).

Rubio, J.E. and Holden, A.V. : The response of a model neurone to a white noise input. Biological Cybernetics $\underline{19}$ 191-195 (1975).

Siebert, W.M. : An extension of a neuron model of ten Hoopen. M.I.T.-Q.P.R.-R.L.E. $\underline{91}$ 231-239 (1968).

Siebert, W.M. : On stochastic neural models of the diffusion type. M.I.T.-Q.P.R.-R.L.E. $\underline{94}$ 281-287 (1969).

Siegert, A.J.F. : On the first passage time probability problem.
Physical Review 81 617-623 (1951).

Sokolnikoff, I.S. and Redheffer, R.M. : Mathematics of Physics and
Modern Engineering. McGraw-Hill, New York (1966).

Stein, R.B. : A theoretical analysis of neuronal variability. Biophys.
J. 5 173-195 (1965).

Stein, R.B. and French, A.S. : Models for the transmission of information by nerve cells. in: Excitatory Synaptic Mechanisms, ed. P.
Andersen and J.K.S. Jansen. Univ. of Oslo Press (1970).

Stein, R.B., French, A.S. and Holden, A.V. : The frequency response,
coherence and information capacity of two neuronal models. Biophys.
J. 12 295-322 (1972).

Stratonovich, R.L. : Topics in the Theory of Random Noise. Vol. I.
Gordon and Breach, New York. (1963).

Sugiyama, H., Moore, G.P. and Perkel, D.H. : Solutions for a stochastic
model of neuronal spike production. Math. Biosciences. 8 323-341 (1970).

Tuckwell, H.C. : Determination of the interspike times of neurons
receiving randomly arriving post-synaptic potentials. Biological
Cybernetics 18 225-237 (1975).

Uhlenbeck, G.E. and Ornstein, L.S. : On the theory of Brownian motion.
Physical Review 36 823-841 (1930). Reprinted in N.Wax: Selected
papers on noise and stochastic processes. Dover, New York (1954).

Wang, M.C. and Uhlenbeck, G.E. : On the theory of Brownian motion II.
Rev. Modern Phys. 17 323-342 (1945). Reprinted in N.Wax : Selected
papers on noise and stochastic processes. Dover, New York, (1954).

Widder, D.V. : The Heat Equation. Academic Press, New York (1975).

Updates: J.Keilson and H.Ross have published tables of the first
passage time distribution for the Ornstein-Uhlenbeck process -
    University of Rochester Centre for System Science CSS 71-08 (1971)
    Selected Tables in Mathematical Statistics. ed. Institute of
Mathematical Statistics, vol. 3 233-327 (1976).

8. SUPERPOSITION MODELS

The models discussed above are all point models in the sense that a spike train is generated at some spatial point whenever a threshold level is exceeded. No account has been explicitly taken of the possibility of a spike train resulting from some combination or interaction of spike trains arising at spatially separate sites. In this section I will consider the simplest kind of interaction, that of superposition : more complex interactions will be considered in sections 9 and 10.

A variety of primary sensory neurones have extensively branched axonal terminations in which spike initiation sites are distributed at widely separate points e.g. Floyd and Morrison (1974) have described slowly adapting mechanoreceptors in the gastrointestinal tract, with a punctate receptive field, the mechanosensitive points being up to 20 cms apart. Thus in neurones with branched processes, when the branch length is much greater than the space constant (see section 3), it is likely that the output spike train results from interaction of spikes generated at separate sites. In central neurones with extensive dendritic arborizations, such the Purkinje cells in the cerebellum (Eccles, Ito and Szentagothai, 1967), the characteristics of the output spike train may be dominated by interaction between dendritic spikes rather than by non-linear summation of post synaptic potentials until a threshold is reached (Sabah and Murphy, 1973). In these cases the interaction is unlikely to be that of simple superposition, but will also involve collision and resetting (see section 9). However, superposition might represent a limiting case of this type of interaction.

When a neurone has several or many synaptic inputs, the sequence of post synaptic potentials will represent the superposition of the activity in each input line. If, as in the neuromuscular junction, the presynaptic fibre is branched and has several 'active spots' or connections with the post synaptic muscle fibre, the sub-threshold post synaptic activity might represent the superposition of activity of the separate active spots (Fatt and Katz, 1952, see section 11.2).

In extracellular recording in the central nervous system, or
from filaments of nerve trunks, the spike activity of several units
may be obtained simultaneously. This composite spike train will
represent the superposition of the spike trains in the individual units.

If ionic current flow across a membrane is through membrane
specializations or channels which are conducting or non-conducting,
the total current flow through an area of membrane is a surface
integral of the current over the membrane area or the superposition
of the activity of the individual channels.

Thus there are a number of situations of interest where the
superposition process may represent a suitable model. If a point
process is considered as a set of points on a line (the time axis),
the superposition of two or more point processes is simply the union
of the sets of points. This process is illustrated in Figure 8.1 in
which two source processes or generator trains $G_1(t)$ and $G_2(t)$

$$G_1(t) \; = \; \Sigma \, \delta \, (t - t_i)$$
$$G_2(t) \; = \; \Sigma \, \delta \, (t - t_j)$$

$$(8.0.1)$$

are superposed or pooled to form the spike train $S(t)$

$$S(t) \; = \; G_1(t) \; + \; G_2(t)$$

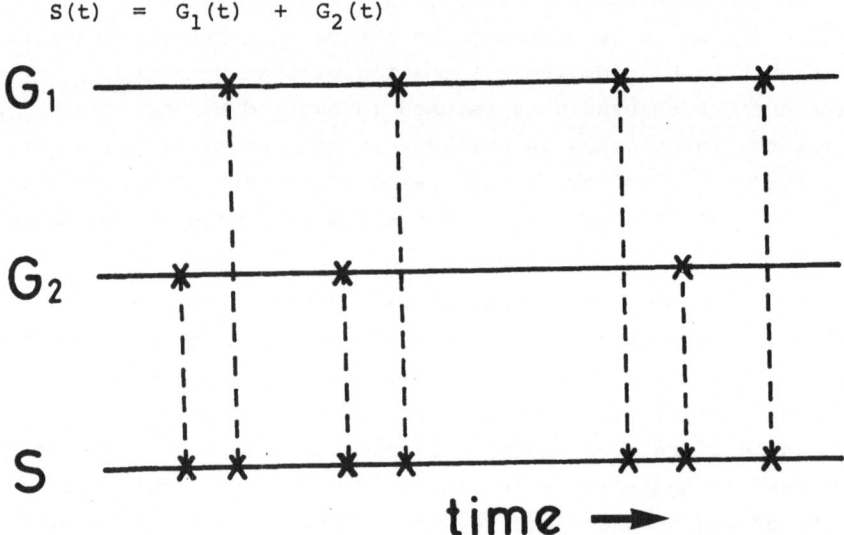

Figure 8.1. Superposition of two generator trains G to form S.

The superposition of renewal process is discussed in Cox (1962); more recent results are reviewed in Cinlar (1972). Problems in superposition theory fall into two classes:

a) given the properties of the generator trains $G_1(t)$, $G_2(t)$, $\cdots$, $G_n(t)$; what are the properties of $S(t)$ given by $S(t) = G_1(t) + \cdots + G_n(t)$ This kind of problem occurs in neural modelling.

b) given the properties of $S(t)$, what can be inferred about the number and properties of the generator trains? This kind of problem occurs in the interpretation of experimentally recorded spike trains.

8.1 Some general properties of superposed processes

If there are p generator trains, each of which has a rate μ, the rate of $S(t)$ is $p\mu$.

The cumulative number of events $N_s(t)$ in $S(t)$ is the sum of the cumulative number of events in the generator trains $N_{g,i}(t)$; $i = 1, 2, \cdots, p$

$$N_s(t) = N_{g,1}(t) + N_{g,2}(t) + \cdots + N_{g,p}(t) \qquad (8.1.1)$$

If the generator trains are independent and have the same density the mean and variance of $N_s(t)$ are p times those of a generator train.

If the generator trains have a Poisson distribution then $S(t)$ has a Poisson distribution. The Poisson process is the only class of point process which is invariant under superposition.

When a large number of independent and uniformly sparse point processes are superposed, $S(t)$ is approximately a Poisson process with a density $f(t)$ given by equation (5.2.4) as

$$f(t) = \mu \exp(-\mu t) \qquad (8.1.2)$$

$$\text{with} \quad \mu = \sum_{i=1}^{p} \mu_i \qquad (8.1.3)$$

where μ_i is the mean rate of the generator train $G_i(t)$, $i = 1,2,\cdots,p$.

The generator trains need not have identical distributions for this convergence to a Poisson distribution as $p \to \infty$ (Khintchine, 1960).

8.2 Superposition of regular generator trains

Cox and Smith (1953) have considered the case when the generator trains $G_1(t)$ are strictly periodic with a period τ_i.

$$G_i(t) \;=\; \sum_{k=0}^{\infty} \delta(t - k\tau_i) \tag{8.2.1}$$

When the periods τ_i and τ_j of the generator trains $G_i(t)$ and $G_j(t)$ are mutually irrational for all $i \neq j$, the superposed train S(t) is not stochastic but is precisely defined. However, Weyl (1916) has given a theorem that if τ_1/τ_2 is irrational, the sequence

$$\{n(\tau_1/\tau_2)\}, \quad n = 1, 2, \cdots \tag{8.2.2}$$

where $\{x\}$ denotes the fractional part of x,
has a uniform distribution over (0,1).

If this is applied to the superposition to two regular pulse trains with periods τ_1 and τ_2, the backward recurrence time $U_T(r)$ from the rth event in $G_1(t)$ is

$$U_T(r) \;=\; \tau_2 \{ r\tau_1/\tau_2 \} \tag{8.2.3}$$

and so by Weyl's theorem the sequence $U_T(r)$ has a uniform distribution over $(0, \tau_2)$. When more generator trains are considered, the sequences of backward recurrence times between any pair of generator trains have independent uniform distributions. When a large number of generator trains are superposed, if the minimum period is large compared with the mean interval of S(t), the inter-event distribution of S(t) tends to an exponential curve. Thus superposition of a large number of strictly periodic generator trains will produce a S(t) which is not stochastic but which will have an interval distribution which would suggest a Poisson process.

This deterministic sequence of intervals with an exponential interval distribution may be distinguished form a Poisson process by its behaviour over $t \gg \tau_i$. Cox and Smith (1953) propose the variance-time curve as a convenient method of distinguishing between the two processes. For a process with a Poisson distribution, the variance as a function of time is given by

$$\text{Var}(t) \quad = \quad \mu\, t \tag{8.2.4}$$

where μ is the parameter of the Poisson process, or the mean rate. For the deterministic process with an exponential distribution, the variance-time curve is given by

$$\text{Var}(t) \quad = \quad \sum_{i=1}^{n} \beta_i (1 - \beta_i) \tag{8.2.5}$$

where β_i is the frequency of intervals of length t containing $r_i = 1$, 2, 3, $\cdots$ events. Thus for τ_i large compared with t

$$\text{Var}(t) \quad \sim \quad \mu t \tag{8.2.6}$$

and for t large compared with τ_i

$$\text{Var}(t) \quad << \quad \mu\, t \tag{8.2.7}$$

and tends to oscillate about n/6. This behaviour of the variance-time curve was illustrated by numerical examples in Cox and Smith (1953) and has been applied to a series of miniature end plate potentials in chapter 8 of Cox and Lewis (1966) - this application showed a deviation from the behaviour of a Poisson distribution at long times.

Proudfoot and Lampard (1973) have obtained the coefficient of variation and serial correlation of the process obtained by taking every kth event of S(t) produced by the superposition of $\underline{n}$ strictly periodic point processes with the same period τ. Such a process would be obtained as the response of a perfect integrator model neurone (see section 3.2) or K-scaler (see section 2.2) to S(t) as an input, when k is the threshold.

For $qn < k < (q + 1)n$, $q = 0, 1, 2, \cdots$
the coefficient of variation of the k-divided, superposed process is

$$\text{C.V.} \quad = \quad \{ k - qn \}\{ (q + 1)\, n - k \} / \{ k^2 (n + 1) \} \tag{8.2.8}$$

and for

$$pn \quad < 2k < \quad (p + 1)n, \qquad p = 0, 1, 2, \cdots$$

the serial correlation coefficient ρ defined by equation (5.2.17) is

$$\rho = - \{k - \tfrac{1}{2} pn\} / \{\tfrac{1}{2} ((p + 2) n - k\}, \text{ for p even} \qquad (8.2.9)$$
$$= - \{k - \tfrac{1}{2} (p+1)n\} / \{\tfrac{1}{2}(p - 1) n - k\}, \text{ for p odd}$$

These equations give a large negative serial correlation coefficient when k is approximately n/2. Such a negative serial correlation co-efficient (of about -0.4 to -0.8) has been observed in the discharges of dorsal spinocerebellar tract neurones (Walloe, 1968, 1970). Thus the negative serial correlation coefficient might be accounted for by k-division of the superposition of approximately periodic, independent muscle spindle afferent inputs to the DSCT neurones.

8.3 Superposition of stochastic, renewal generator trains

In this section I will consider the properties of the spike train S(t) produced when a reasonably small number p of generator trains G_i(t) are superposed. I will assume that the generator trains are realizations of independent renewal processes which need not have the same probability density functions. Since the process of super-position is both associative and commutative the properties of S(t) when p is reasonably small may be obtained by repetition of the method used to obtain the properties of S_2(t) = G_1(t) + G_2(t).

If the two generator trains are independent, stationary realizations of renewal processes they can be characterized by their probability density functions g_1 (τ) and g_2 (τ). Let $\underline{\tau}$ be an inter-spike interval in the superposed train S_2(t), given a spike in G_1(t) or G_2(t) at t = 0. Then

$$\underline{\tau} = \min (\underline{\tau}_1, \underline{\tau}_2) \qquad\qquad (8.3.1)$$

where $\underline{\tau}_1$ and $\underline{\tau}_2$ are the times to the first spike in G_1(t) or G_2(t)

If $\qquad G_1 (\tau) = \int_0^\tau g_1 (\theta) d\theta$

$$G_2 (\tau) = \int_0^\tau g_2(\theta) \, d\theta \tag{8.3.2}$$

are the probability distribution functions of $\underline{\tau}_1$ and $\underline{\tau}_2$ and where

$$\mu_1 = \int_{-\infty}^\infty t \, g_1 (t) \, dt = \int_0^\infty t \, g_1 (t) \, dt$$

$$\mu_2 = \int_{-\infty}^\infty t \, g_2 (t) \, dt = \int_0^\infty t \, g_2 (t) \, dt \tag{8.3.3}$$

The probability distribution function of $\underline{\tau}$, $F (\tau)$, is obtained by using equation (5.2.14) for the recurrence time densities, and is

$$
\begin{aligned}
F (\tau) &= \int_0^\tau \frac{1 - G_1 (\theta)}{\mu_1} \, d\theta + \int_0^\tau \frac{1 - G_2 (\theta)}{\mu_2} \, d\theta \\
&\quad - \int_0^\tau \frac{1 - G_1 (\theta)}{\mu_1} \, d\theta \cdot \int_0^\tau \frac{1 - G_2 (\theta)}{\mu_2} \, d\theta \\
&= 1 - \int_\tau^\infty \frac{1 - G_1 (\theta)}{\mu_1} \, d\theta \int_\tau^\infty \frac{1 - G_2 (\theta)}{\mu_2} \, d\theta \tag{8.3.4}
\end{aligned}
$$

The probability density function $f(\tau)$ of $\underline{\tau}$ is then

$$
\begin{aligned}
f(\tau) &= \int_\tau^\infty \frac{1 - G_1 (\theta)}{\mu_1} \, d\theta \cdot \left\{ \frac{1 - G_2 (\tau)}{\mu_2} \right\} + \\
&\quad \int_\tau^\infty \frac{1 - G_2 (\theta)}{\mu_2} \, d\theta \left\{ \frac{1 - G_1 (\tau)}{\mu_1} \right\} \tag{8.3.5}
\end{aligned}
$$

The probability density function $f (\tau)$ is related to the interspike interval probability distribution function $S_2 (\tau)$ and density function $s_2 (\tau)$

$$f (\tau) = \frac{1 - S_2 (\tau)}{\mu} = \frac{1 - \int_0^\tau s_2 (\theta) \, d\theta}{\mu} \tag{8.3.6}$$

where μ is the mean interval of $S_2(t)$.

Differentiating (8.3.6)

$$s_2(\tau) = \frac{-\mu \frac{df}{d\tau}(\tau)}{d\tau}$$

hence

$$\frac{df}{d\tau} = \frac{-2}{\mu_1 \mu_2} (1 - G_1(\tau))(1 - G_2(\tau))$$

$$+ \int_{\tau}^{\infty} \frac{1 - G_1(\theta)}{\mu_1} d\theta \frac{-g_2(\tau)}{\mu_2}$$

$$+ \int_{\tau}^{\infty} \frac{1 - G_2(\theta)}{\mu_2} d\theta \frac{-g_2(\tau)}{\mu_1} \qquad (8.3.7)$$

and so the interspike interval probability density function $S_2(\tau)$ is given by

$$s_2(\tau) = \frac{\mu}{\mu_1 \mu_2} \{ 2(1 - G_1(\tau))(1 - G_2(\tau))$$

$$+ \int_{\tau}^{\infty} (1 - G_1(\theta) d\theta) g_2(\tau)$$

$$+ \int_{\tau}^{\infty} (1 - G_2(\theta) d\theta) g_1(\tau) \} \qquad (8.3.8)$$

This expression has been obtained in section 4 of ten Hoopen and Reuver (1966) as a special case of their general equation for p independent generator trains. The expression has been used to fit the interspike interval histograms of a thalamic sensory neurone, with $G_1(t)$ being a Poisson process and $G_2(t)$ a process obtained by random deletion of a process with a Gaussian distribution.

For the general case of p generator trains, if an event occurred at t = 0, the probability that the first event of the ith generator train is in (τ,τ + dτ), given the event at t = 0 was not from the ith generator train, is simply

$$\frac{\{ 1 - G_i (\tau)\} \, d\tau}{\mu_i}$$

However, if the event at t = 0 was from the ith generator train, the probability that the next event of S(t) is in $(\tau, \tau + d\tau)$ is the sum of

a) the probability that the next event was from $G_i(t)$

$$g_i (\tau) \, d\tau \prod_{\substack{k=1 \\ k \neq i}}^{p} \int_{\tau}^{\infty} \frac{(1 - G_k (\theta)) \, d\theta}{\mu_k}$$

and b) the probability that the next event was from $G_j(t)$, not $G_i(t)$

$$\sum_{\substack{j=1 \\ j \neq i}}^{p} \frac{(1 - G_j (\tau)) \, d\tau}{\mu_j} \quad (1 - G_i (\tau)) \prod_{\substack{k=1 \\ k \neq i,j}}^{p} \int_{\tau}^{\infty} \frac{(1 - G_k (\theta)) \, d\theta}{\mu_k}$$

The sum of these two probabilities must be multiplied by the probability that the t = 0 event was from the ith train i.e.

$$(1/\mu_i) / \{ \sum_{j=1}^{p} (1/ \mu_j)\}$$

Thus

$$
\begin{aligned}
s (\tau) = \mu \sum_{i=1}^{p} & \left\{ \frac{g_i (\tau)}{\mu_i} \prod_{\substack{k=1 \\ k \neq i}}^{p} \int_{\tau}^{\infty} \frac{\{1 - G_k (\theta)\} \, d\theta}{\mu_k} \right. \\
& \left. + \frac{1}{\mu_i} \{1 - G_i (\tau)\} \sum_{\substack{j=1 \\ j \neq i}}^{p} \frac{\{1 - G_j (\tau)\}}{\mu_j} \prod_{\substack{k=1 \\ k \neq i,j}}^{p} \int_{\tau}^{\infty} \frac{\{1 - G_k (\theta)\} \, d\theta}{\mu_k} \right\}
\end{aligned}
$$

$$(8.3.9)$$

which reduces to (8.3.8) when p = 2. This general expression has also been obtained by Cox and Smith (1954).

An expression for the renewal density h(t) defined by equation (5.2.9) can be readily obtained - this is identical to the expectation density function of ten Hoopen and Reuver (1966) and ten Hoopen (1967).

If S(t) is a renewal process, the renewal density function can be computed by successive convolution from the interspike interval density function s(τ)

$$h(\tau) = s(\tau) + \sum_{n=1}^{\infty} s^{(n)}(\tau) \qquad\qquad (8.3.10)$$

$$s^{(n)}(\tau) = \int_{0}^{\tau} s^{(n-1)}(\theta) \; s(\tau - \theta) \; d\theta$$

$$s^{(0)}(\tau) = s(\tau)$$

S(t) will be a renewal process if p is large (hence S(t) has an approximately Poisson distribution) or if the generator trains have Poisson distributions.

When the generator trains are non-Poisson, the renewal density function of S(t), h(t) can be found by repetitive use of the relation

$$(\mu_1 + \mu_2) \; h(t) = 2 + \mu_2 \; h_1(t) + \mu_1 \; h_2(t)$$

where $h_1(t)$ and $h_2(t)$ are the renewal densities of the two trains.

Lawrence (1973) has investigated the dependency of intervals between events in a superposition process in terms of the joint interval distributions, and has derived an expression for the general joint interval distribution. For the joint distribution of two adjacent intervals, the joint distribution function is given by

$$F(t,\tau) = \text{Prob} \; \{ t_{i+1} < \tau \mid t_i < t \}$$

$$F(t,\tau) = 1 - \sum_{i=1}^{p} M_i \; \{1 - G_i(t,\tau)\} \; \prod_{j=1,\neq i}^{p} \frac{1}{\mu_j} \int_{\theta=t+\tau}^{\infty} (1 - G_i(\theta)) d\theta$$

$$(8.3.11)$$

where M_i $= \dfrac{(1/\mu_i)}{(1/\mu_1) + (1/\mu_2) + \ldots + (1/\mu_p)}$

where $G_i(t,\tau)$ and $G_i(\theta)$ are the joint distribution function and the distribution function of the ith generator train. ten Hoopen and Reuver (1966) have obtained equation (8.3.11) for the special case of p = 2 and give the joint probability density function as

$$
\begin{aligned}
f(t,\tau) \;=\; & \frac{1}{\mu_1 + \mu_2}\; \sum_{i=1}^{2}\; \sum_{\substack{j=1\\ j\neq i}}^{2}\; \left\{ g_i(t,\tau)\int_{t+\tau}^{\infty}\{1 - G_j(\theta)\}\,d\theta \right. \\[2mm]
& + \int_{\tau}^{\infty} g_i(t,\underline{\tau})\,d\underline{\tau}\,\{1 - G_j(t+\tau)\} \\[2mm]
& + \int_{t}^{\infty} g_i(\underline{t},\tau)\,d\underline{t}\,\{1 - G_j(t+\tau)\} \\[2mm]
& \left. + \int_{t}^{\infty}\int_{\tau}^{\infty} g_i(\underline{t},\underline{\tau})\,d\underline{t}\;d\underline{\tau}\;g_j(t+\tau) \right\}
\end{aligned}
\tag{8.3.12}
$$

In principle, one could use Lawrence's general expression to write down complicated expressions for higher order joint interval distributions $F(\tau_1, \tau_2, \cdots \tau_n)$; however, such n-dimensional functions have not been used in analyzing spike trains when n > 2. Rodieck, Kiang and Gerstein (1962) have used a scatter diagram estimation of the joint interval density for adjacent interval pairs.

A more convenient method of characterizing dependency between intervals is the serial correlogram discussed in section 5.2; in particular the first-order serial correlation coefficient ρ_1 defined by equation (5.2.17). ρ_1 may be obtained from the joint probability density function f(t,τ) and the interval density function f(τ) by

$$
\rho_1 \;=\; \frac{\displaystyle\int_{0}^{\infty}\int_{0}^{\infty} t\tau\, f(t,\tau)\, dt\; d\tau \;-\; \mu^2}{\displaystyle\int_{0}^{\infty} \tau^2\, f(\tau)\; d\tau \;-\; \mu^2}
\tag{8.3.13}
$$

If the spike train is a realization of a Markov process, the terms of

the serial correlogram may be obtained from $\rho_j = (\rho_1)^j$. ten Hoopen
and Reuver (1966) used equations (8.3.12) and (8.3.13) to compute ρ_1
for their model of the discharge of a thalamic neurone.

So far I have only discussed the superposition of stationary,
independent generator trains : an example of the superposition of phase-
locked spike trains will be given in section (8.4). Blumenthal et al
(1971) have obtained some limit theorems for the superposition of non-
stationary generator processes; essentially, for large p, the limiting
inter-event interval distribution is exponential with a time varying
mean.

8.4 Some applications of superposition theory

The fluctuations in membrane potential considered in section (1)
are produced by the movement of ions across the membrane. Thus the
current flowing across the membrane can be considered as a surface
integral, over the membrane area, of the membrane current density, or
as the superposition in time of current pulses due to movement of
individual ions or ionic current pulses through discrete membrane
specializations. If each current pulse has an amplitude a, shape
s(t, τ) and the current pulse train

$$I(t) \;=\; \sum_{n=1}^{\infty} a_n \, s_n \, (t - (\sum_{k=1}^{n-1} x_k), \tau) \qquad\qquad (8.4.1)$$

has an inter-pulse interval distribution function p(x), when a, τ and
x are independent, the spectrum is given by equation (1.3.6). The
spectrum depends on

a) the number of pulses/unit time
b) the average value of a^2
c) the square of the average absolute value of the Fourier
    transform of s(t, τ)
d) the real part of $\phi(1 - \phi)$, where

$$\phi \;=\; \int_0^{\infty} p(x) \exp(j\omega x) \, dx$$

Re$\left[\phi(1 - \phi) \right]$ is zero if p(x) is a Poisson distribution.

The experimentally estimated membrane noise spectrum has a region
of $1/\omega$ or flicker behaviour : see section (1.3). Equation (1.3.6) can

generate flicker noise in two ways : if p(x) is Poisson the distribution of time constants could generate flicker noise, or if p(x) is non-Poisson Re $\left[\phi(1 - \phi)\right]$ would contribute to the region of $1/\omega$ behaviour.

If the distribution of time constants is responsible for flicker noise in nerve membranes, since the $1/\omega$ behaviour extends to < 1 Hz., the time constants must be long, and of the order of 1 - 10 secs. Such long time constants have not been experimentally observed, and are thermodynamically extremely unlikely as they imply very high activation energies. This suggests that the $1/\omega$ behaviour is due to a non-Poisson p(x).

However, if each current pulse is associated with the activity of a spatially distinct mechanism, the pulse sequence is the superposition of the pulse sequences representing the activity of each distinct mechanism. Then, from the limit theorems of section 8.2, if the mechanisms are independent p(x) would be Poisson. This suggests that the spatially distinct current mechanisms are not independent, and so there must be some interaction between the ion transit mechanisms.

The flicker noise is predominantly due to a K^+ current, and it is possible to estimate the distance between K^+ transit sites in the membrane. If K^+ transit is through specializations or channels, the channel density is approximately $10-100/\mu m^2$. This gives an average distance between channels of approximately 100 nm, which is greater than the membrane thickness by an order of magnitude, and is much greater than the size of the membrane phospholipid molecules.

Thus the interaction between K^+ transit sites needed to generate a non-Poisson p(x) implies the transmission of a disturbance through the membrane structure over long distances. By use of an argument similar to that proposed by Rushton (1961) for the all or none nature of the propagating action potential, such a disturbance must be an all or none change between two stable states of membrane organisation. This line or argument gives as a hypothesis (Holden, 1976) :

a) K^+ current flow through a localized membrane specialization is associated with a localized structural change.

b) this localized structural change induces a propagative, all or none change or disturbance in membrane organization.

c) this disturbance stabilizes ion transport sites, so that p(x) is non-Poisson.

Another example of superposition of trains of events which are
not independent is given by the analysis of multi-unit electromyograms
(Gath, 1974). The stationary discharge of α-motorneurones is fairly
regular, with a Gaussian distribution having a coefficient of variation
of about 10%. Since α-motorneurone spike trains do not have significantly large first-order serial correlation coefficients (Clamann, 1969)
the discharge of a single motor unit may be modelled by a renewal
process with a Gaussian distribution. For independent generator
trains, when the number of generator trains p is fairly small, the
first-order serial correlation coefficient ρ_1 defined by equation
(5.2.7) may be used to indicate the number p : ρ_1 decreases as p
increases. For fairly regular, independent generator trains the serial
correlogram $\{\rho\}$ will oscillate and show positive peaks at ρ_k, where
k is an integer p. This oscillatory behaviour will be reduced as the
coefficient of variation of the generator trains increases.

Since multi-unit electromyograms do not show an exponential interval distribution, the generator trains are not independent. The
correlation between activity in two generator trains may be specified
in terms of the phase ϕ, which is the ratio of the interval between
an event in $G_1(t)$ and the preceding event in $G_2(t)$, to the corresponding
interval in $G_2(t)$.

$$\phi = (t_{1,m} - t_{2,n})/(t_{2,m+1} - t_{2,n}) \qquad (8.4.2)$$

where $t_{i,j}$ is the time of the jth event in generator train i. When a
small number of phase-locked generator trains are superposed, the resulting train has a unimodal distribution which tends to that of a
Gaussian distribution as the number of phase locked trains increases.
During motor unit recruitment, the multi-unit electromyogram interval
distribution develops through an asymmetric distribution (similar to
a gamma distribution) towards a locally Gaussian distribution - this
suggests the recruitment of correlated motor units.

Sabah and Murphy (1971) have analyzed the stationary discharge of
cerebellar Purkinje cells in terms of a superposition model. The interspike interval histograms of spontaneously active Purkinje cells are
unimodal, with an exponential or slower than exponential tail, and a
mean discharge rate of between 10-70/sec. There is evidence that the
larger dendrites of the flat dendritic tree can generate action poten-

tials, and so the suggestion is that the Purkinje cell discharge
results from the superposition of a sequence of dendritic generator
spikes. The dendritic tree extends for some 400 μm and so with a
dendritic conduction velocity of about 1 m/sec all the generator sites
will be discharged by a somatic action potential within a fraction of
a msec. This is effectively instantaneously compared with the interspike interval, and so the generator sites will probably share a
common refractory period and will not be independent. Whether or not
a centripetal dendritic action potential propagates centrifugally up
other dendrites will depend on the membrane properties and geometry
of the dendritic bifurcations. The threshold conditions at each
branch point will determine the logical operation of the branch point.
For a branch point with daughter branch diameters d_1 and parent branch
diameter d_2, the branch point will act as an OR junction, impulses
from either daughter branching propagating down the parent branch, if

$$d_1/d_2 > \{\alpha/(1 - \alpha)\}^{3/2} \tag{8.4.3}$$

where α is the ratio of the threshold charge : charge carried in the
leading edge of the action potential, which is about 0.5 for a H-H axon.

If the inequality is not satisfied, the branch point will act as
an AND junction, with simultaneous impulses in both daughter fibres
needed to fire the parent branch (Scott, 1973). Thus dendritic interactions in the Purkinje cell might represent a complex pattern of
logical operations.

However, neglecting these complications, an estimate of the number
of generator spikes lost due to the shared refractory period at the
generator sites can be made. If the generator sites are independent
and have a Poisson inter-event distribution, the superposed process is
Poisson, say with parameter μ. If the refractory period is represented
by a dead time ΔT, the fraction of lost spikes f is

$$f = (\Delta T\mu) (1 - \Delta T\mu)$$

An upper bound on f for Purkinje cells is about 0.1: this would be increased if the generator sites were correlated, with a tendency towards

synchronization.

The variance-time curve (see section 8.2) of stationary Purkinje cell discharge shows

a)   for small time the curve is below that of a Poisson process with the same rate. This can be accounted for by the effects of the refractory period

b)   as time increases, the variance-time curve follows that of a Poisson process

c)   at large time, about 5-10 times the mean interval, the variance-time deviates from that of a Poisson process. This behaviour of the variance-time curve is compatible with a simple superposition model, and if the generator trains are assumed to be renewal processes, p, the number of generator sites, was estimated to be 12-150.

A problem with this model is the low spontaneous discharge rate of the Purkinje cells, as a superposition model would imply even lower discharge rates of the generator trains. Since the generator trains are thought to represent the response of dendritic active spots to parallel fibre synaptic input, this would imply a high dendritic threshold or a low discharge rate of the granule cells. A more plausible explanation would be a high degree of correlation between the activity of 'on-beam'granule cells.

8.5 References

Blumenthal, S., Greenwood, J.A. & Herbach, L.: Superimposed non-stationary renewal processes. J. Appl. Prob. 8 184-192 (1971)

Cinlar, E.: Superposition of point processes. In 'Stochastic Point Processes' statistical analysis, theory and applications. ed. P.A.W. Lewis. Wiley, New York (1972).

Clamann, P.: Statistical analysis of motor unit firing patterns in human skeletal muscle. Biophys. J. 9 1233-1251 (1969).

Cox, D.R.: Renewal Theory. Methuen, London (1962).

Cox, D.R. & Lewis, P.A.W.: The statistical analysis of series of events. Methuen: London (1966).

Cox, D.R. & Smith, W.L.: The superposition of several strictly periodic sequences of events. Biometrika 40 1-11 (1953).

Cox, D.R. & Smith, W.L.: On the superposition of renewal processes. Biometrika 41 91-99 (1954).

Eccles, J.C., Ito, M. & Szentagothai : The Cerebellum as a Neuronal Machine. Springer-Verlag, Berlin (1967).

Fatt, P. & Katz, B.: Spontaneous subthreshold activity at motor nerve endings. J. Physiol. 117 109-128 (1952).

Floyd, K.I.W. & Morrison, J.F.B. : Interactions between afferent impulses within a peripheral receptive field. J. Physiol. 238 62-3P (1974)

Gath, J.: Analysis of Point process signals applied to motor unit firing patterns I. Superposition of independent spike trains. Math. Biosci 22 211-222 (1974).

Gath, I.: Analysis of point process signals applied to motor unit firing patterns. II. Superposition of phase-locked spike trains. Math Biosci. 22 223-236 (1974).

Holden, A.V.: Flicker noise and structural changes in nerve membrane. J. of theoretical Biology (1976).

ten Hoopen, M.: Pooling of impulse sequences, with emphasis on applications to neuronal spike data. Kybernetik 4 1-10 (1967).

ten Hoopen, M. & Reuver, H.A.: The superposition of random sequences of events. Biometrika 53 383-389 (1966).

Khintchine, A.Y.: Mathematical methods in the theory of queueing. Griffin: London (1960).

Lawrence, A.J.: Dependency of intervals between events in superposition processes. J. Roy. Stat. Soc. B **35** 306-315 (1973).

Proudfoot, A.D. & Lampard, D.G.: The divided pooled output of several strictly-periodic point processes. J. Appl. Prob. **10** 461-463 (1973).

Rodieck, R.W., Kiang, N. Y-S & Gerstein, G.L.: Some quantitative methods for the study of spontaneous activity of single neurones. Biophys. J. **2** 351-68 (1962).

Rushton, W.A.H.: Peripheral coding in the nervous system. In 'Sensory communication' ed. W.A. Rosenblith, MIT Press, Cambridge (1961).

Sabah, N.H. & Murphy, J.T.: A superposition model of the spontaneous activity of cerebellar purkinje cells. Biophys. J. **11** 414-428 (1971).

Scott, A.C.: Information processing in dendritic trees. Math. Biosci. **18** 153 (1973).

Walloe, L.: Transfer of signals through a second order sensory neurone. Thesis, Univ. of Oslo (1968).

Walloe, L.: On the transmission of information through sensory neurones. Biophys. J. **10** 745-765 (1970).

Weyl, H.: Über die Gleichverteilung von Zahlen mod. Eins. Math. Ann. **77** 313-352 (1916).

9. COLLISION MODELS.

In the models of section 8 there is no interaction between the action potentials of the independent generator trains, and the spatio-temporal pattern of activity in different branches is represented as a simple superposition in time. Here I will consider the type of interaction which occurs when two action potentials collide, or propagate into each other.

9.1. The Mechanism of Collision.

The Hodgkin-Huxley system of equations for a propagating action potential (see section 3.5) in an infinite, non-myelinated axon can be written as:

$$\partial V/\partial x \quad = \quad -r_a i \qquad\qquad\qquad (9.1.1a)$$

$$\partial i/\partial x \; + C \; \partial V/\partial t \quad = \quad -j_i(V,m,n,h) \qquad (9.1.1b)$$

where r_a is the axoplasmic resistance/unit length, C is the membrane capacitance/unit length, i is the axial current and j_i is the membrane ionic current/unit length of axon. m, n and h are the activation and inactivation variables defined in Figure 3.2.

A propagating action potential is represented by a travelling wave solution of equation 9.1.1 in which the dependent variables (V, i, m, n, h) are functions only of a moving spatial variable X where:

$$X \quad = \quad x - \theta t$$

and θ is identified with a conduction velocity. This transformation gives an autonomous system of ordinary differential equations:

$$dV/dX \quad = \quad -r_a i$$

$$di/dX \quad = \quad -r_a C\theta i - j_i$$

$$dn/dX \quad = \quad (n - n_\infty)\,\theta\tau_n$$

$$(9.1.2)$$

$$dh/dX \quad = \quad (h - h_\infty)\,\theta\tau_h$$

$$dm/dX \quad = \quad (m - m_\infty)\,\theta\tau_m$$

The resting condition, $V = i = 0$, with appropriate numerical values
for n, m, and h given by the H-H equations, is a singular point at
which all the derivatives wrt X in equations 9.1.2 are zero. Thus a
propagating action potential is a solution of equations 9.1.2 which
starts from the resting singular point at $X = -\infty$ and returns to it as
$X \to \infty$: this is a homoclinic orbit and its existance depends on the
value chosen for θ (Scott, 1975). Hodgkin and Huxley (1952) found a
value for the conduction velocity of 18.8 m/s which generated a
propagating V(t) which was in good agreement with the experimentally
observed action potential shape and conduction velocity. However, this
is not a unique solution; Huxley (1959) found by numerical computation
a second, smaller pulse which propagated at a conduction velocity of
5.66 m/s. Cooley and Dodge (1966) extended this observation by computing propagating solutions as the maximal sodium and potassium conductances
were reduced. As the conductances are decreased, the amplitude and
conduction velocity of the faster solution decrease until the two
solutions meet. A further decrease in conductance gives decremental
waves as the only solution. Rinzel and Keller (1973) have investigated
travelling wave solutions of a simpler nerve axon equation, the Fitzhugh-
Nagumo equation, which is similar to the V-m reduced system of section
3.6. Here there are two homoclinic orbits, a stable, fast travelling
wave analogous to an action potential and a smaller, unstable, slower
travelling wave. This suggests that the instability of the slower
solution of the H-H axon equations found by Huxley (1959) is not a
numerical artefact. Given that the fast solution of the H-H axon
equations represents a stable, propagating action potential potential, the
problem now is to describe what happens when two such solutions,
propagating in opposite directions, meet and collide.

Experimentally, two propagating action potentials meeting head-on
in a uniform fibre collide and annihilate each other (Tasaki, 1949). If
I assume that at the leading edge of an action potential:

$$|C\,\partial V/\partial t| \quad >> \quad |j_i| \qquad\qquad (9.1.3)$$

i.e. that the predominant membrane current before Na$^+$ activation is a capacitative discharge, equation 9.1.1b gives an approximate conservation of charge:

$$\partial i / \partial x \;\; + \; \partial (CV) / \partial t \;\; \approx \;\; 0 \tag{9.1.4}$$

where the amount of conserved charge is about $V_{max}/\theta r_a$ if V_{max} is the height of the action potential (Scott, 1973). This approximate conservation of charge, together with equation 9.1.1a, gives a linear diffusion equation to represent the interaction of the leading edges:

$$\partial^2 / \partial x^2 (V - V_2) \;\; \approx \;\; r_a C (\partial / \partial t) (V - V_2) \tag{9.1.5}$$

where V_2 is the zero-current, positive slope intercept of the membrane current-voltage relation.(see Figure 1.2d). Thus V will relax towarss V_2, which is more positive than the peak of the propagating action potential. The voltage will then decay to a small negative potential with a time course dominated by the K$^+$ activation time constant τ_n; this trailing edge interaction will slowly relax un il the membrane potential reaches its resting value.

The same sequence can be visualized by thinking of the point of collision as an axis of symmetry, where $\partial V / \partial x = 0$ (Katz and Miledi, 1965). The axial current flow ahead of the leading edge of a freely propagating action potential can leave the axoplasm over a large membrane area at a low density; however, as the two action potentials approach each other, or approach the midpoint where $\partial V / \partial x = 0$, the axial current must leave through a smaller area of membrane at a higher current density. This increase in membrane current density increases the rate of rise of membrane potential and hence both the conduction velocity and the maximal potential reached are increased. This process may be simulated by an action potential propagating into a finite, open circuit cable (see chapter 4 of Jack, Noble and Tsien, 1975): the boundary condition is $\partial V / \partial x = 0$ at the end of the cable and at the point of collision. Goldstein and Rall (1974) have computed the propagation into an open circuit of an action potential given by the empirical equations:

$$\partial^2 V / \partial x^2 \; - \; \partial V / \partial t \; = \; V \; - \; E(1 - V) \; + \; J(V + 0.1)$$

$$\partial E/\partial t \quad = \quad k_1 V^2 + k_2 V^4 + k_3 E - k_4 J \qquad (9.1.6)$$

$$\partial J/\partial t \quad = \quad k_5 E + k_6 EJ - k_7 J$$

where V is a normalized voltage and E and J are auxillary variables. The constants k_1 - k_7 have been chosen to reproduce the shape of the action potential. These equations are a particular case of the general model:

$$\partial^2 V/\partial x^2 - \partial V/\partial t \quad = \quad f(V, W_1, \ldots, W_n)$$

$$\partial W_j/\partial t \quad = \quad f_j(V, W_1, \ldots, W_n) \quad \text{for } j = 1, \ldots, n.$$

where the W_j are auxillary variables, such as m, n and h of the H-H equations. The particular model of equation 9.1.6 was chosen to facilitate numerical computations. The increase in the conduction velocity and the amplitude of the action potential occur within a space-constant λ of the boundary: thus changes occuring with collision will extend over a distance of about 2λ.

However, collision is not just the mutual annihilation of the two action potentials at a point where $\partial V/\partial t = 0$. A propagating action potential extends over a length of axon, given by the product of its duration and conduction velocity. Thus the result of a collision will be a slowly decaying refractory length of axon, which will slow down and distort the next propagating action potential if it reaches the zone of collision before the refractoriness has decayed. Thus a decaying, delay function should be associated with the after-effects of collision. The time course of this decaying delay function could be computed by numerical solution of the H-H axon equations, stimulating both ends of a long cable and then stimulating one end after a variable delay. The introduction of a third, testing action potential means that the solution is no longer symmetric about the point of collision.

9.2. The Geometry of Collision.

A branched dendritic tree or system of axonal terminations may be represented by a graph consisting of a set of points P = { $p_1, \ldots, p_m$ } which represent branch points and terminations, and a set of branches B = { $b_1, \ldots, b_n$ } . Each branch b_i is a line segment joining two adjacent points. Thus the path between two points is an alternating sequence of points and branches. I will assume that there is one, and

only one, path between all pairs of points:: this means that there are
no closed loops and the tree is fully connected.

An action potential will propagate, in either direction, along
a branch. If the branch diameter is constant the antidromic and orthodromic conduction velocities will be the same, but the conduction
velocities will be different if the branch is tapering. If there is
no tapering any branch will be characterized by its length and
conduction velocity, or its conduction time. By appropriately stretching
the branch lengths any conductile tree can be transformed into an
isomorphic tree which has the same point to point conduction times,
but in which all branches have a unit conduction velocity.

If there is only one spike initiation or generator site in a
conductile tree it might be thought that there is no possibility of
collision and so the centripetal spike train is simply the generator
spike train observed after a delay. The tree in this case can be
represented by a single point and branch - see Figure 9.1.a . However,
if there is an abrupt increase in branch diameter, where the propagating
spike is delayed, there is the possibility of the spike antidromically
reinvading the smaller diameter segment as well as propagating
orthodromically (Khodorov et al., 1969; Goldstein and Rall, 1974).
This curious phenomenon of reflection of an action potential is
critically dependent on the ratio of the branch diameters; a 10-fold
increase in diameter produces a failure of conduction and a 3-fold
increase in diameter produces a reflected action potential as well as
the orthodromic action potential. This reflection is caused by
electrotonic spread of depolarization from the larger segment back into
the smaller segment, which has recovered from its refractory period.
If reflection occurs, the reflected spike could collide with the next
generator spike - see Figure 9.2. If the smaller segment is sufficiently
long, or the generator spike rate sufficiently high, it is almost certain
that a collision will occur in the smaller diameter segment. In this
case the centripetal spike train will consist of every other generator
spike. Thus, if changes in conduction velocity due to the after-effects
of collision are neglected, the interspike interval probability density
function $f(\tau)$ of the centripetal spike train will be identical to the
second order interval density (see section 5.2) of the generator spike
train. If the generator train is a renewal process with an interval
density $g(\tau)$, the 2-divided centripetal spike train will have an
an interval density given by:

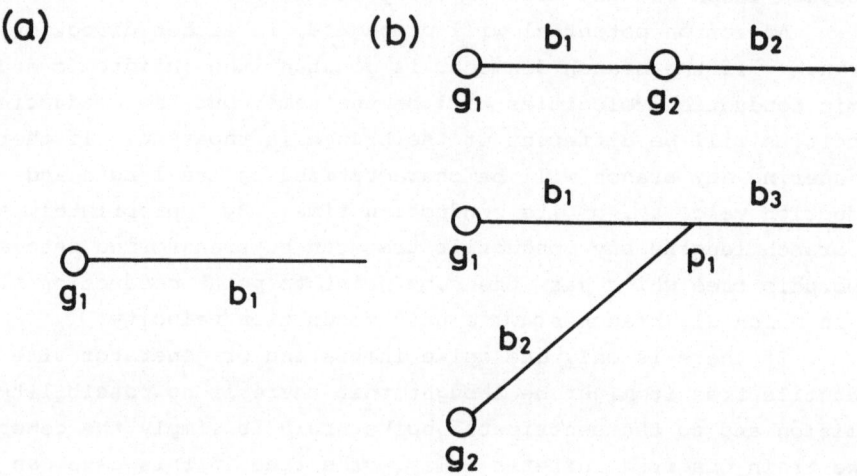

Figure 9.1. Simple conductile trees with generator sites g , branch lengths b and branch point p . See text.

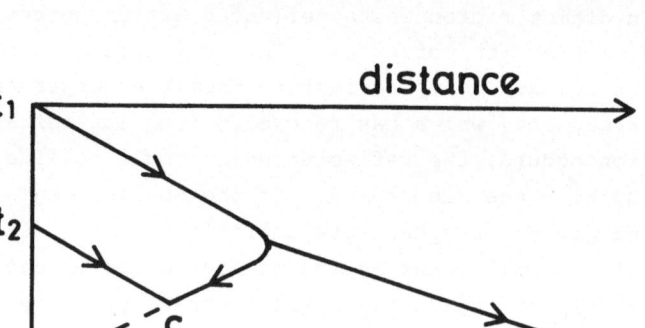

Figure 9.2. Increase in axon diameter can produce a reflected action potential, which can then collide with the next centripetal action potential. See text.

$$f(\tau) \quad = \int_0^\infty g(\tau)g(\tau - t)dt \qquad\qquad (9.2.1)$$

If there are two generator sites in the conductile tree, the
two possible structures are that the generator sites are adjacent, or
are at non-adjacent points - see Figure 9.1.b. When the two generator
sites g_1 and g_2 are at adjacent points, centripetal spikes from g_2 will
not be annihilated by spikes from g_1; however, some spikes from g_1 may
be lost by collision with centrifugal spikes from g_2. In this case, if
the branch length b_1 is associated with a conduction time T, the
centripetal spike train will be the superposition of $g_2(t)$ with the
non-deleted spikes from $g_1(t)$. As an example, if $g_2(t)$ has a Poisson
distribution with parameter σ, the undeleted spikes from $g_1(t)$ will
have an intevval density whose Laplace transform p*(s) is given
by Coleman and Gastwirth (1969) - see section 10.2 - as

$$p^*(s) \quad = \quad q^*(s)/ (1 + q^*(s) - g_1^*(s)) \qquad (9.2.2)$$

where $\qquad g_1^*(s) \quad = \quad$ L.T. $\{g_1(\tau)\}$

$$q^*(s) \quad = \quad \text{L.T. } \{g_1(\tau)\exp(-\sigma\min(t,T))\}$$

Equation 9.2.2 is derived for a modification of the selective interaction model of ten Hoopen and Reuver (1965) which is discussed in section
10.2. Thus the interval density of the centripetal spike train can be
obtained by the superposition of $g_2(t)$ with a process whose interval
density is given by the inverse Laplace transform of p*(s) of equation
9.2.2. Note that these two processes are not independent and so
equation 8.3.8 will not apply.

When the two generator sites are at non-adjacent points, as in
the bottom of Figure 9.1.b, spikes originating at either generator site
may be annihilated by collision with centrifugal spikes originating at
the other generator site. This geometric situation is discussed in
detail in section 9.3.

As the number of generator sites increases, the number of
possible structural forms of the neuronal tree increases. If action
potentials can propagate throughout all the branches of the tree, and
if the activity of all endpoints is observable, the structure of the
tree and the branch conduction times can be deduced from a stimulus-
response time matrix (Murray, 1974). However, in an extensive connected
tree with multiple generator sites the overall possible pattern of

collisions will be complex and it will not be possible to infer the structure of the tree from the observed centripetal spike train.

9.3. Mutual Interaction Between Two Non-adjacent Generator Sites.

The collision model when there are two independent generator sites located at branch terminals has been treated in detail by Clifford and Sudbury (1972): here I will follow their approach. Assuming (a) constant branch diameters, and so the antidromic and orthodromic conduction velocities are identical, and (b) a unit conduction velocity in all branches, the possible paths of an impulse generated at g_1 at a time t_i are:

1) it will propagate towards the branch point p, arriving at p at a time $(t_i + b_1)$, and propagate antidromically towards g_2, reaching g_2 at a time $(t_i + b_1 + b_2)$.

2) it will arrive at the branch point at a time $(t_i + b_1)$, and propagate antidromically towards g_2, but will collide in the branch b_2 at some time T, $(t_i + b_1) < T < (t_i + b_1 + b_2)$.

3) it will collide with an antidromic action potential originating from g_2 before reaching the branch point.

The centripetal spike train will consist of all those action potentials which reach the branch point p. These three possible paths are illustrated in Figure 9.3.a, and there are three analogous paths

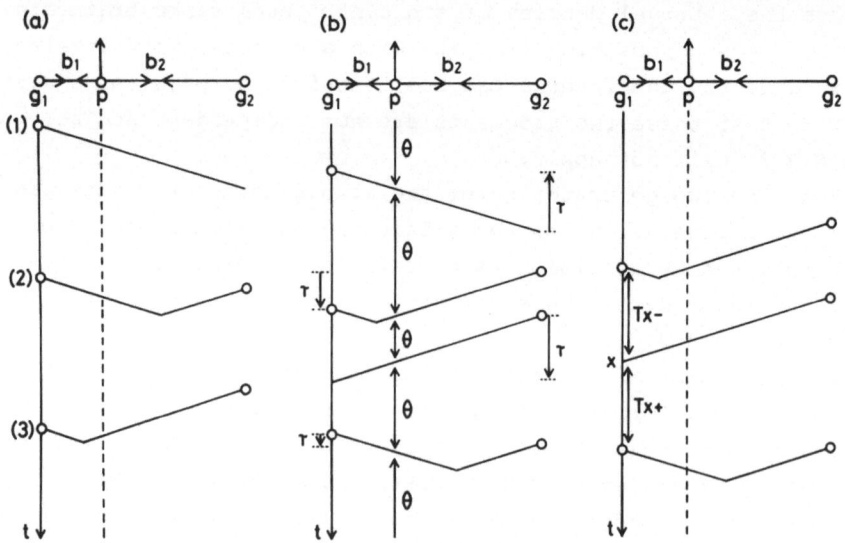

Figure 9.3. Collision patterns when there are two generator sites. See text.

for action potentials originating at g_2.

The process of collision is graphically illustrated by the sequence of linking lines between the generator points g_1 and g_2 shown in Figure 9.3.b. These links may be straight, when there has been no collision, or bent, when a collision has occurred. The intervals θ between intersections of the linking lines with the line from p are the interspike intervals of the centripetal spike train.

Each linking line can be characterized by the time τ between its intersection with g_1 and g_2, where $-(b_1 + b_2) < \tau < (b_1 + b_2)$. If $\tau = -(b_1 + b_2)$, there has been no collision and the spike originated at g_2; if $\tau = (b_1 + b_2)$ there has been no collision and the spike originated at g_1. If $|\tau| < (b_1 + b_2)$ there has been a collision and one generator spike has been lost. The behaviour of the process is characterized by the sequence $\{\tau_i\}$, $i = 1,2,...,$ and the problem is to find the density function of θ_i, the intervals between intersections of the linking lines with the line from the branch point p.

Let $g_1(t)$ and $g_2(t)$ be the generator trains with probability density functions $f_1(T)$ and $f_2(T)$. If τ_n and τ_{n+1} are the durations of two successive links, τ_{n+1} being generated by the last spike of an interval T_1 in $g_1(t)$ and the last spike of an interval T_2 in $g_2(t)$

$$|\tau_{n+1}| < (b_1 + b_2)$$

$$\tau_{n+1} = \tau_n + T_1 - T_2 \qquad (9.3.1)$$

where T_1 and T_2 are random variables with densities $f_1(T)$ and $f_2(T)$. Thus the process $\{\tau_i\}$ appears to be Markov, and Clifford and Sudbury, assuming Poisson distributed generator trains, obtained the transition probabilities of τ. From these transition densities the stationary density of τ and the density function of θ_i, the interspike intervals of the centripetal spike train, were obtained.

However, $\{\tau_i\}$ is not in general a Markov process, as the occurrence of straight links τ_n of length $\pm (b_1 + b_2)$ breaks the recurrence relation of equation 9.3.1. For $\tau_n = -(b_1 + b_2)$, an action potential is initiated at g_2 and propagates to generator site g_1 (point X of Figure 9.3.c). Here the generator potential of g_1 is reset, and if the aftereffects of the antidromic spike are assumed to be identical to the aftereffects of a g_1 spike, the time T_{x+} to the next g_1 spike will have a density $f_1(T)$. However, the time from the preceeding g_1 spike to the point X, (T_{x-}), will have a density given by the backward recurrence time density of $g_1(t)$, which by equation 5.2.14 is:

$$U_1(T) = \frac{1 - F_1(T)}{\mu_1} \qquad (9.3.2)$$

where $F_1(T)$ is the distribution of $g_1(t)$ and has a mean μ_1.

Thus the recurrence relation 9.3.1 will hold, but with T_1 having a density $f_1(T)$ or $U_1(T)$, and T_2 similarly having a density $f_2(T)$ or $U_2(T)$, where $U_2(T)$ is the backward recurrence time density of $g_2(t)$. Thus $\{\tau_i\}$ is not a Markov process, except for the special case considered by Clifford and Sudbury when both the generator trains are Poisson distributed.

Since $\{\tau_i\}$ is not a Markov process, the sequence of intervals $\{\theta_i\}$ of the centripetal spike train will depend on the initial transient behaviour of the process. This can be illustrated by the simple situation of $b_1 = b_2$ when $f_1(T)$ and $f_2(T)$ are strictly regular with intervals T_1 and $T_2 = T_1 + \Delta T$, $\Delta T \ll T_1$. Here the output pulse train will be dominated by the generator site which fires first.

The non-Markovian behaviour of the collision process is further complicated if the aftereffects of collision are considered by introducing some variable delay function, and if the aftereffects of an antidromic action potential resetting the generator site are not the same as the aftereffects of an action potential initiated at that generator site. The forward problem, that of obtaining the characteristics of the centripetal spike train from a known collision geometry and known generator train characteristics, is not very interesting as it can easily be investigated by Monte Carlo methods. However, since the forward problem is so complex analytically the neurophysiologically more interesting problem of infering the generator train characteristics from a known collision geometry and observed centripetal spike train does not appear to be tractable.

9.4 The Biology of Collision.

In the collision models considered above, the axonal branches are long and slowly conducting, and so collision of antidromic and orthodromic action potentials is likely. Although there are examples of neurones with long, thin axonal branches (see Floyd and Morrison, 1974), such a neuronal geometry is rare. A commoner neuronal geometry is one in which the branch lengths are less than a space constant (see equation

3.0.2). In this situation, the predominant interaction between generator trains will not be collision within the branches, but will be antidromic action potentials invading and resetting the generator terminals.

An example of this kind of geometric situation is the amphibian muscle spindle, models of which have been considered in section 4.4. The parent myelinated sensory axon divides outside and inside the spindle capsule into myelinated branches, and these myelinated branches further divide into non-myelinated fibres. Kuroda and Ito (1972) have shown that these branches act as generator sites, and the idea is that an orthodromic action potential from a generator site antidromically invades and resets the other generator sites. Thus the fastest discharging generator site will dominate, and the statistical characteristics of the centripetal spike train will be those of the dominant generator. As the spike rate decreases during adaptation a different generator site may become dominant. If different generator sites have different discharge characteristics, this model suggests that during adaptation as different generators become dominant, there will be discrete changes in the statistical characteristics of the centripetal spike train.

Brokensha and Westbury (1973, 1974) have experimentally investigated and confirmed this prediction. After applying a maintained stretch to the spindle, the rate of discharge adapts in two phases:

a) initially the discharge rate is fairly regular and rapidly adapts.

b) the second phase of adaptation is slower and the discharge rate is more irregular.

If the adapting spike train is divided into runs of 10 intervals, the standard deviation and mean of each run can be estimated . A plot of the standard deviation against the mean interval can be fitted by two straight line segments, with the lower slope at small intervals or rapid discharges. This suggests that a fairly regular, high discharge rate is produced by one generator site and the irregular, slower discharge represents the activity of other generators. Since the faster discharge is more regular, this probably represents the activity of a larger diameter, perhaps myelinated, branch.

Similar processes have been described in other receptors - cold receptors of primates (Duclaux and Kenshalo, 1973), mechano-receptors in frog skin (Lindblom and Tapper, 1966) and sinus hair receptors in the cat (Gottschaldt, Iggo and Young, 1973).

The occurrence of this kind of collision-resetting process in the peripheral nervous system suggests speculations as to the functional significance of such intraneuronal information processing operations. In peripheral sensory neurones the branched axonal tree increases the receptive field area, and if different generator sites have different properties there can be an increase in the dynamic range of sensitivity. The interaction of collision permits a neurone to perform complex logical operations on the generator trains, but for such operations to be utilized by the nervous system the microstructure of the centripetal spike train must contain biologically useable information. There is no experimental evidence that information coded in the microstructure of spike trains from the sensory neurones considered above is used by the nervous system.

9.5 References.

Brokensha, G. and Westbury, D.R.: Evidence from the adaptation of the discharge of the frog muscle spindle for the participation of multiple spike generators. J. Physiol. 232 25-26P (1973).

Brokensha, G. and Westbury, D.R.: Adaptation of the discharge of frog muscle spindles following a stretch. J. Physiol. 242 383-403 (1974).

Clifford, P. and Sudbury, A.: The Markov property of impulse interaction in branching nerve fibres. Math. Biosciences 13 195-203 (1972).

Coleman, R. and Gastwirth, J.L.: Some models for interaction of renewal processes related to neurone firing. J. Appl. Prob. 6 38-58 (1970).

Cooley, J.W. and Dodge, F.A.: Digital computer solutions for excitation and propagation of the nerve impulse. Biophys. J. 6 583-599 (1966).

Duclaux,R. and Kenshalo, D.R.: Cutaneous receptive fields of primate cold fibres. Brain Research 55 437-442 (1973).

Floyd, K. and Morrison, J.F.B.: Interactions between afferent impulses within a peripheral receptive field. J. Physiol. 238 62-63P (1974).

Goldstein, S.S. and Rall, W.: Changes of action potential shape and velocity for changing core conductor geometry. Biophys. J. 14 731-757 (1974).

Gottschaldt, K.M., Iggo, A. and Young, D.W.: Functional characteristics of mechanoreceptors in sinus hair follicles of the cat. J. Physiol.

<u>235</u>   287-315 (1973).

Hodgkin, A.L. and Huxley, A.F.:   A quantitative description of membrane current and its application to conduction and excitation in nerve. J. Physiol. <u>117</u> 500-544 (1952).

ten Hoopen, M. and Reuver, H.A.:   Selective interaction of two independent recurrent processes. J. Appl. Prob. <u>2</u> 286-292 (1965).

Huxley, A.F.: Can a berve propagate a subthreshold disturbance? J. Physiol. <u>148</u> 80P (1959).

Jack, J.J.B., Noble, D. and Tsien, R.W.: Electric current flow in excitable cells. Oxford University Press, London & New York (1975).

Katz, B. and Miledi, R.: Propagation of electric activity in motor nerve terminals. Proc. Roy. Soc. B <u>-61</u> 453-482 (1965).

Khodorov, B.I., Timin, Y.N., Vilenkin, Y. and Gul'ko, F.B.: Theoretical analysis of the mechanisms of nerve impulse propagation along a non-uniform axon.I. Propagation along a region with an increased diameter. Biofizika <u>14</u> 304-314 (1969).

Kuroda, H. and Ito, F.: Functional differentiation of sensory terminal branches in the frog muscle spindle. Proc. Japan Acad. <u>48</u> 206-209 (1972)

Lindblom, U.F. and Tapper, D.N.: Integration of impulse activity in a peripheral sensory unit. Expl. Neurol. <u>15</u> 63-69 (1966).

Murray, M.J.: Conductile trees: obtaining structural and functional information from endpoint measurements. J. Theoretical Biology <u>43</u> 113-132 (1974).

Rinzel, J. and Keller, J.B.: Traveling wave solutions of a nerve conduction equation. Biophys. J. <u>13</u> 1313-1337 (1973).

Scott, A.C.: Strength-duration curves for   threshold excitation of nerves. Math. Biosciences <u>18</u> 137-151 (1973)

Scott, A.C.: The electrophysics of a nerve fibre. Reviews of Modern Physics <u>47</u> 487-533 (1975).

Tasaki, I.: Collision of two nerve impulses in the nerve fibre. Bichim. Biophys. Acta <u>3</u> 494-497 (1949).

10 GATING AND SELECTIVE INTERACTION MODELS

All the models considered above generate spike trains in which
the local behaviour does not change with time, and which have unimodal
interval density functions. Spike trains with these properties are
often found in the peripheral sensory and motor systems; however,
spike trains recorded in the central nervous system often have multimodal interspike interval histograms and exhibit bursting behaviour.
These more complex patterns of behaviour require the development of
more complex neural models.

There seem to be three kinds of approaches: aspects of the
model that represent intrinsic neuronal properties can be modified,
the input pattern can be modified, the discharge pattern of the model
neurone can be ascribed to its connections in some organized neural
network. The first approach would give a neural model which, in the
presence of an unstructured stochastic input, or perhaps even in the
absence of an input, could generate a complex, patterned output. Since
most neurones respond to simple, deterministic inputs by simple, un-
patterned outputs, this seems an unlikely approach. The second approach,
in which complex input patterns produce the complex discharge pattern,
seems more promising, and will be followed in sections 10.1 - 10.3.
However, since the input patterns to central neurones are often not
observable, this approach has the danger of explaining too much by
assumptions which are untestable in practice. The third approach,
postulating feedback connections in small nets, will be introduced in
section 10.4 and further explored in section 12.

There are examples of invertebrate neurones which, when isolated
and in the absence of an input, can generate a patterned, bursting
discharge e.g. the parabolic burster neurone L3 of the abdominal
ganglion of Aplysia (Waziri, Fraser and Kandel, 1965). This neurone
spontaneously generates bursts of impulses, and the burst activity
shows a circadian rhythm (Strumwasser, 1967). The interspike intervals within a burst have a parabolic distribution, and the generation

of bursting discharges in due to an unstable membrane potential. The
silent interburst periods are produced by a hyperpolarizing after-
potential, and so the length of the interburst period depends on the
number of action potentials in the preceding burst. As in cardiac
muscle (see Noble, 1975), the pacemaker potential, or decay of after-
potential, is produced by a time- and voltage-dependent decay of a
K^+ conductance mechanism.

Thus this spontaneous bursting behaviour can be modelled by
introducing slow time constant conductances into the membrane dynamics.

Lewis (1965) has produced bursting discharges with multimodal
interval histograms by varying the simulated K^+ equilibrium potential
in an electronic model of the Hodgkin-Huxley equations.

Complex discharge patterns could then be produced by stochastic
inputs modulating the intrinsic bursting discharge pattern. However
this line of approach seems premature unless suitable slow conductance
changes are observed experimentally in mammalian central neurones.

A simple modification of the input process to the models of
section 3 can produce bursting discharges. If the stochastic input
is continuous and slowly changing (a band-limited white noise), so
that changes in V(t) are faster than changes in the input, there will
be a tendency for bursting discharges (see the discussion following
equation 4.4.6). However, the input to central neurones is a sequence
of synaptic potentials, not a continuous function.

There are a large number of excitatory and inhibitory synapses
on a central neurone, and activity in any one of these synapses produces
only a small change in V(t). The assumption that the activity in
these input lines is independent gives the random walk and diffusion
models of sections 6 and 7. Instead of a large number of independent
input lines, a more realistic model might be a small number of input
lines from the same neural population and with correlated activity.
The simplest realization of this model is to consider only two input
channels, one excitatory and one inhibitory. Here inhibition and
excitation refer to the effect on the firing rate rather than to the
effect on the membrane potential, and so the inhibitory channel could
be pre or post-synaptic.

If the inhibitory effect is very powerful, it can be considered
to act as a gate which when open (no inhibition) permits the excitatory
input channel to generate spikes: see section 10.1. If the inhibitory

channel is less powerful there can be different kinds of selective
interaction between the input channels - see sections 10.2 - 3.

10.1 Compound Exponential Distributions and Gating Models

Cortical neurones often show irregular, bursting discharges in
the absence of a controlled experimental input - see the monograph by
Burns (1968) for example. This background activity probably represents
the response of the neurone to ongoing synaptic inputs, rather than a
true, spontaneous discharge produced by some slow conductance mechanism.
Thus an analysis of this background activity might give insights into
the background activity of the input channels to the neurone. Whatever the mechanism producing this background discharge, the response
of the neurone to a controlled experimental input will be a change in
this background discharge, perhaps produced by changes in the activity
of the mechanism generating the background discharge.

Smith and Smith (1965) have investigated the statistical characteristics of this background discharge in unidentified neurones in
the cortex of unanaesthetized cats. These neurones showed a bursting
discharge pattern: this background discharge is reduced or abolished
by anaesthesia.

Analysis of the interspike interval histograms showed a group
of units whose interval distributions appeared to be a mixture of two
exponentials; a semilogarithmic plot of an estimate of the survivor
function could be separated into two exponentials by peeling. By
segmenting the data it was apparent that these two exponentials processes were associated with exponentially distributed intervals within
exponentially distributed bursts. The faster exponential process was
Poisson distributed.

Smith and Smith accounted for these observations by a model in
which a Poisson distributed process is gated by a slower, exponentially
distributed process. The formal model is:
    (a)   excitation arrives as a Poisson shower, with parameter C
    (b)   during each OFF period the neurone is inactive and the
excitatory input pulses have no effect; during each ON period each
excitatory input pulse produces an output pulse
    (c)   the duration of the ON periods has a probability density

$$g(t) = \exp(-Bt)$$

(d)   the duration of the OFF periods has a probability density function

$$h(t) = \exp(-At)$$

(e)   each onset of an ON period initiates an action potential; this device avoids the possibility of empty bursts

(f)   the random variables in (a), (c) and (d) are mutually independent. If the process is started at $t = 0$, and the probability

that there is a second action potential in $(t, t + \delta t)$ and that it is in the same burst i.e. the gate has not turned off is given by:

$$p_1(t) = C \exp(-Bt) \exp(-Ct)\ \delta t \qquad (10.1.1)$$

The probability that the gate has turned OFF is

$$p_2(t) = B \exp\{-(B + C)t\}\ \delta t \qquad (10.1.2)$$

Starting when the gate goes off, the probability that it opens for t the first time in the interval $(t, t + \delta t)$ is:

$$p_3(t) = A \exp(-At)\ \delta t \qquad (10.1.3)$$

Taking the one-sided Fourier transforms of $p_1(t)$, $p_2(t)$ and $p_3(t)$ gives the characteristic functions $M_1(j\omega)$, $M_2(j\omega)$ and $M_3(j\omega)$, from which the cumulative characteristic function $M(j\omega)$ can be obtained:

$$M(j\omega) = M_1(j\omega) + M_2(j\omega)\ M_3(j\omega) \qquad (10.1.4)$$

and contour integration in the complex plane gives the interval distribution function:

$$F(t) = \frac{BA}{(B + C - A)} \exp(-At) +$$

$$+ \quad \frac{(C - A)(B + C)}{B + C - A} \quad \exp\{-(B + C)t\} \qquad (10.1.5)$$

which is the distribution of a renewal process as the sum of two exponentials. Thus this model accounts for the observed interval distribution.

However, there is the problem of identifying the proposed gating mechanism with a known biophysical process. The effect of closing the gate is to completely suppress the effect of the excitatory, Poisson distributed input. This could be achieved by a large, maintained increase in membrane conductance tending to clamp the membrane potential close to the resting potential: a burst of post-synaptic inhibition could do this. Alternatively, the excitatory input channels could be suppressed, either by a strong, post-synaptic inhibition of their cell bodies or by a pre-synaptic inhibition of their axonal processes.

Whatever biophysical mechanism is proposed for the gating mechanism, a criticism of this kind of model for bursting discharges is that the observed bursting is accounted for by some unobserved, intermittent or bursting process. Thus the problem of the generation of the slow, intermittent component of bursting discharges is pushed one neurone upstream.

Another kind of model for bursting discharge is to generate the burst activity intraneuronally. Thomas (1966) has discussed a model for bursting discharges based on an intraneuronal mechanism proposed by Burns (1955). This model proposes that the generation of after-bursts following an evoked action potential is due to different rates of recovery of the membrane potential at opposite ends of the neurone, the deeper end repolarizing more slowly than the superficial end. The cell membrane is assumed to produce an action potential followed by a slow negative after-potential, which decays to a resting value V_r. These changes in potential have the same amplitude at different ends of the neurone, but are slower at the deeper end: this accounts for the observed surface-positivity observed during an evoked burst. After the rapid phase of repolarization (with the membrane potential at V_1 which is more positive than V_r) the membrane potential decays exponentially at the deep and superficial ends of the neurone, with time constants τ_1 and τ_2, $\tau_1 > \tau_2$. Thus the resulting potential difference between the two ends of the neurone, $v(t)$, is given by

$$v(t) = (V_1 - V_r) \{ \exp(-t/\tau_1) - \exp(-t/\tau_2) \} \qquad (10.1.6)$$

which has a maximum $v(\theta)$ at

$$\theta = \{ \log(1/\tau_1) - \log(1/\tau_2) \} / (1/\tau_1 - 1/\tau_2) \qquad (10.1.7)$$

and so if $v(\theta) > V_o$, the threshold, a second action potential is generated at a time $t \leqslant \theta$. This subsidiary process continues recursively and generates action potentials so long as $v(\theta) > V_o$. Burns illustrated this model by an electronic neural naalog.

The model of Thomas assumes:

(a) externally evoked discharges are produced by a Poisson process with parameter α.

(b) the afterpotentials following an evoked action potential, v_x, are independently and identically distributed random variables. If $v_x > V_o$, a subsidiary process starts immediately after an evoked action potential: let $r = \text{Prob} \{ v_x > V_o \}$ be the probability of starting a subsidiary process

(c) the subsidiary process generates the first action potential of the afterburst at a time y_o, and this action potential has an afterpotential v_y leading to the second subsidiary action potential of the afterburst at a time $(y_o + y)$ if $v_y > V_o$. This process continues iteratively until $v_y < V_o$. The v_y's are identically and independently distributed random variables, with an exponential distribution with parameter ω. Let $p = \text{Prob} \{ v_y > V_o \}$.

(d) an evoked discharge produced by the external Poisson input process terminates the current subsidiary process and initiates a new subsidiary process with probability r.

This model is formally similar to the branching Poisson process described by Lewis (1964), except that Lewis's model permits over-lapping of subsidiary processes. The asymptotic properties of branching Poisson processes are discussed in Lewis (1969, 1970).

Using the method of generating functions Thomas derives the interspike interval probability density function of the model as

$$f(t) = \frac{\omega(1 - p) + \alpha(1 - r)}{\omega + \alpha + \omega(r - p)} \; \alpha . \exp(-\alpha t)$$

$$+ \quad \frac{r(\omega + \alpha)}{\omega + \alpha \ (r - p)} \quad (\alpha + \omega) \quad \exp -(\alpha + \omega)t \qquad (10.1.8)$$

which is a mixture of two exponentials as desired.

This model is interesting in that it allows some expression of the complexity of intraneuronal processes; however, a biophysically more satisfactory model, based on the known geometry and modelled electrical characteristics of branched, tapering dendritic trees with non-linear current-voltage relations (see chapters 7 and 9 of Jack, Noble and Tsien, 1975) would be prohibitively complex.

Thomas's model is two-state semi-Markov with two types of interval, the two states corresponding to gate ON and gate OFF (Cox, 1963; see also section 7.5 of Cox and Lewis, 1966), and predicts that the serial correlogram (equation 5.2.18) should decrease exponentially with increasing lag. In an extensive analysis of the background discharge of neurones in the foetal rabbit brain, Hyvarinen (1966) found few units with such serial correlograms. Ekholm and Hyvarinen (1970) proposed a pseudo-Markov model for these spike trains, and this abstract model has been further explored by Ekholm (1967, 1968, 1971, 1972).

The train of action potentials generated by Ekholm's two-state pseudo-Markov model consists of an alternating sequence of runs of state 1 or state 2 action potentials. These action potentials are indistinguishable, and the state refers to the interval probability distribution functions $F_1(x)$ and $F_2(x)$ from which the interspike intervals are drawn. The length $Y_{1,1}$ of the first run of state 1 action potentials is a positive integer drawn from a discrete probability distribution $p_1(k)$: say $Y_{1,1} = k$. Then k intervals x_1, x_2, ... x_k are independently drawn from the interval distribution $F_1(x)$. Thus the first k action potentials (numbered from 0 at t = 0 to (k - 1) at t $= x_1 + x_2 + ... + x_k$) are from the state 1 process. The (k + 1)th action potential is from the state 2 process and starts a run of length $Y_{2,1}$ of state 2 action potentials. The length $Y_{2,1}$ of this first run of state 2 action potentials is drawn from a discrete probability distribution $p_2(k)$; say $Y_{2,1} = j$. Then j intervals are drawn independently from $F_2(x)$ taking the process to a time $t = x_1 + x_2 + ... x_{k+j}$, when the length $Y_{1,2}$ of the second run of state 1 action potentials is drawn from $p_1(k)$. Thus the process alter-

nates between runs of state 1 and state 2 action   potentials, and
the process is characterized by the discrete distributions $p_1(k)$
and $p_2(k)$ and the interval distributions $F_1(x)$ and $F_2(x)$.

The type of interval is chosen at the preceding action potential, and
so there is a matrix of transition probabilities:

type of the (k + 1)th event

		1	2
type of the	1	θ_1	$1-\theta_1$
kth event	2	$1-\theta_2$	θ_2

where the discrete distributions are geometric:

$$p_1(k) = (1 - \theta_1)\,\theta_1^{k-1}$$
$$p_2(k) = (1 - \theta_2)\,\theta_2^{k-1} \qquad (10.1.9)$$

The two-state pseudo-Markov process can be mapped on to an alternating
renewal process (see chapter 7 of Cox, 1962) which has the action
potential number as a discrete axis. The discrete alternating renewal
process is the sequence $Y_{1,1}, Y_{2,1}, Y_{1,2}, \ldots$, which is not
stationary but can be modified (see section 3.2 of Ekholm, 1971) to
form an equilibrium discrete alternating renewal process. The full
process is thus a set of random functions defined on an alternating
renewal process. Ekholm has obtained the second order properties of
the stationary interspike intervals and the stationary process of
counts (see equation 5.3.6). If m_1 and m_2 are the means of the run
distributions $p_1(k)$ and $p_2(k)$, and μ_1, μ_2 and σ_1^2, σ_2^2 the means and
variances of the interval distributions $F_1(x)$ and $F_2(x)$, the marginal
interval distribution of the full process is:

$$F(x) = \frac{m_1}{m_1 + m_2}\,F_1(x) + \frac{m_2}{m_1 + m_2}\,F_2(x) \qquad (10.1.10)$$

with a mean

$$\mu \; = \; m_1\mu_1/(m_1 + m_2) \quad + \quad m_2\mu_2/(m_1 + m_2)$$

and variance

$$= \frac{m_1 m_2}{(m_1 + m_2)^2} \; (\mu_1 - \mu_2)^2 \quad + \quad \{ \; m_1\sigma_1^2 \; + \; m_2\sigma_2^2 \} \; /(m_1 + m_2)$$

and Ekholm obtains the serial correlogram $\rho(k)$, $k = 1, 2, \ldots$ (defined in equation 5.2.18) as

$$\rho(k) \; = \; \frac{m_1 m_2}{(m_1 + m_2)^2} \; (\mu_1 - \mu_2)^2 \; \frac{1}{\sigma^2} \left\{ 1 - \frac{m_1 + m_2}{m_1 m_2} \; t(k - 1) \right\}$$

$$(10.1.11)$$

where $t(k)$ is a recursive sequence defined by

$$t(k) \; = \; \sum_{j=0}^{k} \left\{ \sum_{n=j+1}^{\infty} P_1(n) \quad \sum_{m=k-j+1}^{\infty} P_2(m) \right\} +$$

$$\sum_{j=0}^{k} \left\{ \sum_{q=0}^{j} P_1(q) \; P_2(j-q) \quad t(k-j) \right\}$$

$$(10.1.12)$$

This sequence permits numerical computation of the serial correlogram once the run distributions are known.

Ekholm and Hyvarinen (1970) applied this model to analyse the data of Hyvarinen (1966). Units with bimodal interspike interval histograms where chosen, and a histogram could be divided to obtain estimates of the two probability distributions $F_1(x)$ and $F_2(x)$. The distributions of runs $p_1(k)$ and $p_2(k)$ could then be estimated. The computed serial correlogram provided a good fit to the estimated serial correlograms of those units which had geometric run distributions. This explicit method of analysis presupposes that the spike train can be divided into a series of runs, and that the intervals of the different distributions $F_1(x)$ and $F_2(x)$ are clearly distinguishable. An implicit analysis could start with assumptions of the form of the run and interval distributions, and would obtain estimates of the parameters of these distributions from the behaviour of the full process.

The two-state pseudo-Markov model is an abstract model for the generation of point processes and has been applied in other contexts e.g. Ekholm (1972) applies it to the analysis of vehicle flow data. In the neurophysiological context, a problem is to identify possible biophysical mechanisms underlying the two states, and the run distributions. Ekholm and Hyvarinen (1970) suggest two possible interpretations. Thomas's model may be modified to produce a two-state pseudo-Markov model, if the distribution of intervals between subsidiary spikes is not Poisson, and if the process controlling the length of the subsidiary burst has some kind of memory. Another path to a pseudo-Markov model is to let the ratio of the rates of inhibitory to excitatory input pulses to the random walk and diffusion models change, so that short intervals tend to be generated by inputs when $r_i:r_e$ is small and longer intervals tend to be generated when $r_i:r_e$ is large.

A different approach to the multiple exponential interval distributions of Smith and Smith (1965) is taken by Reuver and ten Hoopen (1972), who propose two threshold models for clustered or bursting discharges. In their first model, the excitatory input process is Poisson with parameter λ, and an excitatory input pulse causes the instantaneous transition:

$$V \rightarrow (V + 1) \quad \text{if } V < k$$
$$V=k \rightarrow V=k$$

$$(10.1.13)$$

The potential V decays in a stepwise manner with the duration of the effect of an input pulse having an exponential distribution, with parameter μ. Thus V(t) is a birth and death process with a reflecting barrier at k (see section 6.6)

An action potential is generated whenever an input pulse arrives if $V = k$ or $(k - 1)$. There is no resetting of V(t) after an action potential; V(t) represents the effect of the input rather than the membrane potential, and so this model can be considered to separate the integrative locus of the model neurone from the spike generating locus. The interval distribution generated by this model is a mixture of two distributions, one obtained by the process during intervals when $V(t) \geqslant (k - 1)$ and one obtained during intervals when $V(t) < (k - 1)$. These two distributions correspond to the two states of intraburst and interburst activity.

The Laplace transform of the interval probability density was obtained as:

$$P_1{}^*(s) = \frac{\lambda}{s+\lambda}\left\{1 - \frac{k(k-1)\mu^2}{\{(s+\lambda)^2+(2k-1)\mu(s+\lambda) + k(k-1)\mu^2\}}\right\}$$

$$+ \frac{k(k-1)\mu^2 \lambda^2 A_{k-2}(s)}{(s+\lambda)^2 + (2k-1)\mu(s+\lambda) + k(k-1)\mu^2 A_k(s)} \qquad (10.1.14)$$

where $A_k(s)$ is obtained from the recurrence relation:

$$A_0(s) = 1$$

$$A_1(s) = -s - \lambda$$

$$A_k(s) = -\{s + \lambda + (k-1)\mu\} A_{k-1}(s) - \{(k-1)\lambda\mu\} A_{k-2}(s)$$

This model generates an interval distribution which is a mixture of eight exponentials, and on a semilogarithmic plot can be approximated by a mixture of two exponentials. However, this first model does not provide a good fit to the data of Smith and Smith, and so Reuver and ten Hoopen modified it to produce their second model in which action potentials which terminate intervals during which $V(t) \geqslant (k-1)$ are replaced by a pair of action potentials separated by a small interval η. This increases the mean rate of discharge without altering the shape of the distribution at long intervals. The Laplace transform of the interval probability density of this model was obtained as:

$$p_2^*(s) = \exp(-\eta s)\left\{1 - \frac{k(k-1)\mu^2}{\{s^2 + (2k-1)\mu s + k(k-1)\mu^2\}}\right\}$$

$$\left\{2 - \frac{k(k-1)\mu^2}{\{s^2 + (2k-1)\mu s + k(k-1)\mu^2\}}\right\}^{-1}$$

$$+ \quad p_1^*(s)\left\{2 - \frac{k(k-1)\mu^2}{\{s^2 + (2k-1)\mu s + k(k-1)\mu^2\}}\right\}^{-1} \qquad (10.1.15)$$

The interval probability density function of this second model gave a reasonable fit to the date of Smith and Smith, although the interval density is a mixture of ten exponentials. The density is not sensitive to the value of η, as long as η is much smaller than the mean interval. The details of these two models are not all that interesting; what is of general interest is that a separation of the integrative mechanisms of the model (the summed effect $V(t)$) from the spike generation-reset mechanism permits a simple Poisson input to produce bursting discharges without needing to postulate a gating mechanism.

10.2   Selective Interaction Models

The models of section 10.1 can account for bursting discharges, which may have interspike interval densities which are mixtures of several exponentials, or are bimodal; in this case the mean intra-burst and interburst intervals differ widely. However, the inter-spike interval histograms of central neurones are often even more complex, and can be multimodal. In this section I will consider some models which generate multimodal interval densities; these are reviewed in section 4 of Fienberg (1974), and in section 9.4 of Srinivasan (1974).

Bishop, Levick and Williams (1964) have presented interval histograms of the background, dark discharge of lateral geniculate neurones of the anaesthetized cat. All the interval histograms showed a predominance of short (< 8 msecs) intervals, due to bursting probably produced by the barbiturate anaesthesia. Semi-logarithmic plots of the interval histograms could be classified into four groups:

a)   histograms with a deep notch between the short and long interval ranges

b)   histograms with a not so pronounced notch

c)   histograms with no notch, but a change of slope between the short and long interval ranges

d)   multimodal histograms

Groups (a) - (c) can be considered to form a continuum, and may be accounted for by the models of section 10.1. The multimodal histograms of (d) cannot be accounted for by the models of section 10.1, and had 3-9 modes, with the property that the intervals corresponding to the second and subsequent modes form an arithmetic progression in which the modes are at integer multiples of a shorter mode. Histogram types (a) - (c) were accounted for by a simple superposition model (see section 8). The input fibres to the lateral geniculate neurones are the axons of the retinal ganglion cells and in the dark the discharge of retinal ganglion cells is stationary and has an interval distribution which may be fitted by a Gamma distribution (Kuffler, Fitzhugh and Barlow, 1957). If the background activity of an excitatory input fibre is assumed to have a Gamma distribution defined by equation 5.2.6, with an order $\kappa = 2$, the probability density of the intervals of an input line is simply

$$p(t) \; = \; t. \exp (-t) \tag{10.2.1}$$

if κ / μ of (5.2.6) are set to unity.

If there are n active input lines to the lateral geniculate neurone, which are mutually independent, and an action potential in any one line produces an output spike from the lateral geniculate neurone, the lateral geniculate output will simply be the superposition of the input spikes, and will have an interval density f(t) given by

$$f(t) \; = \; \tfrac{1}{2} \left[1 + \frac{t}{2n} \right]^{n-2} \left[1 - \frac{1}{n} + \frac{2t}{n} + \frac{t^2}{n^2} \right] \exp (-t) \tag{10.2.2}$$

which can be obtained from equation 10.2.1 by use of equation 8.3.9. As the number n of independent input lines increases the distribution intersects the vertical probability axis at higher values. As it

stands, the distribution f(t) does not reproduce the shape of the
histogram types (a) - (c); however, if this distribution is mixed
with a narrow distribution of very short intervals the observed histograms can be reproduced. Thus the model has an input of independent
Gamma distributed spikes, every input spike producing an action
potential in the lateral geniculate neurone, but some input spikes
producing a high frequency burst.

This model ignores the subthreshold integrative properties of
the lateral geniculate neurone, and predicts that the discharge rate of
the lateral geniculate neurone will be greater than the mean discharge
rate of the ganglion cell input lines. The converse is found experimentally; models which explicitly deal with the subthreshold interactions will be discussed in section 10.3.

Multimodal interval histograms may be produced by introducing
inhibitory lines from the retinal ganglion cells to the lateral geniculate neurone. If the effect of an action potential in an inhibitory
line is to delete an excitatory input spike, or to block the effect of
an excitatory input spike, the effect of the inhibitory input will be
to randomly delete some of the output spikes, and so the spike train
from the lateral geniculate neurone will contain intervals which are
sums of adjacent intervals of the superposed input spike train. This
will generate a multimodal interval histogram, with the mode intervals
being integer multiples of the mean interval of the excitatory inputs.

This simple idea of inhibitory deletions has generated a
family of selective-interaction models (ten Hoopen and Reuver, 1965b,
1967, 1968; ten Hoopen, 1966; Dietz, 1968; Coleman and Gastwirth,
1969; Srinivasan and Rajamanur, 1970, 1971; Basawa, 1971; Smith, 1971,
1973; Rade, 1972a,b). The type of selective-interaction model depends
on the kind of effect produced by an inhibitory input, and the statistical properties of the excitatory and inhibitory inputs.

In the first model of ten Hoopen and Reuver (1965b) the input
spike trains to the neurone are two independent, stationary renewal
processes, the excitatory process having a density $\phi(t)$ and the
inhibitory process a density $\psi(t)$. The arrival of an excitatory
input spike produces an output spike, and the arrival of one or more
inhibitory input spikes deletes or blocks the next excitatory spike.
There is no decay of the inhibitory effect, and so this model could be
thought of as a perfect integrator with reset model, with a threshold
$V_o = 1$ and excitatory inputs producing a change in potential of +1,
and inhibitory inputs producing a change of -1.

Two versions of this model were considered by ten Hoopen and Reuver (1965), when $\phi(t)$ is arbitrary and $\psi(t)$ Poisson distributed, which is their model A, and when $\phi(t)$ is Poisson distributed and $\psi(t)$ has an arbitrary distribution, which is their model B. In model A, when the inhibitory process has a Poisson distribution $\psi(t)$ with parameter μ, the Laplace transform of the output interval distribution function is obtained as:

$$P^*(s) \quad = \quad \phi^*(s + \mu)/\{1 + \phi^*(s + \mu) - \phi^*(s)\} \qquad (10.2.3)$$

and so if $\phi(t)$ is a Gamma density:

$$\phi(t) \quad = \quad \frac{m(mt)^{k-1}\exp(-mt)}{\Gamma(k)} \qquad k > 1 \qquad (10.2.4)$$

and $k > 20$, $p(t)$ is multimodal (ten Hoopen, 1966; Dietz, 1968). The first mode is produced by the mean interval of the excitatory input process, the second mode is due to a deletion leaving second-order intervals of the excitatory input, and similarly for higher modes. Thus the modes will occur as an arithmetic progression, with a separation $(k+1)/m$, with the height of the peak decreasing and its spread increasing as the mode number increases. The mean of $p(t)$, $E\{p(t)\}$, and the variance σ^2 are given by Coleman and Gastwirth (1969) as:

$$E\{p(t)\} \quad = \quad \frac{k}{m} (1 + \frac{\mu}{m})^k$$

$$(10.2.5)$$

$$\sigma^2 \quad = \quad \frac{k}{m^2} (1 + \frac{\mu}{m})^k \{k + 1 + k(1 + \mu/m) - \frac{2km}{\mu + m}\}$$

and Coleman and Gastwirth also obtain expressions for the mean and variance when $\phi(t)$ is arbitrary.

Model B, when $\phi(t)$ is Poisson distributed with parameter λ and $\psi(t)$ has an arbitrary distribution, is more complex since the resultant spike train is not a renewal process. The Laplace transform of the output interval density is obtained by ten Hoopen and Reuver (1965b) and Lawrence (1970a,1971), who also obtains the Laplace transforms of the interval densities of the deleted excitatory spikes, the inhibitory spikes which produce deletions and the inhibitory spikes which do not produce deletions.

This basic model has been generalized to cover the situation when $\phi(t)$ and $\psi(t)$ are arbitrary renewal processes (ten Hoopen and Reuver, 1967a; Srinivasan and Rajamannur, 1970), when $\phi(t)$ is any stationary point process (Lawrence, 1970b) and when $\phi(t)$ and $\psi(t)$ are any independent Markov processes (Basawa, 1971).

A more interesting type of generalization has been developed by Coleman and Gastwirth, who permit a decay time to be associated with the inhibitory effect. Thus this takes the selective-interaction model in the direction of the leaky integrator type of model, even though there is still no subthreshold integration. The three models investigated by Coleman and Gastwirth (1969) has $\phi(t)$ arbitrary, with a mean λ, and $\psi(t)$ Poisson distributed with parameter μ, and:

(a) an inhibitory input deletes the next excitatory pulse arriving within a time T, where T is a constant

(b) an inhibitory input deletes the next excitatory pulse arriving in a time θ, where θ is a positive random variable with a distribution function H(t)

(c) an inhibitory pulse deletes all subsequent excitatory input pulses arriving within a time T, which is a constant. Thus an inhibitory pulse is followed by a dead time.

The first model, in which the inhibitory effects last for a constant time T, gives an interspike interval density whose Laplace transform is:

$$P_a{}^*(s) \quad = \quad q^*(s) / \{1 + q^*(s) - \phi^*(s)\} \qquad (10.2.6)$$

where $q^*(s)$ is the Laplace transform of a function $q(t)$ given by:

$$q(t) \quad = \quad \phi(t)\exp(-\mu\min(t,T))$$

This model has been applied in section 9.2.

The second model, where the duration of the inhibitory effect has a distribution H(t), has an interval density whose Laplace transform is

$$P_b{}^*(s) \quad = \quad \frac{\lambda(\lambda+s)}{\lambda(\lambda+s) + s\{\int_0^\infty \exp\{-(\lambda+s)t\}\exp\{-\mu\int_0^t (1-H(x))dx\}dt\}^{-1}}$$

$$(10.2.7)$$

The third model, in which an inhibitory input deletes <u>all</u> excitatory inputs which arrive within the following time T, provides a link with the theory of queues (Takacs, 1962). The time T, which is a positive random variable with a distribution H(t), can be thought of as a dead time. There are two models for the dead period:

(i) the dead period can be that of a type II counter (see chapter 6 of Takacs), which is a counter registering Poisson distributed events which is locked for a dead time following every input event. While the counter is locked no input events are registered.

(ii) the arrival of an inhibitory input removes any remaining effect of any previous inhibitory inputs, and so the dead period persists until the dead time produced by an inhibitory input ends before the next inhibitory input arrives.

For model (i), if $\phi(t)$ is Poisson distributed with parameter λ, the Laplace transform of the interval   density   is given by

$$P_{c,i}^*(s) \; = \; \left\{ 1 + \alpha.\exp{(\mu/\alpha)} \left(\frac{\mu}{\alpha}\right)^{s/\alpha} \left[\int_0^{\mu/\alpha} x^{s/\alpha \, - \, 1} \exp{(-x)} \, dx \right]^{-1} \right\}^{-1}$$

$$(10.2.8)$$

if H(t) has an exponential density h(t) with parameter α.

For model (ii), Fienberg (1974) points out that the expression for the Laplace transform of the interval distribution given by Coleman and Gastwirth (their equation 5.9) is incorrect, and gives the Laplace transform of the interval   density   obtained by Hochman (unpublished thesis) as:

$$P_{c,ii}^*(s) \; = \; \frac{\lambda}{\lambda + \mu + s - \mu(s + \mu)h^*(s + \mu)/\{s + \mu h^*(s + \mu)\}}$$

$$(10.2.9)$$

Coleman and Gastwirth obtain the conditions under which the output spike train is a Poisson process. A number of generalizations and developments have been made of these models. The type II counter version has been generalized by Srinivasan and Rajamannur (1971). Smith (1971, 1973) treats these selective interaction models in terms of a general shot noise model generated by a semi-Markov process, as when $\phi(t)$ is non-Poisson the output spike train is not a renewal process. The output spikes are taken to occur at the transition points of a semi-Markov process. If I(t) is a continuous time,

finite state Markov process, an inhibitory input occurs every time
there is a transition to an initial state in $\{I(t)\}$. This generates
a large class of inhibitory process distributions $\psi(t)$. The output
process is then analyzed as a semi-Markov process whose embedded
Markov chain is $I(t_n)$, where t_n is the time of the nth output spike.
A different kind of generalization is given by Rade (1972), who
permits the inhibitory input to delete the next k excitatory inputs
with probabilities p_k, where $\sum_{k=0}^{\infty} p_k = 1$.

However, all these variants of the basic selective-interaction
model do not take any account of subthreshold interactions between
excitation and inhibition. Models in which subthreshold interactions
occur are considered in section 10.3.

10.3   Models with Subthreshold Interactions.

The major defect of the models of section 10.2 is that no account
is taken of subthreshold interactions, and so the idea of the integrative functions of a neurone is replaced by a gated relay device.
Subthreshold summation of excitatory inputs was introduced to the
selective interaction class of models by ten Hoopen and Reuver (1967b).
In this modification of their earlier models (ten Hoopen, 1966;
ten Hoopen and Reuver, 1965), excitatory inputs arrive with a
distribution $\phi(t)$ and an output action potential is generated when
k excitatory inputs have arrived. The summation mechanism is then
reset. The inhibitory inputs arrive independently with a distribution
$\psi(t)$, and the effect of an inhibitory input is to wipe out any subthreshold summated excitation. Thus the model is that of a perfect
integrator which is reset either by an action potential or by an
inhibitory input. If $\phi(t)$ is Poisson distributed, the output spike
train will have a Gamma distribution (equation 5.2.6) if there is no
inhibitory input. When $\phi(t)$ and $\psi(t)$ are independent recurrent
processes with arbitrary distributions the output spike train will
not be a renewal process, except for the case when both $\phi(t)$ and
$\psi(t)$ are exponentially distributed. Thus the expression for the

output spike train distribution function, p(t), is cumbersome:
however, if ϕ (t) has a small coefficient of variation the function
p (t) is multimodal. ten Hoopen and Reuver discuss this model in
terms of queueing theory, and introduce a further modification where-
by an inhibitory input resets the excitatory input process. This
generates the output spike train as a renewal process and simplifies
the mathematics; however this modification is not appropriate in
the neurophysiological context. The non-Markovian, non renewal
process generated by the basic model has been further explored by
Srinivasan and Rajamannur (1970a, 1970b) and Srinivasan, Rajamannur
and Rangan (1971). Osaki and Vasudevan (1972) have introduced the
modification of letting the effect of an excitatory input be a step
towards the threshold level, where the step size is a positive
random variable with an arbitrary distribution.

These modifications all produce cumbersome expressions for p (t),
and the behaviour of p (t) could be systematically investigated by
numerical methods. However, in the neurophysiological context, this
is not very useful since it seems inappropriate to combine the over-
simplified perfect integrator model with complicated interactions
between arbitrary input distributions. The kind of approach required
is to allow subthreshold interactions within some kind of single time
constant model, without ending up with the diffusion approximations
of section 7.

There are a number of ways of introducing some kind of decay
time constant into the subthreshold interaction models. A simple
approach is to let the effect of an excitatory input have a finite
lifetime: for the generation of an output action potential k excitatory
inputs must arrive within some time θ (Rapoport, 1950; ten Hoopen and
Reuver, 1965a). This is identical to the coincidence detector neural
machine of section 2.2. The output interval density function is
available for k = 2 and for larger values of k when $t < \theta$ (van der
Velden, 1944). ten Hoopen and Reuver approach this model by an
approximation in which the lifetimes of the effects of excitatory
inputs are exponentially distributed, with a mean lifetime of θ. This
approach is taken in terms of a birth and death process in ten Hoopen
(1966b), who compares the behaviour of this model with that of the
leaky integrator, and the model is applied to problems of photon
detection by Srinivasan and Rangan (1970).

A different approach has been adopted by Leslie (1969) as a

general model for clusters of Poisson distributed points, in which
the condition is that k excitatory inputs arrive with no interval
between these excitatory inputs being greater than an interval γ.
The Laplace transform of the interval density of the output spike
train generated by this model is obtained by Leslie as:

$$f*(s) \; = \; \frac{A^k \, B^{k-1}}{1 - A(1 - B)(1 - A^{k-1}B^{k-1})(1 - AB)^{-1}} \tag{10.3.1}$$

where $\quad A \; = \; \lambda/(\lambda + s)$

$\qquad\quad B \; = \; 1 - \exp\{-(\lambda + s)\gamma\}$

which by numerical inversion generates unimodal interval distributions.
Inhibitory inputs have been added to Leslie's model by Hochman
and Fienberg (1971). Their first model has independent excitatory
and inhibitory inputs which are Poisson distributed with parameters
λ and μ. An action potential is generated at the kth excitatory
input of a sequence of excitatory inputs in which no interval between
inputs is greater than γ and which is not interupted by an inhibitory
input. This model generates a unimodal interval density (Fienberg
and Hochman, 1972) which has a Laplace transform given by:

$$g_1*(s) \; = \; \frac{(s + \mu) \, f*(s + \mu)}{s + \mu f*(s + \mu)} \tag{10.3.2}$$

where $f*(s)$ is given by equation 10.3.1.
The second model of Hochman and Fienberg adds a dead time
period after each inhibitory input: the dead time is a positive
random variable T with a probability density function $h(t)$. During
the dead time neither excitatory nor inhibitory inputs are registered.
The Laplace transform of the interval density is:

$$g_2*(s) \; = \; \frac{(s + \mu) f*(s + \mu)}{s + \mu\{1 - h*(s)\} + \mu h*(s) f*(s + \mu)}$$

$$\tag{10.3.3}$$

Fienberg and Hochman (1972) show that this model can generate multimodal interval density functions, with the first mode occurring at t_z
which is the single mode of $Z(t) = f(t) \exp(-\mu t)$, where $f(t)$ is the
inverse transform of (10.3.1) and for a constant dead time T the
higher modes occur at t_n given by

$$t_n = t_z + (n - 1)T + (n - 2)/\mu \text{ for } n = 2, 3, \cdots$$

Fienberg and Hochman (1972) present multimodal computed interval
densities, and by suitable choice of the parameters the histograms of
Bichop et al (1964) and the interval densities of the model of Smith
and Smith (1965) can be reproduced. The interval density is clearly
multimodal when T and μ are large and $1/\mu \ll T$. Note that the distance
between the second and third mode, and subsequent pairs of modes,
depends on $(T + 1/\mu)$, whereas in the selective interaction model of
ten Hoopen (model A of section 10.2) the distance between adjacent
modes depended only on the mean rate of the excitatory input, λ, and
the order of the Gamma density which represented the interval density
of the input.

The third model of Hochman and Fienberg modifies the effect of
the dead time T. During the dead time, any excitatory input fails to
register, but an inhibitory input starts a new dead time period. If
$h(t)$ is the probability density function of a single dead time period,
with distribution $H(t)$, the density function of the concatenated
dead time period, $h_c(t)$, is given by the renewal equation:

$$h_c(t) = \exp(-\mu t) h(t) + \mu \int_0^t \exp(-\mu x) \{ 1 - H(x) \} h(t-x) \, dx$$

and so

$$h_c*(s) = \frac{(s + \mu) h*(s)}{s + \mu h*(s)}$$

Thus if $h(t)$ is unimodal, $h_c(t)$ is also unimodal with its mode occurring earlier than the mode of $h(t)$. This suggests that the interval
density of the output spike train from this model is also unimodal.
The Laplace transform of the interval density is obtained as

$$g_3^*(s) \quad = \quad \frac{s + h^*(s + \mu) \; f^*(s + \mu)}{s + \mu h^*(s + \mu) \; f^*(s + \mu)} \qquad\qquad (10.3.4)$$

and numerical inversion shows this to be unimodal.

Thus whether or not these modifications of Leslie's model of clusters in a Poisson point process generate unimodal or multimodal interval densities depends on the law governing the behaviour of the process during the dead time period. The concept of a dead time is rather an abstract representation of the effects of inhibitory action potentials, and so perhaps it might be biophysically more satisfying to return to models in which an inhibitory input produces a change in the state variable $V(t)$ representing the membrane potential.

Tuckwell (1975a) has developed the differential-difference equation model of Stain (1965) - see equation 7.7.7, and has obtained exact analytic results for the interspike interval moments when the model is subject to a purely excitatory input. A SILIT model (section 7.7) receives excitatory input pulses which are Poisson distributed with parameter λ; the effect of an excitatory input is a positive jump discontinuity in the membrane potential $V(t)$, with a magnitude $\underline{a}$. If $V(0) = x < V_o$, where V_o is the threshold level, the problem is to find the distribution of times θ when $V(t)$ first reaches V_o: this has the density

$$f(\tau, V_o|x)\,d\tau \quad = \quad \text{Prob } \{\theta \in (\tau, \; \tau + d\tau) \mid V(0) = x\} \qquad (10.3.5)$$

The transition density of $V(t)$ satisfies the backward Kolmogorov equation:

$$\frac{-\partial f(x, t)}{\partial t} \quad = \quad -\lambda f(x, t) \quad \frac{-\sigma x \; \partial f(x, t)}{\partial x} \quad + \quad \lambda f(x+a, t) \quad (10.3.6)$$

which is obtained from equation 7.7.7, and σ is the reciprocal of the time constant of the leaky integrator.

For a temporally homogenous Markov process, which may have jump discontinuities in its sample path, Tuckwell (1975b) has shown that if $T_A(x)$ is the first exit time of the process $V(t)$ from the set A $\equiv (v_1, v_2)$, $x \in A$, and defining the moments of $T_A(x)$ by

$$t_{A, (n)}(x) = E\{T_A(x)^n\} \qquad (10.3.7)$$

then $t_{A, (0)}(x)$ satisfies:

$$-\sigma x \frac{dt_{A, (0)}(x)}{dx} + \lambda t_{A, (0)}(x+a) - \lambda t_{A (0)}(x) = 0 \quad (10.3.8)$$

for $x \in A$ and $t_{A, (0)}(x) = 1$ for $x \notin A$

If $t_{A, (0)}(x) = 1$ is a solution of (10.3.8) then $V(t)$ leaves the set A in a finite time and the first exit moments, $t_{A, (n)}(x)$ satisfy the recurrence relation:

$$-\sigma x \frac{dt_{A, (n)}(x)}{dx} + \lambda t_{A, (n)}(x+a) - \lambda t_{A, (n)}(x) = -nt_{A, (n-1)}(x)$$

$$(10.3.9)$$

$$\text{for } x \in A, \quad n = 1, 2, \ldots$$

Tuckwell (1975a) shows that when there is no inhibitory input, the moments of the interspike interval density are all finite - the first two moments are given by equations 7.7.8 and 7.7.9. Since the moments are all finite the recurrence relation (10.3.9) holds with $t_{A, (0)}(x) = 1$ for n=1. Use of this recurrence relation for the first moment of the time for $V(t)$ to leave the interval $A = (\Delta, 0)$ gives

$$-\sigma x \frac{dt_{A, (1)}(x)}{dx} + \lambda t_{A, (1)}(x+a) - \lambda t_{A, (1)}(x) = -1$$

$$\text{for } x \in A \qquad (10.3.10)$$

with $t_{A, (1)}(x) = 0$ for $x \in A$, and $t_{A. (1)}(x)$ has to be continuous on A and continuous at the lower boundary. The moments of the interspike interval density depend on the ratio $k = V_0/a$, and analytic results are simplified when the ratio k is a positive integer. If k is not an integer non-standard integrals result and numerical methods must

be used. Tuckwell (1975a) obtains expressions for the expected value
of the interspike interval θ as a function of λ for $k = 2$ i.e.
$V_o = 2a$.   If $\lambda = n\sigma$

$$E \{\theta\} = T_r + \frac{1}{\lambda} \{ 2 + \frac{1}{K(n)} \} \tag{10.3.11}$$

$$K(n) = 1 - n \{ \log 2 + \sum_{s=1}^{n-1} k_{ns}(1-2^{-s}) \}$$

$$k_{ns} = \frac{(-1)^s}{s} \binom{n-1}{s}$$

where T_r is the absolute refractory period.

By computing $E \{\theta\}$ for different values of λ and σ the relationship
between the first exit time from $(0, 2a)$ for an initial value of
$x = 0$ was obtained as a function of λ/σ. As $\lambda \to \infty$ the first exit
time asymptotically approaches $2/\lambda$: this is the result obtained for
the perfect integrator model subject to a Poisson excitatory input.
When λ/σ approaches zero the decay process predominates and the exit
time tends to infinity.

The relationship between the mean discharge rate and the excitatory input rate was computed using values of σ, V_o and T_r appropriate
for α-motoneurones, in order to compare the theoretical relationship
with the experimental results of Redman and Lampard (1968) and Redman,
Lampard and Annals (1968). Redman et al stimulated four muscle
afferent nerve filaments asynchronously, to produce, by superposition,
a Poisson distributed group Ia excitatory input to an α-motoneurone,
with the input rate directly controlled. Over a mean input rate
range of 25 - 120 impulses/sec the relationship between discharge
rate and input rate was approximately exponential. The theoretical
relationship and experimental results agree fairly well, the
theoretical mean discharge rate being of the right order of magnitude
but less than the observed discharge rate. A major cause of the
underestimate could be that the maximum amplitude of the evoked
excitatory postsynaptic potential is dependent on the stimulus rate,
the evoked excitatory postsynaptic potential being largest for stimulus
repetition rates of about 55 - 60/sec.

The inclusion of a Poisson distributed inhibitory input with
parameter μ to this differential-difference equation model gives the

forward Kolmogorov equation 7.7.7, and if the moments of the interspike interval density were finite a similar method could be used. It appears that the moments are not finite and so (7.7.7) cannot be solved by Tuckwell's method.

However, Tuckwell (1975a) introduces the assumption that the inhibitory input is weaker than the excitatory input, and takes a linear approximation

$$t_{A, (1)}(x-b) \simeq t_{A, (1)}(x) - b \frac{dt_{A, (1)}(x)}{dx}$$

where b is the amplitude of the inhibitory postsynaptic potential. This is equivalent to adding a constant, hyperpolarizing current to the model, which will produce a linear, hyperpolarizing drift in $V(t)$ and confine $V(t)$ to the interval $(\mu b/\lambda, \infty)$. The moments of the inter-spike interval density are now finite and the mean discharge rate is sensitive to the level of the inhibitory input; the introduction of an inhibitory input half as strong as the excitatory input ($\mu b = 1/2. \lambda a$) halves the mean discharge rate obtained using this approximation. The analytic problems produced by introducing inhibition to the differential-difference equation model suggest that numerical simulation, similar to the extensive computations of Segundo et al (1968) for excitatory inputs, may be a fruitful approach.

10.4 Neural Micronets with Reciprocal Inhibition.

An alternative approach to bursting discharge patterns is that the bursting behaviour is produced by the pattern of connections in a network of neurones. I will call a neural network consisting of only a few neurones a micronet; the behaviour of such a micronet is specified in terms of the activity of its constituent neurones, whereas the behaviour of the larger networks considered in section 12 can only be described in terms of the average activity of its constituents. The simplest micronet consists of only two neurones.

The rhythmic, bursting discharge of vertebrate and invertebrate motor efferents is alternating, in that the discharge of flexor

motoneurones alternates with the discharge of extensor motoneurones.
This alternating rhythmic discharge may be modelled by reciprocally
inhibitory connections between flexor and extensor motoneurones
pools (Brown, 1911).

Reiss (1962) has investigated the discharge patterns produced by
interactions between two neurones, using digital and analog simulations.
If an almost regular input excites a pair of reciprocally inhibiting
neurones, rhythmic, alternating bursts are generated. By appropriate
manipulation of the parameters of the neurone model and the strength
of the inhibitory cross-connections, one unit only can be made to fire
at any one time: its discharge inhibits the other unit. If the units
have some slow time constant mechanism, representing the effects of
fatigue, the discharge rate of the active unit will decrease with time.
Thus the silent unit is subject to a maintained excitatory input and
a decreasing inhibitory input, and is released from inhibition. The
silent unit now becomes active and dominant, and inhibits the first.
The rate of alternation of dominance between the two units is an almost
monotonic function of the rate of the common excitatory input. The
pattern of dominance is not necessarily complete, as when one unit is
active the other may fire in rapid bursts. This is similar to the
discharge patterns of the reciprocally innervated extensor and flexor
motoneurones of the locust leg (Hoyle, 1966).

Similar analog simulations have been performed by Harmon (1964),
using shorter time constants to mimic the flight control system of
the locust (Wilson, 1964). Alternating burst discharges can also be
produced by excitatory cross-connections between two neural models
(Pavlidis, 1965; see also section 2.6 of Pavlidis, 1973). The generation of periodic burst discharges by micronets of reciprocally
inhibiting neurones has been further modelled by Kling and Szekely
(1968), Wilson and Waldron (1968) and Suzuki et al. (1971). Morishita
and Yajima (1972) have investigated the conditions for the stability
of a network of reciprocally inhibiting neurones. A continuous
variable neural model was used, consisting of a multi-input summation
device, a first-order, low-pass filter and a diode type nonlinearity
in cascade. For a mutually inhibiting network, with each element
inhibiting all the other elements, the network dynamics can be
described by an autonomous system of nonlinear differential equations.
The steady-state solutions show three modes of behaviour, depending
on the weighting function associated with the inhibitory connections:
a unique, equilibrium solution, multiple equilibrium solutions with

hysteresis, and periodic solutions. These correspond to contrast enhancement by lateral inhibition, short term information storage and pattern generation.

Kinouchi and Inai (1974) have investigated in detail the alternating output pattern generated by a micronet of two reciprocally inhibiting leaky integrator model neurones, subject to a pulse train or constant input. The possible stable patterns for a regular pulse train input are when one unit is completely dominant, when both units are synchronously active or when there is an alternating output pattern. The stable patterns for a constant input are only the two cases of one unit being completely dominant, and when there is an alternating output pattern.

Thus one unit of the reciprocally inhibitory neurone pair can exhibit deterministic burst discharges when subject to either a pulse train input or a constant input. This would suggest that stochastic burst discharge patterns with multimodal interval histograms could be modelled by a stochastic input to a reciprocally inhibitory neurone pair. However, there are a number of micronet structures which can generate burst discharge patterns, and so if complex discharge patterns are to be modelled in terms of a micronet there is a choice of the neural model used and the net structure. In cases where histological evidence does not suggest a specific net structure it might be possible to model a given discharge pattern by complex neural models in a simple net or by simple neural models in a complex net.

This idea is illustrated by Oguztoreli (1972), Leung et al (1974a, b) and Stein et al (1974a, b). Stein et al (1974a) have developed a third order neural model incorporating adaptation: for a perfect integrator the time t_i to reach threshold V_o in response to a constant input $f(u) = f$ is

$$V_o = \int_0^{t_i} f(u)\ du = ft_i \qquad\qquad (10.4.1)$$

and so the discharge rate r is simply

$$r = f/V_o$$

Introduction of an absolute refractory period t_o gives

$$r \quad = \quad f/(ft_o + V_o) \tag{10.4.2}$$

and so the discharge rate saturates at large positive inputs. However, negative inputs do not give negative discharge rates and so r cannot be negative when f is negative - setting r = 0 when f is negative introduces a sharp discontinuity in slope at f = 0. This sharp discontinuity can be avoided by an arbitrary function of the form (Cowan, 1970)

$$x \quad = \quad rt_o$$
$$= \quad 1/\{1 + \exp(-f)\} \tag{10.4.3}$$

where $0 < x < 1$ is a normalized output discharge rate.

Let the discharge rate respond to a step change in input by an exponential change from its initial value to a steady state value, with a rate constant $\underline{a}$

$$\frac{1}{a} \frac{dx}{dt} + x = \frac{1}{\{1 + \exp(-f)\}} \tag{10.4.4}$$

and let the discharge rate adapt exponentially with a rate constant p:

$$\frac{1}{a} \frac{dx}{dt} + x = \frac{1}{1 + \exp\{-f - b \int_0^t x(t-u) \exp(-pu)\, du\}} \tag{10.4.5}$$

and so when the constant b is negative the response adapts with a rate p, the extent of the adaptation dependent on b and the convolution of the rate at earlier times with the exponential decay. This adaptation process will build up with time, and so rate constants for the onset and decay might be required:

$$\frac{1}{a} \frac{dx}{dt} + x = \frac{1}{1 + \exp\{-f - b \int_0^t x(t-u)\{\exp(-pu) - \exp(-qu)\}\, du\}} \tag{10.4.6}$$

and p < q.

Leung et al (1974b) discuss equation 10.4.6 as a member of a general class of integro-differential equations which can be represented as a system of first-order differential equations:

$$\frac{1}{a} \frac{dx_1}{dt} = -x_1 + \frac{1}{1 + \exp(-f -bx_2 + bx_3)}$$

$$dx_2/dt = x_1 - px_2 \tag{10.4.7}$$

$$dx_3/dt = x_1 - qx_3$$

with appropriate initial conditions, and x_1 is identical with x of (10.4.6). Equation 10.4.5, obtained from (10.4.6) by letting $q \to \infty$, can be represented by two coupled first order differential equations, and if $b \to 0$ equation 10.4.4 results which can be represented by a single differential equation.

Thus a micronet of N third order model neurones can be represented by the system of equations:

$$\frac{1}{a_i} \frac{dx_{i,1}}{dt} = -x_{i,1} + \frac{1}{1 + \exp\left\{-f_i - \sum_{j=1}^{N} c_{i,j} x_{i,2} + b_i x_{i,3}\right\}}$$

$$\tag{10.4.8}$$

$$dx_{i,2}/dt = x_{i,1} - p_i x_{i,2}$$

$$dx_{i,3}/dt = x_{i,1} - q_i x_{i,3}$$

where $i = 1, 2, \ldots N$ and $c_{i,j}$ is the strength of the connection from neurone i to j. This is negative for an inhibitory connection. The properties of this set of equations are explored in Oguztoreli (1972), Leung et al (1974) and Stein et al (1974b).

Stein et al (1974b) and Leung et al (1974b) discuss the conditions under which the linear behaviour of a micronet of N model neurones is equivalent to the linear behaviour of a single neurone of order K, for K = N. A single neurone of order 2 (i.e. equation 10.4.5) has three possible modes of behaviour in response to a small perturbation; damped oscillations, a decaying response and an unstable, non-oscillatory response. A micronet of two first order coupled neurones has the same range of linear responses, and so cannot generate an oscillatory response, except when the coupling coefficients $c_{1,2}$ and $c_{2,1}$ have the opposite sign. Thus reciprocal inhibition between two first order neurones will not generate burst discharges.

However, two coupled second order model neurones, or a single fourth-order model neurone, can generate oscillatory responses. For weak interconnections damped oscillations are produced, stronger interconnections give maintained oscillations and powerful inter-connections give damped oscillations. Thus linear analysis shows that a fourth-order system can generate oscillatory responses, and this system can be interpreted either as a multiple time constant neural model, or a coupled micronet.

Matsuyama, Shirai and Akizuki (1974) have investigated numeric-ally and by simulation the discharge pattern produced by two leaky integrator model neurones with mutual inhibitory coupling. Two micronet structures are treated; when the two neural models receive independent excitatory inputs and when they receive the same input sequence, but one model neurone receives the excitation after a delay. When the inputs are Poisson processes the discharge of the neural models has a multimodal interval density, and shows an irregular rhythm, the periodicity of the rhythmic bursts depending on the time constants of the leaky integrator. More complex discharge patterns are produced when the excitatory pulse train input is a sinusoidally modulated point process.

10.5. References.

Basawa, I.V.: Some models based on the interaction of two independent Markovian point processes. J. Appl. Prob. 8 193-197 (1971).

Brown, T.G.: The intrinsic factors in the act of progression in the mammal. Proc. Roy. Soc. (London) B 308-319 (1911).

Burns, D.D.: The mechanism of afterbursts in cerebral cortex. J. Physiol. 127 168-188 (1955).

Bishop. P.O., Levick, W.R. and Williams, W.O.: Statistical analysis of the dark discharge of lateral geniculate neurones. J. Physiol 170 598-612 (1964).

Burns, B.D.: The Uncertain Nervous System. Edward Arnold, London (1968).

Coleman, R. and Gastwirth, J.L.: Some models for interaction of renewal processes related to neuron firing. J. Appl. Prob. $\underline{6}$ 38-58 (1970).

Cox, D.R.: Renewal Theory. Methuen, London (1962).

Cox, D.R.: Some models for series of events, Bull. I.S.I. $\underline{40}$ 737-46 (1963)

Cox, D.R. and Lewis, P.A.W.: The Statistical Analysis of Series of Events. Methuen, London (1966).

Dietz, K.: Erzeugung multimodaler intervallverteilungen durch ausdunnung von ennenerungsprozessen. Kybernetik $\underline{4}$ 131-136 (1968).

Ekholm, A.: A generalization of the two-state semi-Markov model. Social Research Institute of Alcohol Studies, Helsinki. Report 27 (1967).

Ekholm, A.: A pseudo-Markov model for stationary series of events. Social Research Institute of Alcohol Studies, Helsinki. Report 31 (1968).

Ekholm. A.: A pseudo-Markov model for stationary series of events. Commentationes Physico-Mathematicae $\underline{41}$ 73-120 (1971).

Ekholm, A.: A generalization of the two-state, two interval semi-Markov model. In: Stochastic Point Processes: Statistical Analysis, Theory and Applications. ed. P.A.W.Lewis. Wiley-Interscience, N.Y. (1972).

Ekholm, A. and Hyvarinen, J.: A pseudo-Markov model for series of neuronal spike events. Biophysical J. $\underline{10}$ 773-796 (1970).

Fienberg, S.E.: Stochastic models for single neurone firing trains: a survey. Biometrics $\underline{30}$ 399-427 (1974).

Fienberg, S.E. and Hochman, H.G.: Modal analysis of renewal models for single neurone discharge. Kybernetik $\underline{11}$ 292-297 (1972).

Harmon. L.D.: Neuromimes : action of a reciprocally inhibitory pair. Science $\underline{146}$ 1323-1325 (1964).

Hochman, H.G.: Some renewal process models relevant to neural discharge. Doctoral dissertation, Dept. of Theoretical Biology, Univ. of Chicago (1971).

Hochman, H.G. and Fienberg, S.E.: Some renewal models for single neuron discharge. J. Appl. Prob. $\underline{8}$ 802-808 (1971).

ten Hoopen, M.: Multimodal distributions. Kybernetik $\underline{3}$ 17-24 (1966a).

ten Hoopen, M.: Probabilistic firing of neurons considered as a first passage problem. Biophysical J. $\underline{6}$ 435-451 (1966b).

ten Hoopen, M. and Reuver, H.A. : On a waiting time problem in physiology. Statistica Neerlandica 19 27-34 (1965a).

ten Hoopen, M. and Reuver, H.A. : Selective interaction of two independent recurrent processes. J. Appl. Prob. 2 286-292 (1965b).

ten Hoopen, M. and Reuver, H.A. : Interaction between two independent recurrent time series. Information and Control 10 149-158 (1967a).

ten Hoopen, M. and Reuver, H.A. : On a first passage problem in stochastic storage systems with total release. J. Appl. Prob. 4 409-412 (1967b).

ten Hoopen, M. and Reuver, H.A. : Recurrent point processes with dependent interference with reference to neuronal spike trains. Math. Biosci. 2 1-10 (1968).

Hoyle, G.: Exploration of neuronal mechanisms underlying behaviour in insects. In: Neural Theory and Modeling. Ed. R.F.Reiss. Stanford Univ. Press, Stanford. (1964).

Hyvarinen, J.: Analysis of spontaneous spike potential activity in developing rabbit diencephalon. Acta Physiol. Scandinavica 68 Supplementum 278 (1966).

Jack, J.J.B., Noble, D. and Tsien, R.W.: Electric Current Flow in Excitable Cells. Clarendon Press, Oxford. (1975).

Kinouchi, Y. and Inai, Y.: On the output pattern of a reciprocally inhibited two neurone system. Bull. of Faculty of Engineering, Tokushima University 10 55-65 (1974).

Kling, U. and Szekely, G.: Simulation of rhythmic nervous activities. Kybernetik 5 89-103 (1968).

Kuffler, S.W., Fitzhugh, R. and Barlow, H.B.: Maintained activity in the cat's retina in light and darkness. J. Gen. Physiol. 40 683-702 (1957).

Lawrence, A.J.: Selective interaction of a Poisson and renewal process: first order stationary results. J. Appl. Prob 7 359-72 (1970a).

Lawrence, A.J.: Selective interaction of a stationary point process and a renewal process. J. Appl. Prob. 7 483-489 (1970b).

Lawrence, A.J.: Selective interaction of a Poisson and renewal process: the dependency structure of the intervals between responses. J. Appl. Prob. 8 170-183 (1971).

Leung, K.V., Mangeron, D., Oguztorelli, M.N. and Stein, R.B.: On the stability and numerical solutions of two neural models.Util. Math.5 167 (1974)

Leung, K.V., Mangeron, D., Oguztorelli, M.N. and Stein, R.B.: On a class of nonlinear integro-differential equations. I and II. Rend. Accad. Naz. Lincei Series VIII, vol 54 342-348 & 699-705 (1973).

Lewis, P.A.W.: A branching Poisson process for the analysis of computer failure patterns. J. Roy. Stat. Soc. B 26 398-441 (1964).

Lewis, P.A.W.: Asymptotic properties and equilibrium conditions for branching Poisson processes. J. Appl. Prob. 6 355-371 (1969).

Lewis, P.A.W.: Asymptotic properties of branching renewal p ocesses. I.B.M. R.C. 2878 (No. 13535) May 1970.

Morishita, I. and Yajima, A.: Analysis and simulation of networks of mutually inhibiting neurones. Kybernetik 11 154-165 (1972).

Noble, D.: The Initiation of the Heartbeat. Clarendon Press, Oxford (1975).

Oguztorelli, M.N.: On the neural equations of Cowan and Stein(I). Utilitas Mathematica 2 305-317 (1972).

Pavlidis, T.: A new model for simple neural nets and its application in the design of a neural oscillator. Bull. Math. Biophys. 27 215-229 (1965).

Pavlidis, T.: Biological Oscillators : Their Mathematical Analysis. Academic Press, New York (1973).

Rade, L.: A mode of interaction of a Poisson and a renewal process and its relation with queuing theory. J. Appl. Prob. 9 451-456 (1972a).

Rade, L.: Thinning of renewal point processes. Goteborg, Sweden (1972b).

Rapoport, A.: Contribution to the probabilistic theory of neural nets II. Facilitation and threshold phenomena. Bull. Math. Biophys. 12 187-197 (1950).

Redman, S.J. and Lampard, D.G.: Monosynaptic stochastic stimulation of cat spinal motoneurones. I. Repponse of motoneurones to sustained stimulation. J. Neurophysiol. 31 485-498 (1968).

Redman, S.J., Lampard, D.G. and Annal, P.: Monosynaptic stochastic stimulation of cat spinal motoneurones. J. Neurophysiol. 31 499-508 (1968).

Reiss, R.F.: A theory and simulation of rhythmic behaviour due to reciprocal inhibition in small nerve nets. Proc. AFIPS Spring Joint Computer Conference 21 171-194 (1962).

Smith, D.R. and Smith, G.K.: A statistical analysis of the continual activity of single cortical neurones in the cat unanaesthetized isolated forebrain. Biophysical J. 5 47-74 (1965).

Smith, W.: A general shot noise model applied to neural spike trains. Proc. of the 3rd Annual Southeastern Symposium on System Theory, Vol.2. (1971).

Smith, W.: Shot noise generated by a semi-Markov process. J. Appl. Prob. 10 685-690 (1973).

Srinivasan, S.K.: Stochastic Point Processes and Their Application. Griffin, London (1974).

Srinivasan, S.K. and Rajamannar, G.: Selective interaction between two independent stationary recurrent point processes. J. Appl. Prob. 7 476-482 (1970a).

Srinivasan, S.K. and Rajamannar, G.: Counter models and dependent renewal point processes related to neurone firing. Math. Biosci. 7 27-39 (1970b).

Srinivasan, S.K., Rajamannar, G. and Rangan, A.: Stochastic models for neurone firing. Kybernetik 8 188-193 (1971).

Srinivasan, S.K. and Rangan, A.: A stochastic model for the quantum theory in vision. Math. Biosci. 9 31-36 (1970).

Stein, R.B.: A theoretical analysis of neuronal variability. Biophysical J. 5 173-194 (1965).

Stein, R.B., Leung, K.V., Mangeron, D. and Oguztorelli, M.N.: Improved neuronal models for studying neural networks. Kybernetik 15 1-9 (1974).

Stein, R.B., Leung, K.V., Oguztorelli, M.N. and Williams, D.W.: Properties of small neural networks. Kybernetik 14 223-230 (1974).

Strumwasser, F.: Neurophysiological aspects of rhythm.In : The Neurosciences - a study program. Ed. G.C. Quarton, T. Melnechuk and F.O. Smith. Rockefeller Univ. Press, New York. (1967).

Suzuki, R.: Dynamics of 'Neuron Ring'. Kybernetik 8 39-45 (1971).

Takacs, L.: Introduction to the Theory of Queues. Clarendon Press, Oxford (1962).

Thomas, E.A.C.: Mathematical models for the clustered firing of single cortical neurones. British J. of Mathematical and Statistical Psychology 19 151-162 (1966).

Tuckwell, H.C.: Determination of the inter-spike times of neurones receiving randomly arriving post-synaptic potentials. Biological Cybernetics 18 225-237 (1975a).

Tuckwell, H.C.: On the first exit time problem for temporally
homogenous Markov processes. J. Appl. Prob.

van der Velden, H.A.: Over het aantal lichtquanta dat nodig is
voor een lichtprikkel bij het menselijk oog. Physica 11 179-189(1944).

Waziri, R., Frazier, W.T. and Kandel, E.R.: Analysis of pacemaker
activity in an identifiable burst generator in Aplysia. Physiologist
8 190 (1965).

Wilson, D.M.: Relative refractoriness and patterned discharge of
locust flight motor neurones. J. Expl. Biol. 41 191-205 (1964).

Wilson, D.W. and Waldron, I.: Models for the generation of motor
output patterns in flying locusts. Proc. I.E.E.E. 56 1058-1064 (1968).

11. MODELS OF SYNAPTIC TRANSMISSION

The models discussed in sections 3-10 all deal with stochastic trains of action potentials : in this section I will deal with models of the processes which occur when an action potential reaches the axonal termination and produces electrical changes in a post-synaptic cell. There are two types of specific mechanism for the transmission of electrical activity between excitable cells - these are electrical transmission and chemical transmission. Eccles (1964) has given an extensive review of the basic physiology of these transmission mechanisms in the nervous system.

In electrical transmission between excitable cells, there is a low resistance pathway between the two cells, and so potentials in one cell can spread with little decrement to the second cell. This effect-ive electrical continuity need not imply a structural, cytoplasmic continuity, but could be produced by a favourable geometric relation-ship between processes of the two cells, or by a structural modification, perhaps a fusion, of the cell membranes. In some cases the voltage decrement produced by the low resistance pathway is less than the safety factor for propagation of action potentials, and so action potentials can propagate freely in both directions : the best known example of this is the low resistance pathway between adjacent cardiac muscle cells. In this case the tissue behaves as a functional syncytial unit, in which the partial differential equation for the spread of excitation is a three dimensional analog of the cable equation (see chapter 5 of Jack, Noble and Tsien, 1975).

It is possible to have a uni-directional, or rectifying, low resistance pathway between two excitable cells e.g. Furshpan and Potter (1959) have demonstrated a low resistance connection which acts as a unidirectional excitatory electrical synapse between the giant axon and flexor motor axons in the crayfish abdominal cord. Specific, inhibitory electrical synapses also exist e.g. the inhibitory, electrical synapse on the axon cap of the Mauthner cell of the goldfish (Furukawa and Furshman, 1963). This curious effect is produced by a large extra-cellular resistance funnelling the external local circuit current of

the pre-synaptic action potential across the post-synaptic membrane,
producing a hyperpolarization or decrease in excitability of the spike
producing area.

Whatever the functions or importance of electrical synapses in
the activity of the nervous system (see Bennett, 1966), electrical
synapses can be modelled by a simple, passive electrical circuit, and
so are not all that interesting from the point of view of stochastic
modelling. French and DiCaprio (1975) have used the methods of
section 5.4 to estimate the frequency response and coherence functions
of an electronic junction between two neurones in the leech : this
junction is a nonrectifying electrical synapse and can be represented
by a simple linear resistance. The low attenuation of this junction
tends to synchronize the action potentials of both neurones.

In chemical transmission between excitable cells, the depolarization of the pre-synaptic terminals produces an increase in the
rate of release of a chemical transmitter, which diffuses across the
synaptic cleft. At all chemical synapses the transmitter appears to
be released in discrete multi-molecular packets or quanta. At the
post-synaptic membrane the transmitter binds with receptor sites and
causes a change in the conductance of the post-synaptic membrane.
The transmitter is inactivated enzymatically. The ionic currents
flowing through the changed conductance of the post-synaptic membrane
produce local, post-synaptic potentials, which, if depolarizing, may
initiate an action potential. Thus there are several stages in this
process which may be modelled by stochastic processes:
     (a) the nature of the spontaneous release of quanta of
transmitter (section 11.2)
     (b) the effect of pre-synaptic depolarization or action potentials
on the release of transmitter (section 11.3)
     (c) the effects of prolonged stimulation on the release of
transmitter : fatigue and facilitation (section 11.4)
     (d) the nature of the fluctuations in post-synaptic potential
produced by quantal conductance changes (section 11.5)
     (e) the dynamics of information transfer across a synapse
(section 11.6)

The neuro-muscular junction provides a model preparation for
investigating chemical transmission : the main advantage of this
preparation is the accessibility of the junction and the controllabil-
ity of the extracellular fluid composition. It is assumed that

chemical transmission in the central nervous system is by mechanisms
which are qualitatively similar to transmission at the neuro-muscular
junction, but which can use different excitatory and inhibitory
transmitters. The most apparent difference is purely quantitative:
at the neuro-muscular junction a pre-synaptic action potential
releases enough transmitter to almost invariably evoke a post-synaptic
muscle action potential; at central synapses much less transmitter
is released and so post-synaptic potentials must summate before a
post-synaptic action potential is initiated.

11.1   The Neuro-muscular junction

At the vertebrate neuro-muscular junction, the myelinated
motor axon loses its myelin sheath and branches, running in troughs
in the muscle fibre and forming discrete contacts with a specialized
part of the muscle fibre, the motor endplate. At the endplate the
muscle fibre membrane is thrown into folds, which increase the
surface area and hence conductance of the post-junctional membrane.
The membranes of the motor axon terminations and the muscle fibre
are separated by a cleft of about 500 A, the contents of which are
contiguous with extracellular fluid. If the specific resistivity of
the fluid or gel within the cleft is approximately that of extracellular fluid (100 ohm-cms), Katz (1966) calculates that, assuming
normal values of membrane resistance, and given the geometry of a
1.5 μm diameter axon terminal running for 1 mm along a muscle fibre
of 150 μm diameter, the electrotonic spread of current from a pre-
junctional action potential would produce a negligible (< 1 μV)
change in the post-junctional membrane potential. This argues
strongly against electrical transmission at the neuro-muscular
junction.

In fact, transmission has been shown to be chemical, the
transmitter at the vertebrate skeletal neuro-muscular junction being
acetylcholine. The inactivating enzyme, acetylcholine esterase, is
located on the muscle membrane surface. Electron micrographs show
numerous vesicles, about 500 A in diameter, in the axon terminal, and
del Castillo and Katz (1955) suggest that these vesicles may be
identified with quanta of transmitter. This attractive idea need
not be taken too seriously : however, there is a tendency to identify
any collection of vesicles beneath a neuronal membrane with a
chemically transmitting presynaptic terminal.

PRESYNAPTIC   SYNAPTIC CLEFT   POST SYNAPTIC

acetylcholine
synthesis

    storage in vesicles

                     spontaneous Ca^{2+}
                     entry

                     spontaneous   diffusion → receptor → spontaneous, Ca^{2+}
                     quantal release   site   dependent miniature
                                                                   endplate potentials

action potential

                     depolarization of
                     terminal
                     Ca^{2+} entry

                     simultaneous release → diffusion → receptor → endplate potential
                                                          site

                                                                   propagated muscle
                                                                   action potential

                     choline   hydrolysis by
                     reabsorption   acetylcholine
                                          esterase

For a chemical to be identified as a synaptic transmitter it must fulfil the following criteria:

(a)   it must be located and synthesized intraneuronally, and highly concentrated in the axon terminals, bound to protein granules or enclosed within vesicles

(b)   a local inactivating enzyme must be present

(c)   it must be released from the pre-synaptic terminals by pre-synaptic action potentials

(d)   the post-synaptic membrane must respond to the applied transmitter in the same way, by the same conductance changes, as to pre-synaptic action potentials

(e)   pharmacological agents which interfere with the synthesis, storage, release and inactivation of the proposed transmitter should appropriately modify the response of the post-synaptic cell to pre-synaptic action potentials.

Acetylcholine fulfils all these criteria at the vertebrate skeletal neuro-muscular junction, and is the excitatory transmitter at this junction. There is no peripheral inhibition of vertebrate skeletal muscle. Acetylcholine produces its excitatory effect, the endplate potential, by producing a local increase in sodium and potassium conductance : this acetylcholine activated ionic current mechanism is voltage insensitive and has a reversal potential of around -15 mV.

The presence of calcium ions is important in the release of transmitter; a reduction of external calcium ion concentration reduces the amplitude of the evoked endplate potential (an index of the amount of transmitter released). Magnesium ions competitively inhibit this effect of calcium; thus a low calcium, high magnesium external solution can be used to reduce the number of quanta released by a pre-synaptic action potential.

Figure 11.1 schematically illustrates the essential steps in chemical transmission at the vertebrate neuro-muscular junction.

11.2 Spontaneous Miniature Endplate Potentials

Fatt and Katz (1950, 1952) first described the occurrence of
small (< 1 mV) spontaneous depolarizations under the endplates of the
neuromuscular junction, the depolarizations having the same time course
as the endplate potential. These miniature endplate potentials (m-e.p.ps.)
are believed to represent the post-junctional membrane response to the
spontaneous pre-junctional release of packets or quanta of the transmitter Acetylcholine : this idea is supported by the localization
of the m-e.p.ps. to the neuromuscular junction and the similarity of
the pharmacological sensitivity of the m-e.p.ps. and endplate
potentials to curarine and prostigmine. The evidence for this quantal
hypothesis is reviewed by Martin (1966), Stevens (1968) and van der
Kloot, Kita and Cohen (1975).

The intervals between the spontaneous m-e.p.ps. are irregular,
although visual inspection of the records suggests some infrequent
bursts. Fatt and Katz analyzed records in which spontaneous bursts
of m-e.p.ps. were absent, and the mean rate of m-e.p.ps was about
5/sec. The spontaneous rate varied in different preparations over a
3 decade range, from 0.1 - 100/sec., but high and low discharge rates
were not chosen for analysis as (a) at high rates there would be
summation and coincidences giving inaccuracies in interval
measurement and (b) at low discharge rates the recording of a large
sample of intervals would take too long, and there would be problems
in maintaining the stability of the preparation and the stationarity
of the discharge.

The inter-event interval histogram could be fitted a simple
exponential curve:

$$n \; = \; N\Delta t/T \, \exp \, (-t/T) \qquad\qquad (11.2.1)$$

where N is the total number of intervals and n is the number of
intervals whose length lies in (t, t+Δt), and T is the mean interval.
This simple exponential suggests that the m-e.p.ps. have a Poisson
interval distribution, and the m-e.p.ps. occur randomly in time
with a probability which is constant and independent of time and
the occurrence of previous m-e.p.ps.

If m-e.p.ps. occur randomly, with a probability $\mu\Delta t$ that in an interval $(t, t+\Delta t)$ there is a m-e.p.p., and Δt is sufficiently small so that the probability of more than one m-e.p.p. is negligible, the probability that no m-e.p.ps. occur in $(0, t)$ is

$$P_o(t) \quad = \quad (1 - \mu t/m)^m$$

where $m = t/\Delta t$. As $\Delta t \to 0$ and $m \to \infty$

$$P_o(t) \quad = \quad \exp(-\mu t) \tag{11.2.2}$$

and so the probability that the interval τ between two adjacent m-e.p.ps. is in the range $(t, t+\Delta t)$ is

$$\text{Prob } (t < \tau < t+\Delta t) \quad = \quad \exp(-\mu t)\mu\Delta t$$

and so

$$p(\tau)\Delta t \quad = \quad \mu \exp(-\mu\tau)\Delta t \tag{11.2.3}$$

where $p(\tau)$ is the probability density function for the intervals, with a mean interval of $1/\mu$. Thus μ is the average rate of m-e.p.ps.

The probability density function for the number r of m-e.p.ps. occurring per unit time is then

$$p(r) \quad = \quad \exp(-\mu) \; \mu^r/r! \tag{11.2.4}$$

which is the Poisson distribution

Thus, if the m-e.p.ps. are Poisson distributed:

(a) the interval distribution will be exponential

(b) the intervals between spontaneous m-e.p.ps. will form a realization of a stationary process

(c) the intervals will be identically and independently distributed

(d) the simultaneous occurrence of two or more m-e.p.ps. is not permitted.

There are a number of statistical tests which can be used to determine whether or not a sequence of events forms a realization of a Poisson point process: these are discussed in Cox and Lewis (1966) and are reviewed in van der Kloot et al. (1975). The problems I will

in this section are whether or not the spontaneous m-e.p.ps. recorded
at the frog neuro-muscular junction show any deviations from a Poisson
process, and whether or not a post-junctional Poisson distributed
sequence of m-e.p.ps. necessarily implies a Poisson release mechanism.

Fatt and Katz (1952) recorded m-e.p.ps. using extracellular and
intracellular recording methods; in both cases the m-e.p.ps. were
localized to the endplate region of the muscle fibre. With intracellular recording, the m-e.p.ps. appeared as small depolarizations on
the resting membrane potential, and could be obtained 1 - 1.5 mm from
the neuro-muscular junction, the amplitude of the m-e.p.ps. increasing
as the microelectrode approached the junction. With extracellular
recording, the m-e.p.ps. appeared as brief (time to peak < 1 msec) negative deflections of about 1 - 2 mV, with an extremely restricted
localization - moving the microelectrode a few μm caused the m-e.p.ps.
to vanish. The rate of the extracellularly recorded m-e.p.ps. was less
than the rate of the intracellularly recorded m-e.p.ps. This suggested
the interpretation that there are a number of active spots distributed
close together on the endplate surface, the active spots spontaneously
releasing quanta of transmitter. Intracellular recording will not
discriminate between the activity of different active spots, and so will
record the superposition of the activity of all the active spots.
However, extracellular recording will only pick up the activity of a few
active spots close to the tip of the recording microelectrode, and so
the rate of extracellularly recorded me.p.ps. will be less than the rate
of intracellularly recorded m-e.p.ps. in the same preparation. Further,
if the number of active spots is large and their activity is mutually
independent, the intracellularly recorded m-e.p.ps. will have a Poisson
interval distribution.

Fatt and Katz compared the cumulative distribution function
obtained from an intracellularly recorded, burst-free series of m-e.p.ps
with the cumulative distribution function:

$$P(\tau) = 1 - \exp(-\mu\tau) \tag{11.2.5}$$

computed for a Poisson distribution with the same mean rate μ. The
fit appeared good, and this simple method of testing whether or not the
intervals are exponentially distributed has been applied to a variety
of invertebrate and vertebrate neuro-muscular junctions: see van der
Kloot et al. for references.

Lewis (1965) and Cox and Lewis (1966) analyzed Fatt and

Katz's data in greater detail, and some of the statistics used showed a deviation, significant at the 5% level, from the Poisson distribution. However, this might give a spurious rejection of the Poisson hypothesis as the deviations were due to a lack of short intervals, and this might be due to quantization errors as the intervals were measured to the nearest 10 msec.

Cohen et al (1973) question the Poisson hypothesis for m-e.p.ps. recorded at the frog neuro-muscular junction. The rate of spontaneous m-e.p.ps. is increased in hypertonic solutions, and in such preparations there is a poor fit with the Poisson distribution. Since the number of active spots at a neuro-muscular junction is fairly large (del Castillo and Katz, 1956), superposition will tend to produce a Poisson distribution of intervals of the intracellularly recorded m-e.p.ps. if the active spots are independent, whatever the stochastic properties of the release mechanism. Thus any deviations from the Poisson hypothesis should be more apparent in extracellularly recorded m-e.p.ps. trains, in which the activity of only a few active spots is observed.

Extracellular recording in a hypertonic solution of 1000 intervals between spontaneous m-e.p.ps. gave a cumulative distribution function which was singificantly different from the Poisson distribution at the 0.005% confidence level. Further, the autocorrelation co-efficients of the intervals were positive for the first 100 lags, 99 of the first 100 lags exceeding the 5% confidence limits. Thus the externally recorded m-e.p.ps. train appears to be more ordered than the simple Poisson hypothesis would predict.

Cohen et al (1974a, b) confirm and extend these results. Series of spontaneous m-e.p.ps. were recorded in normal Ringer, high K^+ Ringer, low Na^+ Ringer, hypertonic Ringer and hypertonic Ringer at low temper-atures, and five goodness of fit tests were applied to determine if the interval distribution was exponential. These tests showed that it was extremely unlikely that the intervals were generated by a process having an exponential distribution. Further, adjacent intervals were not independent, as the autocorrelogram showed a greater than chance number of autocorrelation coefficients which exceeded that 5% confidence limits. There was a tendency for runs of positive or negative autocorrelation coefficients to be clustered. The unsmoothed power spectrum of the intervals also showed deviations from the Poisson hypothesis.

If the m-e.p.ps. were generated by a Poisson process there would be no coincidences, when two or more m-e.p.ps. occur simultaneously. However, occasionally 'giant' m-e.p.ps. occur, with an amplitude which is an integer multiple of the normal m-e.p.p. amplitude. Under the Poisson hypothesis, these 'giant' m-e.p.ps. would represent two or more m-e.p.ps. occurring within the resolution time, which is approximately 0.5 msec. The number n_m of apparent coincidences, when $\underline{m}$ m-e.p.ps. occur within the resolution time T would be:

$$n_m = (T\mu)^{m-1} N \qquad \text{if} \qquad T \ll 1/\mu \qquad (11.2.6)$$

for a series of N m-e.p.ps. Cohen et al found that the number of 'giant' m-e.p.ps. was consistently greater than the number of apparent coincidences predicted by (11.2.6).

Thus it appears that at the frog neuro-muscular junction the simple Poisson hypothesis for the spontaneous m-e.p.p. interval distribution does not hold : This is also found at invertebrate (e.g. see Usherwood, 1972, Cohen et al 1974c) and mammalian (e.g. see Liley, 1957) neuro-muscular junctions. Deviations from the Poisson hypothesis are significant at the 5% level even when the data is preselected to avoid m-e.p.p. bursts. However, the nearly exponential nature of the m-e.p.p. interval distribution is striking : this is probably due to the effects of superposition. In their analysis of Fatt and Katz's data, Cox and Lewis (1966) suggested that the observed m-e.p.p. sequence could be generated by the superposition of about 170 periodic processes : however an insufficient number of intervals were available for testing this simple model.

Cohen et al (1974b) analyze the deviations from the Poisson hypothesis in greater detail. The deviations could not be accounted for by a non-homogenous Poisson process model, in which the parameter μ of the Poisson process changed with time. The intensity function, which is defined by equation 5.2.9 as the renewal density, is the limit

$$\lim_{\Delta t \to 0} \frac{\text{Prob. of one, or more, events in } (t, t + \Delta t)}{\Delta t} ,$$

and estimates of the intensity function deviated from the Poisson hypothesis by having significantly more events at short intervals, and

there was no significant periodicity. Estimates of the power spectrum
showed an excessive concentration of variance at low frequencies :
this is the frequency domain analog of the significant positive autocorrelations. The variance-time curve (see section 8.2) was consistently and significantly above that predicted by the Poisson hypothesis.
For a Poisson process, the variance would increase linearly with
time, with a slope given by the parameter μ. A more regular process
would have a variance-time curve which fell beneath this straight line.
A process produced by the superposition of p independent periodic
processes would have a variance-time curve which starts off along
the line predicted by the Poisson hypothesis and then levels off
asymptotically to a value p/6. Thus the variance-time curves of
Cohen et al do not support the simple superposition model of Cox
and Lewis.

All these deviations from the Poisson hypothesis are consistent
with a clustered point process model, similar to the models considered
in section 10.1 for bursting neuronal discharge patterns. Cohen et al
consider the branching renewal process introduced by Lewis (1964) to
account for the distribution of failure times of a computer. This
process consists of a series of primary events Z_1, Z_2 ...... Z_n, each
of which generates a random number S of subsidiary events Y_1, Y_2 .... Y_S,
where S = 0, 1, 2, ... The subsidiary processes are independently and
identically distributed, and are independent of the primary series.
The complete branching process is the superposition of the primary
and subsidiary events, which are assumed to be indistinguishable.
If the primary process is Poisson a branching Poisson process is
obtained : the properties of such a process are discussed in Lewis
(1969, 1970).

If $P(\tau)$ is the m-e.p.p. interval distribution, the survivor
function $G(\tau)$ defined by (5.2.8) is, for a Poisson distribution
with parameter μ :

$$G(\tau) = 1 - P(\tau)$$
$$= 1 - \{1 - \exp(-\mu\tau)\}$$
$$= \exp(-\mu\tau) \qquad (11.2.7)$$

and a plot of the ln-survivor function is a straight line of slope $-\mu$
and intercept 0. Cohen et al's plots of the ln-survivor function
deviate from this straight line, and show an initial concavity

(produced by the excess of short intervals) and an approximately
linear tail. Assuming that P (τ) is a branching Poisson process,
the expected mean interval between primary events can be estimated
from the slope of linear tail. If E {Z } is the expected mean interval
of the primary process, and μ is the mean rate of the complete process,
the mean number of secondary releases E {S} is given by

$$E \{S\} = \mu \, E \{Z\} - 1$$

and the expected mean interval between subsidiary events, E {Y } can
be found from:

$$E \{Y\} \; = \; \frac{-E \{Z\} \; (b + \ln(1 + E \{S\})}{E \{S\}}$$

where b is a constant, given by the intercept with the vertical axis
of the regression line fitting the tail of the ln-survivor function.
Since the estimated values of E {S} were consistently small, it is
difficult to make any inferences about the distribution of the
subsidiary intervals : the simplest assumption is that they have a
Poisson distribution. In this case, the survivor function of the
complete process can be computed as :

$$\frac{1 + E\{S\}\exp(-t/E\{Y\})}{1 + E\{S\}} \quad \exp(-t/E\{\tau\}) \; + \; \frac{E\{S\} \; E\{Y\}}{E\{\tau\}} \quad \exp(-t/E\{Y\}) \quad -1$$

and the computed ln-survivor functions provide a good fit to the
experimental ln-survivor functions. Computer simulations of branching
Poisson processes show the qualitative features of m.e.p.p. trains,
the large frequency range obtained from different realizations of the
same process and the bursting behaviour. The branching Poisson
process model provides a good fit to both intracellularly and extracellularly recorded m-e.p.p. trains : if the process observed intracellularly is the superposition of many similar processes, only one
of which is observed extracellularly, this good fit to both kinds of
record probably reflects the invariance of the branching Poisson
process under superposition.

An alternative model which could account for the experimental
results is one in which there is an exponential interval distribution

with a time-varying, periodic mean. Such a model is difficult to
distinguish from the branching Poisson process model, but Cohen et al's
observation of an excess of significant positive autocorrelations in
the autocorrelogram of the m-e.p.p. amplitude suggests that subsequent
m-e.p.ps. have similar amplitudes, and so are perhaps released at
the same site i.e. there is some kind of branching process.

A similar analysis of spontaneous m-e.p.ps. at an invertebrate
neuro-muscular junction (Cohen et al 1974c) supports a branching
Poisson process model.

11.3 Pre-synaptic depolarization enhanced release

A presynaptic action potential is the normal stimulus to
enhanced transmitter release, and so, given that transmitter is
released in quanta of uniform size, it is reasonable to investigate
the relation between the rate of release of quanta, r(t), and the
potential across the pre-synaptic terminal. From this relation the
time course of transmitter release produced by a presynaptic action
potential can be computed.

Liley (1956) has investigated the dependence of the rate of
m-e.p.ps. on the presynaptic membrane potential in the rat diaphragm
neuro-muscular junction. The presynaptic potential was altered in
two ways, either by the application of rectangular current pulses
with external electrodes (cathodal current depolarizing the terminals)
or by changing the potassium ion concentration of the external bathing
fluid (high external K^+ depolarizes). The rate of release of quanta,
measured by the rate of spontaneous m-e.p.ps., varied over two
decades as an exponential function of applied current. If the presynaptic membrane current-voltage relation was linear this means that
the rate of release varies exponentially with the change of presynaptic potential. A similar exponential relation was found between
the rate of release of quanta and the logarithm of the external K^+
concentration for K^+ concentrations greater than 7.5 mM/L. If the
presynaptic membrane potential is given by the Nernst equation, the
potential is proportional to the logarithm of the external K^+ concentration. A 10-fold increase in the rate of release of quanta was
produced by a 15 mV depolarization, and this effect was abolished by

low Ca^{++}-high Mg^{++} solutions. Extrapolation of this relation to the effect of a 1 msec action potential with an amplitude of 120 mV on a background release rate of 1 quantum/sec would give about 2-300 quanta released/action potential, which is sufficient to account for the size of the endplate potential.

This exponential relation between the steady-state rate of transmitter release and presynaptic voltage V

$$r(v, t = \infty) = r_o(1 + a \exp(bV)) \qquad\qquad (11.3.1)$$

is also found at the frog neuromuscular junction (del Castillo and Katz, 1954) : the constants a and b differ in the two preparations.

However, an increase in the rate of quantal release does not immediately follow depolarization : there is a lag between depolarization and the release of quanta. If a hyperpolarizing pulse is applied during this lag the transmitter release can be blocked. Further, transmitter release is not a simple function of depolarization : for strong depolarizations the transmitter is released when the depolarizing pulse is terminated. These experimental findings, and related experiments on the calcium dependence of transmitter release, suggest that depolarization produces a delayed increase in the conductance of the presynaptic membrane to Ca^{++}, or a delayed increase in the permeability of the presynaptic membrane to a charged calcium compound, and that it is the entry of this calcium current which causes the increase in the rate of release of quanta (Katz and Miledi, 1967). This suggests that the distribution of synaptic delays, which are produced by the delay in transmitter release and the diffusion and postsynaptic reaction delays, might provide information about the time course of transmitter release.

Katz and Miledi (1964) have measured the transmission delay at the neuromuscular junction, where the delay is the time between the peak depolarization of the presynaptic action potential in the axon terminal and the first detectable postsynaptic current. External, focal recordings were made with a $CaCl_2$ filled micropipette in Ca^{++}-free solutions: thus transmitter release was blocked except at the active spot under the micropipette, and so delays at a single active spot could be measured. The transmission delay was variable, the histogram of delay times being unimodal and positively skewed,

with a minimum delay of 0.4 to 0.5 msec and a modal delay of 0.75 msec.
This delay is the sum of the delay in release, the diffusion delay
and any delay in receptor reaction. However, the diffusion delay
over a synaptic cleft of 500 A is only about 1 µsec if the diffusion
coefficient is similar to that in free solution. Also the response
to an iontophoretically applied acetylcholine pulse begins in less
than 150 µsecs and so the transmission delay is mostly due to the
delay in transmitter release.

Given the quantal release of transmitter, the synaptic delay
histogram reflects the time course of the enhanced probability of
release of quanta: if S(t) is the cumulative distribution function
of the synaptic delay, or the probability that the synaptic delay is
less than t, the rate of release of quanta r(t) in response to a
presynaptic action potential is

$$r(t) = \frac{\frac{dS}{dt}}{1 - S(t)} \qquad (11.3.2)$$

Stevens (1968) has computed r(t) from the histogram of synaptic delays
of Katz and Miledi (1964), and the computed r(t) shows a delay of
about 0.5 msec and a very rapid, transient 1000 fold increase in the
release rate.

The presynaptic action potential causes a large, transient
increase in the rate of quanta release, and so further confirmation
of the quantal hypothesis can be obtained by estimating the average
number of quanta released by an action potential (the quantum content)
and the total number of quanta which are present and available for
release in the presynaptic terminal. Extrapolation of the steady
state relation between quantal release rate and depolarization suggests
that several hundred quanta will be released by a presynaptic action
potential when the preparation is in a normal bathing solution.
Thus the discrete, quantal nature of the components of the endplate
potential will not be apparent. Further, the ratio of the amplitude
of the endplate potential to the mean amplitude of spontaneous,
unitary m-e.p.ps. will not measure the number of quantal components
making up the endplate potential. A single quantum of transmitter
will produce a discrete change in cation conductance, of about 10^{-7} S,
and this conductance channel has a reversal potential of about -15 mV.

Thus the change in postsynaptic voltage will only be proportional to the number of quantal components at potentials far from the reversal potential, or when the number of quanta released is small.

This situation is produced in a low Ca^{++}-high Mg^{++} bathing solution, where the amplitude of the endplate potential evoked by a presynaptic action potential is greatly reduced. In section 11.2 I have shown that a detailed examination of the intervals between spontaneous m-e.p.ps. shows deviation from a simple Poisson release process. However, the effect of these deviations on the predicted number of quanta released by a presynaptic action potential will be small, and so from equation 11.2.4 the probability of k releases in a time t when the mean release rate if μ is given by:

$$p(k,t) = \frac{(\mu t)^k}{k!} \quad \exp(-\mu t) \qquad (11.3.3)$$

In response to a presynaptic action potential, the release rate $r(t)$ undergoes a large, transient increase. Irrespective of the form of this increase in $r(t)$, the number of quanta released in a time interval t_o will follow the Poisson distribution (11.3.3) and the average quantum content q will be μt_o and the standard deviation of the quantum content $\sqrt{(\mu t_o)}$. Thus the coefficient of variation will be inversely proportional to the square root of the average quantum content:

$$c.v. = \frac{\sqrt{(\mu t_o)}}{\mu t_o} = \frac{1}{\sqrt{q}} \qquad (11.3.4)$$

del Castillo and Katz (1954) showed that a plot of the logarithm of c.v. against q is linear with a slope of 0.5 for low values of q, and passes through zero at m = 1. The deviations at high q are probably almost entirely due to errors in estimating the quantum content from the ratio of the amplitude of the endplate potential to the mean amplitude of spontaneous m-e.p.ps.

A more complete test of the Poisson hypothesis is by a comparison of the theoretically expected and experimentally observed amplitude distributions of the evoked endplate potentials. In a series of trials, the quantum content should fluctuate as predicted by the

binomial distribution, which is approximated by the Poisson distribution
when the probability of release of a single quantum is small. Thus
the number of endplate potentials with a quantum content of x, n_x,
occurring in N trials should be:

$$n_x = \frac{N \exp{(-q)} \, q^x}{x!}$$

(11.3.5)

Thus in N trials one would expect $n_o = N \exp(-q)$ failures, when no
quanta are released and no endplate potential is observed. The mean
quantum content can be estimated from the mean amplitude of the evoked
endplate potential divided by the mean amplitude of spontaneous
m-e.p.ps. or independently from the number of failures n_o which occur
in N trials:

$$q = \ln{(N/n_o)}$$

Boyd and Martin (1956) have shown that these two independent estimates
of q agree for the mammalian neuromuscular junction when the quantum
content is sufficiently small (< 4) for nonlinear summation to be
negligible.

From the mean and variance of the amplitude distribution of
spontaneous unit m-e.p.ps., and equation 11.3.5, the amplitude
distribution of the evoked m-e.p.ps. can be computed by adding the
expected distributions of single, double, triple, ... unit responses
which have means and variances 1, 2, 3, .. times that of the spontaneous
m-e.p.ps. The predicted amplitude distribution computed using this
method fits the experimental amplitude distribution of evoked endplate
potentials in low Ca^{++} solutions at the frog (del Castillo and Katz,
1954) and mammalian (Boyd and Martin, 1956) neuromuscular junctions.

Thus the simple Poisson release hypothesis gives an adequate
fit to the experimental results of evoked endplate potentials in
low Ca^{++} solutions, this is in spite of the fact that deviations from
the simple Poisson hypothesis are apparent in sequences of spontaneous
m-e.p.ps., and that no account has been taken of the effect of latency
fluctuations on the endplate potential. Soucek (1971) has obtained a
general expression for the evoked endplate potential amplitude
distribution as a function of the mean quantum content, the pulse
shape and amplitude distribution of spontaneous m-e.p.ps. and the latency

distribution. The effect of latency fluctuations is to make the
peaks of the multimodal amplitude distribution asymmetric.

del Catillo and Katz (1954) point out that the release rate
r(t) is the product of the total number of available quanta Q and
the probability p that any given quantum is released in a second

$$r\ (t)\ =\ p\ (t)\ Q\ (t) \tag{11.3.6}$$

This assumes that the release of quanta is mutually independent. The
problem is to estimate p and Q, and to see if there is any relation
between the number of available quanta and the number of presynaptic
vesicles.

Equation 11.3.5 is the Poisson distribution, which is an
approximation to the binomial distribution

$$n_x\ =\ \frac{N\ Q!}{(Q-x)!\ x!}\ \ p^x\ (1\ -\ p)^{(Q\ -\ x)} \tag{11.3.7}$$

for the expected number of responses n_x containing x quanta in N
trials. The coefficient of variation of the binomial distribution is
$\sqrt{(1/q\ -\ 1/Q)}$, which is less than the coefficient of variation of the
Poisson approximation. Thus when the quantum content q is not negligibly
small compared to the number of quanta available for release, there
should be deviations from the Poisson prediction of equation 11.3.4.
There are such deviations in normal Ca^{++} solutions (Martin, 1955),
but these are removed by allowing for nonlinear summation of potentials
or by curarizing to reduce the amplitude of the endplate potential.
The plot of the logarithm of the coefficient of variation against the
corrected quantum content agrees with the Poisson prediction for
quantum contents up to 200, and so the total number of available
quanta must be greater than 400. If Q < 400 the reduced coefficient
of variation of the binomial distribution would have been detectable.
This minimal figure of 400 for the number of available quanta gives
an upper limit of 0.25 for p.

A different method of estimating p, and hence, given r(t), of
estimating Q, is by using twin stimuli. If initially there are Q_o
quanta available for release, and the first conditioning stimulus
releases q quanta, the number of quanta available for release will be

reduced to $Q_1 = Q - q$. The second, test stimulus evokes a reduced
endplate potential, the amplitude of this second test endplate potential
returning exponentially to its control value as the interval between
the two stimuli is increased. If the reduction in the amplitude of
the test endplate potential is due to a reduced quantum content q, and
not to some desensitization of the postsynaptic membrane, and this
reduced quantum content is due to the reduction in Q and not to a change
in p, then

$$\frac{V - V_1(T)}{V} = p \exp(-kT) \tag{11.3.8}$$

where V is the amplitude of the conditioning endplate potential, V_1 (T)
is the peak amplitude of the second, test endplate potential when
the interval between the two stimuli is T, and k is the rate constant
of recovery of the amplitude of the test endplate potential. This
method of estimating p has been applied to the frog (Takeuchi, 1958)
and mammalian (Liley and North, 1953) neuromuscular junctions, and a
problem is that at short intervals there appears to be less depression
or even a facilitation. However, for intervals greater than 500 msec
a plot of the logarithm of the RHS of (11.3.8) against the interval
T is linear, and extrapolation of the linear segment back to a zero
interval gives an estimate of p. Since nonlinear summation has been
neglected this will be an underestimate; however, p was estimated as
0.14 at the frog and 0.45 at the mammalian junctions in normal bathing
solutions. Since the quantum content in normal solutions is about
100 for frog junctions and about 300 for mammalian junctions, in both
cases the total number of quanta available for release is about 1000.
The acetylcholine content of these quanta is only 1-2% of the total
acetylcholine content of the presynaptic terminal, which is found in
the presynaptic vesicles. Thus even if one quantum corresponds to the
contents of one vesicle, the total number of quanta available for
release does not correspond to the total number of presynaptic vesicles.

In a recent review of the quantum hypothesis Wernig (1975)
points out that estimates of Q at different synapses, even when
obtained using the binomial distribution rather than the Poisson
approximation, can be very low e.g. between 2 and 10 at the crayfish
neuromuscular junction. This is difficult to reconcile with the idea
that Q represents the number of vesicles which are available for
release, and Wernig suggests that Q might in fact be an estimate of
the number of active release sites where quanta can be released.

This fits with the observation that presynaptic vesicles appear to be clustered around sites on the presynaptic membrane.

11.4 Models of facilitation and depression

The quantal release rate r(t) depends on the presynaptic voltage, and if the presynaptic quanta available for release form a homogenous population the release rate is the product of the total number of quanta available for release, Q, and their probability of release, p :

$$r(t) \; = \; Q(t) \; p(t) \tag{11.4.1}$$

In this section I will consider some models which account for changes in r(t) as measured by changes in quantum content q: these changes are produced by repetitive stimulation of the presynaptic axon. If, during a series of presynaptic action potentials, there are changes in the amplitude of the evoked endplate potential, these changes might be produced by changes in the quantum content or by a change in the sensitivity of the postsynaptic membrane to acetylcholine. Iontophoretic application of acetylcholine to the endplate does produce a desensitization of the postsynaptic membrane, and the response to subsequent test applications is reduced. However, this occurs at concentrations of acetylcholine much greater than those produced by the release of a few hundred quanta.

After a conditioning stimulus, or train of conditioning stimuli, the amplitude of the endplate potential evoked by a test stimulus is increased. Using low Ca^{++}-high Mg^{++} solutions to reduce the quantum content, del Castillo and Katz (1954) showed that the number of failures of the test response was less than the number of failures of the conditioning response; thus, using equation 11.3.5, the quantum content was increased. Thus facilitation is produced by an increase in quantum content and not in the amplitude of the post-synaptic response to a quantum.

Mallart and Martin (1967) have investigated the time course of facilitation at the frog neuromuscular junction in solutions where

the average quantum content of the control endplate potential was
about 7. This low quantum content meant that there were considerable
fluctuations in the responses between trials; however, averaging
the responses in about 50 trials gave a smooth time course of facilitation. In order to measure facilitation a dimensionless index f was
defined by

$$f = \frac{v - v_c}{v_c} \qquad\qquad (11.4.2)$$

where v is the mean amplitude of the test endplate potential and v_c
the mean amplitude of the control endplate potential. A graph of
f against time, obtained in response to a single conditioning stimulus,
suggests that there are at least two components in the time course of
facilitation. The first 100 msec could be fitted by a single exponential

$$f = f_1 \exp(-bt) \qquad\qquad (11.4.3)$$

where the zero intercept f_1 was 0.95 and the time constant 1/b about
45 msec. There also appeared to be a slow component of facilitation.
The components of facilitation could be separated by using a conditioning stimulus train to precede the test stimulus. If the only component
of facilitation is that given by equation 11.4.3 the time course of
the facilitation produced by a conditioning stimulus train can be
predicted. The first stimulus of the conditioning train will produce
a facilitation f_1 which will decay towards zero with a time constant
1/b. After a time ΔT the second condioning stimulus will add an
increment f_1 to the residual facilitation $f_1 \exp(-b\Delta T)$, and this
new level of facilitation will decay until the arrival of the third
conditioning stimulus. Thus, during the conditioning stimulus train
the facilitation will rise in a sawtooth curve, the inside envelope
of the curve increasing exponentially towards a steady state level k_1

$$F_1 = k_1 (1 - \exp(-bt)) \qquad\qquad (11.4.4)$$

where the steady state level k_1 depends on f_1 and b

$$k_1 = f_1 / (\exp(b\Delta T) - 1)$$

After the last conditioning stimulus the facilitation, given by the sum of the accumulated facilitation (given by equation 11.4.4) and the increment in facilitation produced by the last stimulus, will decay exponentially towards zero with a time constant $1/b$. Thus any difference between this theoretical curve and the experimentally observed curve of facilitation will give the time course of the second component of facilitation.

The second component of facilitation extracted by this method is qualitatively similar to the first component in that its magnitude and time course are the same for each conditioning stimulus, and the effect of a conditioning stimulus sums linearly to the residual effect of preceeding stimuli. However the time scale and amplitude are markedly different from those of the first: the second component of facilitation begins after a delay of about 60 - 80 msec, and then slowly decays with a time constant $1/g$ of about 250 msec. Thus if the delay is t_2, the facilitation due to this second component is given by

$$F_2 = k_2 (1 - \exp \{-g (t - t_2)\})$$
$$(11.4.5)$$

where k_2 is given by:

$$k_2 = f_2 \{ 1 + \frac{1}{\exp (g\Delta T) - 1} \}$$

and after the last stimulus of the conditioning train the second component of facilitation will decay towards zero with a time constant $1/g$. The overall time course of facilitation will be the sum of the two components. Values of the parameters of this two component model obtained by fitting the time course of facilitation produced by a short conditioning train could be used to predict the time course of facilitation produced by different conditioning trains.

This demonstration of two separate kinetic processes determining the level of facilitation suggests that there are two separate mechanisms responsible for facilitation or primary potentiation. The first component seems to be produced by an increase in the probability of release of quanta, perhaps due to some transient change in the release mechanism. The delayed, second stage of facilitation might be due to an increase in the number of quanta available for release (transmitter 'mobilization') or to an increase in the probability of release of transmitter produced by some different mechanism. Braun and Schmidt (1966) observed that the second component of facilitation appeared to be coincident with an increase in the amplitude of the pre-synaptic action potential measured extracellularly - this suggests that the second component is concomitant with an accumulated pre-synaptic hyperpolarization.

Both these two components can be accounted for by an increase in the probability of release of quanta, and so in equation 11.4.1 $p(t)$ has a transient increase while $Q(t)$ is constant. These two components correspond to the facilitation produced by short trains of conditioning stimuli, or the primary potentiation described by Hubbard (1963) at the mammalian neuromuscular junction. After long trains of conditioning stimuli, this primary potentiation is followed by a phase of reduced transmitter release, or depression, and then a second-ary potentiation or post-tetanic potentiation. This post-tetanic potentiation is not due to an increased sensitivity of the post-synaptic membrane to acetylcholine but to an increased transmitter release. The mechanism producing post-tetanic potentiation is obscure: a simple explanation would be an increased transmitter mobilization giving an increase in the number of quanta available for release. However, if post-tetanic potentiation was produced by an increase in $Q(t)$, the amplitudes of endplate potentials evoked by a short train of stimuli during the period of post-tetanic potentiation would be increased - in fact only the first 2-3 responses are potentiated. Since the duration of post-tetanic potentiation is sensitive to the number and frequency of stimuli in the conditioning train perhaps changes in the ionic composition of the axonal terminals may be responsible for the change in quantum content.

The depression in the endplate potential amplitude which precedes post-tetanic potentiation appears to be produced by a depletion in the number of quanta available for release (Thies, 1965). This depression begins after the primary potentiation and lasts up to several seconds, and is maximal about 200 - 400 msecs after the conditioning stimulus. The depression is enhanced in high Ca^{++} solutions and reduced in low Ca^{++}-high Mg^{++} solutions, suggesting that it is produced by the release of transmitter. The quantum content of the evoked endplate potential is reduced proportionally to the evoked endplate potential amplitude at low rates of stimulation

(2 - 20/sec for guinea pig preparations and 2-5/sec for frog
preparations), and so the depression is not due to a change in the
sensitivity of the postsynaptic membrane. There is a linear relationship between the maximal depression of the test endplate potential
amplitude and the quantal content of the conditioning endplate
potential - the simplest explanation of this is that the depression
is due to a depletion of presynaptic transmitter. In the first 200
msecs after the conditioning stimulus there is less depression or
even a primary potentiation as although $Q(t)$ is reduced the probability
of release is enhanced.

If depression is produced by a depletion of the store of
acetylcholine available for release, the recovery from depression
represents a refilling of the store of available acetylcholine from
the store of reserve acetylcholine. The rate of refilling is
independent of the magnitude of depletion, as the rate of recovery
is not correlated with the magnitude of maximal depression and is
approximately constant. Extrapolation of the recovery curve back to
the conditioning stimulus gives the time course of the presumed
transmitter depletion: the time course of depression is the product
of this curve and the curve for primary potentiation.

11.5 Transmitter induced fluctuations in conductance

A spontaneous, unit m-e.p.p., with an amplitude of about 1 mV,
is produced by the interaction of a quantum of acetylcholine with
receptor sites on the postsynaptic membrane, which produces a cation-
selective increase in conductance of about 10^{-7}S with a reversal
potential of -15 mV. A single quantum of acetylcholine contains
perhaps thousands of molecules, and so the postsynaptic response to
a single quantum represents the approximately simultaneous response
of a large number of acetylcholine sensitive ionic current mechanisms.
Thus fluctuations in potential produced by the interaction of single
acetylcholine molecules with receptor sites will not be apparent.

If the interaction of a molecule of acetylcholine with a
receptor site causes a discrete ionic channel to open, the application
of acetylcholine would change the average number of open ionic channels,
and fluctuations in conductance produced by fluctuations in the number
of open channels might be detectable. Katz and Miledi (1970, 1971,

1972, 1973a, b) have shown that when acetylcholine is applied to an endplate, the resultant depolarization is accompanied by an increase in the voltage noise. This fluctuating acetylcholine potential is produced as a result of the interaction of acetylcholine molecules with receptor sites, and can also be produced by the acetylcholine substitute carbacol.

The spectrum of this acetylcholine voltage noise is not the spectrum of the conductance fluctuations, but is dominated by the low-frequency cutoff produced by the passive RC cable properties of the muscle cell. The voltage noise spectrum, $S_V(\omega)$, current noise spectrum $S_I(\omega)$, conductance noise spectrum $S_G(\omega)$ and current to voltage transfer function $Z(\omega)$ are related by:

$$S_V(\omega) = S_I(\omega) \, |Z(\omega)|^2$$

$$S_I(\omega) = S_G(\omega) \quad (<V> - V_{eq})^2$$

where V_{eq} is the reversal potential for the conductance and $<V>$ is the mean value of the membrane potential, and so

$$S_G(\omega) = \frac{S_V(\omega)}{(<V> - V_{eq})^2 \, |Z(\omega)|^2} \tag{11.5.1}$$

Thus the conductance spectrum can be computed from the voltage noise spectrum if the current to voltage transfer function $Z(\omega)$ is also measured using white noise. However, this method of obtaining $S_G(\omega)$ has the disadvantage in practice that the division of spectral estimates severely limits the accuracy, especially for potentials near V_{eq}.

Katz and Miledi (1972) avoided this problem by extracellular recording of the voltage change produced by local circuit current flow. This is assumed to be proportional to the endplate current noise: however, Katz and Miledi (1973b) note that when this method is used to record the endplate current during spontaneous m-e.p.ps. there is a distortion of the time course of the falling phase of the endplate current. With such focal, extracellular recording, the amplitude of the observed potential fluctuations is critically dependent on the position of the extracellular microelectrode, and so one can only investigate the time course of the conductance changes,

and not their amplitude. The power spectrum of the extracellular
voltage noise, which is assumed to be proportional to the conductance
noise, was estimated over the frequency range of 3-200 Hz, and since
this is less than two decades the shape of the spectrum is not known.
However, assuming that the conductance spectra are relaxation spectra
with a Lorentzian shape

$$S(\omega) \quad \propto \quad \frac{\tau^2}{1 + \omega^2 \tau^2} \tag{11.5.2}$$

the time constant τ can be calculated from the half-power frequency
or from the ratio of the asymptotic value of $S_I(\omega)$ as $\omega \to 0$ to the
total noise variance. These two methods gave different values for τ,
with means of 1.1 msec and 0.7 msec: this suggests that the spectrum
is more complex than a simple Lorentzian, but that if the acetylcholine
noise is produced by channels opening and closing the average length
of time a channel is in the open state is approximately 1 msec.

If the acetylcholine potential noise is made up of a shot process
of brief pulses with a shape $h(t)$ occurring at a mean rate r, the
mean value $< V >$ and variance $< E^2 > \; = \; (V(t) - < V >)^2$ are
given by Campbell's theorem as:

$$< V > \quad = \quad r \int h(t) \; dt \tag{11.5.3}$$

$$< E^2 > \quad = \quad r \int h^2(t) \; dt \tag{11.5.4}$$

The shape $h(t)$ is unknown, but a plausible guess is an instantaneous
rise to an amplitude a and an exponential decay with a time constant τ

$$h(t) \quad = \quad a \exp (-t/\tau) \tag{11.5.5}$$

and so

$$< E^2 > \quad = \quad < V > a/2 \tag{11.5.6}$$

This analysis ignores the effects of nonlinear summation, and so is
only an approximate, order of magnitude, calculation. However, using
this method Katz and Miledi (1972) calculated the amplitude of the
elementary shot effect to be about 0.5 μV, which is about three

orders of magnitude less than the amplitude of spontaneous m-e.p.ps.
This analysis is not very sensitive to the shape h(t) assumed for the
elementary shot effect, as long as it is a brief, monophasic pulse.
However, as the mean acetylcholine depolarization $< V >$ increases,
the amplitude $\underline{a}$ of the elementary shot effect will be underestimated
if the effects of nonlinear summation are ignored. A simple model
for nonlinear summation of the acetylcholine potential is to consider
the acetylcholine channel, with a conductance g and reversal potential
V_{eq}, shunting the muscle fibre input resistance R.   The acetylcholine
potential is related to the acetylcholine conductance g by

$$\frac{< V >}{(V_{eq} - V_r)} = \frac{gR}{(1 + gR)} \tag{11.5.7}$$

where V_r is the resting membrane potential.   It is reasonable to
assume that at low concentrations the acetylcholine conductance g is
proportional to the concentration of acetylcholine, or to the number
of open acetylcholine channels. A small fluctuation in conductance,
δg, produced by a fluctuation in the number of open acetylcholine
channels, will produce a voltage fluctuation δV;   that is,

$$\frac{dV}{dg} = \frac{(V_{eq} - V_r) R}{2(1 + gR)^2} \tag{11.5.8}$$

Thus equation 11.5.6 corrected for nonlinear summation becomes

$$< E^2 > = \frac{(V_{eq} - V_r) gRa}{2(1 + gR)^4}$$

$$= \frac{< V > a}{2(1 + gR)^3} \tag{11.5.9}$$

Katz and Miledi (1972) used this correction factor in their estimates
of $\underline{a}$:   this method of correction applies to steady voltages produced
by steady changes in conductance, and will tend to overcorrect (and
hence overestimate $\underline{a}$) when rapid transients such as the elementary
shot effects are considered.

A plot of the corrected values of E against $\sqrt{< V >}$ is linear,
with a slope of $\sqrt{(a/2)}$, as expected. This supports the model;

howver, estimates of a ranched from 0.064 to 1.54 μV, with a mean of
0.42 μV. This wide range is due in part to different fibre input
resistances produced by different fibre diameters, and to experiments
being performed at different temperatures. However, after allowing for
these effects there is still a considerable scatter in the estimates of
a, over a three-fold range. This might be within the limits of the
experimental technique, or might suggest that the elementary shot
effects have a wide amplitude distribution.

Stevens (1972) points out that Katz and Miledi's results are
compatible with a different microscopic model: instead of treating
the conductance change as a rapid increase followed by an exponential
decay, or some similar monophasic waveform with a deterministic
duration, an alternative model is to have the channel open at random
and remain open for a random period. If an acetylcholine channel can
have a conductance of either zero (closed) or G (open) and the open-
close transition of the acetylcholine-receptor complex occurs as a
Poisson process with a closing rate α and opening rate β, then the
probability p(k, t) that, given the channel was in state k at time
t = 0, the channel is open at time t, is given by:

$$\tau \frac{dp(k \mid t)}{dt} + p(k \mid t) = \alpha/(\alpha + \beta) \tag{11.5.10}$$

where τ is now $1/(\alpha + \beta)$. The state k can be either open (o) or
closed (c). At low acetylcholine concentrations the probability of the
channel being open is small and the fraction of channels with n
acetylcholine molecules bound to their receptors (n is the number of
acetylcholine molecules which have to be bound to a receptor before
the channel can open) is $Kc^n(t)$, where K is the binding constant and
c(t) the acetylcholine concentration. If there is no interaction
between channels

$$\frac{dp(k \mid t)}{dt} = -(\alpha + \beta Kc^n(t)) \, p(k \mid t) + \beta Kc^n(t) \tag{11.5.11}$$

but since only a small fraction of the channels are open $\beta Kc^n(t) << \alpha$
and so equation 11.5.11 approximates to

$$\frac{dp(k \mid t)}{dt} = -\alpha p(k \mid t) + \beta Kc^n(t) \tag{11.5.12}$$

If the initial acetylcholine conductance is zero, all the
channels are closed at time t = 0, and so $p(c \mid t)$ is the probability
that a channel is open at a time t. This is the same as the fraction
of channels that are open, and so the conductance is

$$g(t) = p(c \mid t) \, G \, N$$

when N is the total number of channels and G is the conductance of
a single, open channel.

The covariance function $C(t)$ of the current $I(t)$ through a
single channel is

$$C(t) = < I(0) I(t) > - < I(\infty) >^2$$
$$= < (V - V_{eq})^2 \, g(0) \, g(t) > - < (V - V_{eq}) \, g(\infty) >^2$$

which for steady state fluctuations in the presence of a constant
acetylcholine concentration c gives:

$$C(t) = G^2 (V - V_{eq})^2 \, p(o \mid \infty) \, p(o \mid t) - G^2 (V - V_{eq}) \, p^2(o \mid \infty)$$
$$(11.5.13)$$

Substituting for $p(k \mid t)$ obtained by solving equation 11.5.12 gives
the covariance function as

$$C(t) = G^2 (V - V_{eq})^2 \, \beta \, Kc^n \, \exp(-\alpha t), \quad t \geqslant 0 \qquad (11.5.14)$$

and the spectral density of the total current noise is obtained from
the real part of the Fourier transform of $NC(t)$

$$S_I(\omega) = \frac{2 \, NG^2 (V - V_{eq})^2 \, \beta Kc^n / \alpha}{1 + (\omega/\alpha)^2} \qquad (11.5.15)$$

Thus this model also generates a spectrum with a Lorentzian shape.
Since both the exponential pulse and the two-state, variable duration
models both generate relaxation spectra the results of Katz and
Miledi (1972) cannot distinguish between them: what is needed is
a direct measurement of the current noise spectrum over a wider
frequency range, so that the shape of the spectrum is clearly
defined and the absolute values of the noise power are obtained
without use of perhaps misleading correction factors. This has

been achieved by Anderson and Stevens (1973) using a voltage clamp method.

Excitation-contraction coupling in the muscle fibres was broken by ethylene glycol pretreatment: this appeared to lower the internal K^+ concentration, as the resting membrane potential was less negative than normal and the reversal potential of the endplate current was about 0 mV rather than -15 mV. However, this did not appear to affect the acetylcholine-receptor interaction. The experiments were performed on frog muscle at $10^{\circ}C$: This low temperature reduces the rate of spontaneous m-e.p.ps. while optimizing the acetylcholine noise spectrum with respect to the bandwidth of the voltage clamp circuit. Two microelectrodes were inserted into the endplate region, one for current injection and the other for recording membrane potential: the separation of the microelectrodes was less than 50 μm so that there was little voltage decrement between the two electrodes. The localization of the voltage clamp electrodes at the endplate was confirmed by the time course of spontaneous miniature endplate currents (m-e.p.c.) and evoked endplate currents. M-e.p.c. rise times of < 0.5 msec and amplitudes >1 nA, and evoked e.p.c. rise times of < 1.5 msec and amplitudes >100 nA were taken to indicate location at an endplate. Since the endplate in frog muscle can extend for up to 400 μm along a muscle fibre (or up to 0.18 space constants along a muscle fibre 2.5 mm long) this voltage clamp system does not ensure spatial uniformity of voltage over all the endplate: the maximum voltage error would be about 15%.

Before this system could be used to measure the spectrum of acetylcholine current noise, the spectrum of the background noise arising in the equipment and muscle membrane had to be measured. For a frequency range of 1-500 Hz, the background noise was 2-4 orders of magnitude less than the acetylcholine noise. At frequencies less than 200 Hz a flicker noise predominates (see section 1.3): this flicker noise had an equilibrium potential which did not appear to be the K^+ equilibrium potential, and so perhaps flicker noise in frog muscle membrane is produced by a less selective ionic current mechanism than the K^+ current mechanism which generates flicker noise in nerve membranes. There was also a Lorentzian component in the background noise, with a corner frequency determined by the membrane potential: this might be associated with fluctuations in H-H K^+ channels. However, whatever the origin of the background noises, these noises limit the resolution of the system to about 1×10^{-24} A^2 sec at 100 Hz, which is about three orders of magnitude less than the acetylcholine noise. The acetylcholine noise spectrum can be obtained by subtracting the background noise spectrum from the total noise spectrum measured in the presence of acetylcholine. This correction is at most a few per cent, and assumes that the background and acetylcholine noises are mutually independent.

When acetylcholine is iontophoretically applied to the endplate by an extracellular microelectrode the mean and the variance of the endplate current, measured under a voltage clamp, both increase. From the measurements of the mean and variance of the acetylcholine current, μ_I and σ_i^2, the mean and variance of the conductance fluctuations can be calculated

$$\mu_g = \frac{\mu_I}{(V - V_{eq})}$$

$$\sigma_g^2 = \frac{\sigma_I^2}{(V - V_{eq})^2}$$

The computed acetylcholine conductance variance was found to be proportional to the mean acetylcholine conductance, the constant of proportionality being 0.205×10^{-10} S. This is equivalent to Katz and Miledi's (1972) result that the standard deviation of the corrected voltage noise was proportional to the square root of the mean of the corrected voltage noise, with a constant of proportionality of $\sqrt{(a/2)}$. In terms of Katz and Miledi's model these results mean

$$\sigma_g^2 = G \mu_g / 2 \qquad\qquad (11.5.16)$$

while in terms of the two-state, variable duration model the interpretation is

$$\sigma_g^2 = G \mu_g \qquad\qquad (11.5.17)$$

where G is the conductance of a single open channel.

The spectrum of the acetylcholine noise has a clear Lorentzian shape, being constant at low frequencies and approaching $1/\omega^2$ at

high frequencies (see equation 11.5.15). However, as Verveen and
DeFelice (1975) point out, such a Lorenztian spectrum is generated
by a wide range of microscopically distinct models, and so although
Anderson and Stevens (1973) obtain the Lorentzian shape of the
acetylcholine conductance noise spectrum over 1-500 Hz at different
clamp potentials this does not distinguish between their two-state,
variable duration model and Katz and Miledi's model.

If the acetylcholine-receptor interaction in the presence of
a constant acetylcholine concentration can be represented by

$$nACh + R \underset{}{\overset{K}{\rightleftharpoons}} ACh_nR \underset{\alpha}{\overset{\beta}{\rightleftharpoons}} ACh_nR* \qquad (11.5.18)$$

where n acetylcholine molecules bind very rapidly with a receptor R,
and the acetylcholine-receptor complex fluctuates between the closed
state ACh_nR and the open state ACh_nR* with rate constants α and β,
then these rate constants not only underly the acetylcholine noise
(equations 11.5.10-15) but also underly the kinetics of endplate
currents produced by transmitter release evoked by nerve action
potentials.

Magleby and Stevens (1972) used a similar voltage clamp method
to measure the effects of membrane potential on the time course of
evoked endplate currents. The decay of endplate current follows a
simple exponential, with the rate constant α that characterizes the
exponential decay being weakly voltage dependent:

$$\alpha (V) = B \exp (AV) \qquad (11.5.19)$$

where A and B are constants, A = 0.008 /mV and B = 1.7 /msec. The
form of this empirical relation was also derived theoretically from
the theory of rate processes, with A reflecting a change in dipole
moment associated with the conformational change of the gating
molecule, and B being related to the free energy barrier for this
change in the absence of an electric field. The dependence of α
on voltage can be measured from the decay constants of evoked endplate
currents and m-e.p.cs, and also from the cutoff frequency of the
acetylcholine conductance noise spectrum. From equation 11.5.15,
the cutoff frequency ω_c, which is the frequency at which the spectral
power has declined to a half of its asymptotic value at $\omega = 0$, is
given by

$$\omega_c = \alpha$$

and estimates of α obtained from ω_c at different clamp potentials agree with the values computed from equation 11.5.19.

The decay rate constant of evoked endplate currents increases as temperature is lowered and has a Q_{10} of approximately 3, and so α estimated from ω_c should have a similar temperature dependence, and an increase in temperature should shift the conductance noise spectrum toward higher frequencies. These effects are observed.

The parameter G of equation 11.5.15 represents the conductance of one open channel. Magleby and Stevens estimated the instantaneous current voltage relations of the acetylcholine conductance mechanism by imposing step changes in clamp potential during the exponential decay phase of evoked endplate currents. Extrapolation of the exponentially decaying current back to the time of the step change in potential gives the current flowing immediately after the step. The instantaneous current-voltage relation measured in this way is linear, and so the acetylcholine conductance is independent of membrane potential. Thus values of G obtained from equation 11.5.15 should be the same, irrespective of the clamp potential. The conductance of a single channel can be obtained from either

$$G = \frac{S_I(0)\,\alpha}{2\,\mu_I\,(V - V_{eq})}$$

or from the total fluctuation variance

$$\sigma_I^2 = \int_0^\infty S(\omega)\,d\omega = G\,\mu_I\,(V - V_{eq})$$

A problem with this method is that imperfect voltage clamping will tend to underestimate G, as the space constant will be longer for steady currents than for rapid fluctuations and hence $S(0)/\mu_I$ will be underestimated. However, in experiments where the voltage clamp appeared to be spatially uniform over the endplate, estimates of G were not dependent on the value of the clamp potential.

Thus the experimental results of Magleby and Stevens (1972) and Anderson and Stevens (1973) are fully self-consistent with the

two state, variable duration model of equations 11.5.10-15. However,
they are also consistent with the exponentially decaying pulse model
of conductance, and in fact are consistent with any model which is
compatible with the fluctuation-dissipation theorem (Kubo, 1957,
1959). The fluctuation-dissipation theorem states that the autocovariance function (defined by equation 5.4.8) of reversible
fluctuations of a system about its mean equilibrium position describes
the relaxations of the system when the system is subject to small
perturbations. This assumes that the behaviour of the system is
describable by a Markov process, in that its future behaviour depends
only on its present state.

However, although the experimental results are compatible
with a wide range of models, they can be interpreted in detail in
terms of the two state, variable duration model. The conductance of
a single channel, obtained from equation 11.5.17, is about 3×10^{-11}s
when the channel is open. The channel is controlled by a gating
molecule which can be in two distinguishable conformations, only
one of which permits Na^+ and K^+ to flow through the channel. The
binding of acetylcholine to its receptor lowers the energy barrier
between the open and closed conformations, and so a channel opens
and closes randomly. The mean duration of the open, conducting
state is voltage sensitive: hyperpolarization increases the energy
barrier as the gating molecule has a dipole moment. At 8°C the
channel is open for about 11 msec, during which about 2×10^5 ions
pass through the channel at a rate of about 10^7/sec: this suggests
that the channel is a gated, aqueous pore. A single quantum opens
about 1700 channels.

11.6 Dynamics of Signal Transfer Across a Synapse.

A presynaptic action potential produces a postsynaptic
potential, which, as the endplate potential at the neuromuscular
junction, almost invariably produces a postsynaptic muscle action
potential. In the central nervous system the magnitude of the
excitatory postsynaptic potential is much less than the change in
potential required to reach threshold : this is the basis of the
diffusion models of section 7. Thus for a postsynaptic action
potential to be generated there must be spatial or temporal integration
of a number of excitatory postsynaptic potentials before the threshold
of the postsynaptic cell is reached. The relationships between the

postsynaptic and presynaptic action potential trains will define the
dynamics of signal transfer across the synapse.

One method of characterizing the steady-state signal transfer
characteristics is by the relation between the output rate of action
potentials and the rate of input action potentials. If an excitatory
pulse train with a constant rate x is the input to a perfect integrator model (see section 3.2), which requires k input pulses to
reach threshold, the steady-state input-output relation will be that
of a simple K-divider (see section 2.2) and the output rate y will be:

$$y \quad = \quad x/k$$

or, if there is an absolute refractory period t_o

$$y \quad = \quad (x/k) \: / \: (1 + xt_o/k) \qquad\qquad (11.6.1)$$

which is a rectangular hyperbola. For the same input to a leaky integrator model, with a threshold k, time constant τ and each input pulse
producing a unit depolarization, the steady-state input-output relation
is approximately:

$$y \quad = \quad \frac{1}{\ln \{x\tau/(x\tau - k)\} \: + t_o} \qquad\qquad (11.6.2)$$

for $x\tau \gg k$. At low input rates no output pulses are generated, and
so there is a critical input rate below which there is no output.
Above this critical input rate the leaky integrator shows a smaller
sensitivity to low rates than to high rates: this is in contrast to
the greater sensitivity of the leaky integrator to smaller amplitude
inputs (see equation 3.3.3).

These simple input-output curves do not show an important
feature observed at synapses: the saturation of the output discharge at
high input rates and the inability of the postsynaptic cells to follow
very high input rates. Bullock (1964) comments on the variety of
experimentally obtained synaptic input-output relations, and suggests
that a behavioural classification might be appropriate - e.g. a synapse
that had a sharp critical input rate threshold, above which it responds
maximally, might be considered to act as a switch. However, this
diversity of synaptic input-output relations might be a quantitative
diversity rather than a qualitative diversity, in that the experimental
input-output relations can all be considered to fall on a suitably
scaled segment of a monotone increasing sigmoid curve. For synaptic
junctions where the postsynaptic neurone is not spontaneously active,

most experimentally obtained input-output relations can be fitted by
a function of the form:

$$y(x) \quad = \quad \frac{1}{1 + \exp\ (-a\ (x - x_o)\ )} \qquad (11.6.3)$$

if this function is suitably scaled and shifted along the x-axis.
The value of this function increases from 0 at $x = 0$ to 1 at $x = +\infty$,
with a single inflection point at x_o, and the parameter a determining
the slope of the relation. In fact this is a purely empirical function,
used by Wilson and Cowan (1972) to describe the response function of
population of neurones, and also occurs as the logistic function of
population growth (see Crow and Kimura, 1970).

However, there are circumstances in which the input-output
relation need not be monotonic. Perkel et al (1964) have investigated
the effects of regular excitatory or inhibitory pulse train inputs
to a digitally simulated model neurone which had a regular pacemaker
discharge. The neural model had an absolute refractory period,
during which presynaptic events had no effect, and after which the
membrane potential recovered exponentially from a hyperpolarized
value towards the resting potential. After the generation of an
impulse, the threshold decayed exponentially, with the asymptotic
value of the threshold more negative than the resting potential, thus
generating a regular pacemaker or carrier discharge rate. When
subject to regular excitatory or inhibitory input trains, each input
pulse producing a postsynaptic potential, the overall trend of the
input-output relation was positive, with an increased rate of
excitatory inputs tending to increase the output rate, and an increased
rate of inhibitory inputs tending to decrease the output rate. However,
some portions of the input output curve had a paradoxical negative
slope, where an increase in the output discharge rate was produced by
a decrease in the excitatory input rate or an increase in the
inhibitory input rate. This effect has been observed experimentally
at the crayfish stretch receptor neurone and is produced by a
'beating' between the regular input and the regular pacemaker discharge.
Since this effect is only found with regular inputs it is perhaps no
more than a curiosity with little or no functional significance.

When an irregular input train is used to avoid the effect of

beating, the two possibilities are a periodic input, in which the
pulse rate is modulated, or a nonperiodic and stochastic input.
The input-output relation can then be described in either the time
or the frequency domain.

Terzuolo, McKean, Poppele and Rosenthal (1969) consider the
problem of whether the impulse rate or interval is the relevant
parameter in a periodically modulated pulse train. The motor nerve
to a mammalian muscle was electrically stimulated by a frequency
modulated pulse train generated by a relaxation oscillator, and the
muscle tension was modulated. The Bode plot of the muscle tension
modulation against modulation frequency was characteristic of a
simple, first-order low-pass filter with transfer function

$$G(j\omega) = \frac{1}{1 + j\omega\tau}$$

The modulation of the muscle tension was independent of the carrier
discharge rate, and so the impulse rate rather than the impulse
interval appears to be the functionally significant parameter of
the input train.

This first-order low pass filter characteristic of the modulated
pulse train-tension transduction is produced by the relaxation time
constant of the muscle tension. Terzuolo and Bayly (1968) have
performed a similar experiment on the inhibitory synapse of the
crayfish stretch receptor neurone: this preparation has a single
spike train input and single spike train output, and so avoids the
problems of averaging the effects of many approximately synchronous
units which are inherent in the Terzuolo et al (1969) nerve-muscle
preparation. The large inhibitory fibre was electrically stimulated
by the output from a relaxation oscillator and the modulated sensory
spike train from the stretch receptor recorded; the output carrier
rate was adjusted by applying maintained stretch to the receptor.
Bode plots of the cell-to-cell transfer function between the averaged
instantaneous rates of the input and output pulse trains, when
normalized with respect to the carrier rate, were characteristic of
a low pass filter with a phase shift of 180° produced by the inhibitory
nature of the synaptic junction. This transfer function is of a
cascade of the inhibitory synapse and the spike generation mechanism

encoder of the stretch receptor neurone. Terzuolo, Purple, Bayly and
Handelman (1968) determined the transfer function of the encoder by
injecting sinusoidally modulated current into the cell body, and
so the transfer function of the inhibitory synapse can be obtained
by subtracting, on a Bode plot, the encoder transfer function from
the cell-to-cell transfer function.

The transfer function of the inhibitory synapse obtained by
this method is once again that of a low pass filter. Such a low pass
characteristic might be found at all synapses, as the shape of the
postsynaptic potential produced by a single presynaptic input spike
can be considered to represent the impulse response of the synapse.
The corner frequency of the transfer function would be determined by
the time course of the decay of the postsynaptic potential; this
would not necessarily be the membrane time constant but would be
dependent on the geometry and cable properties of the postsynaptic
cell.

The idea that a synapse operates as a low pass filter seems
reasonable when the spectral density of a sinusoidally modulated
pulse train is considered. The cosine expansion of such a pulse
train has been derived by Bayly (1968, 1969) and is given by equation
5.3.16: a sinusoidally modulated pulse train has a line spectrum
with components at D.C., the modulating frequency, the carrier
frequency and integer multiples of the carrier frequency, and at
sidebands above and below each carrier band. Thus if the sinusoidal
modulation frequency is considered as a signal, the sinusoidally
modulated pulse train contains power at the signal frequency and at
other frequencies. This power at other frequencies may be considered
to be produced by distortion of the signal, and when the modulation
frequency is less than the carrier frequency the signal can be
recovered from the modulated pulse train by a low pass filter.

Watanabe (1962) estimated the transfer functions of reflexes
in the abdominal ganglion of the crayfish using a sinusoidally
modulated pulse train input, and found two kinds of transfer function:
(a) reflexes which had a bursting discharge after the cessation of
a regular pulse train input had a high order transfer function
similar to that of a proportional-integral-derivative system. These
were presumed to be polysynaptic reflexes. (b) reflexes without an
afterburst had a first order transfer function similar to that of
a proportional derivative system. These were presumed to be mono-

synaptic reflexes. The increased gain at high (20 Hz) frequencies might be due to the dynamic properties of the postsynaptic spike generation mechanism. A major problem in these experiments was that the reflex response to a maintained, sinusoidally modulated input decreased with time, and so there is some habituation or fatigue occurring.

Watanabe (1969) extended this analysis of the transfer functions of reflexes in the crayfish by using stationary, random inputs to avoid the problem of habituation. The pre-ganglionic nerve was electrically stimulated by an exponentially distributed stochastic pulse train with a dead time of 5 msec - this approximates a Poisson input. The frequency response function and coherence function between the input and output spike trains were computed by methods similar to those described in section 5.4. In response to regular pulse train inputs, two types of postsynaptic discharge were obtained: the one-to-one type, with a postsynaptic spike following most of the presynaptic input pulses, and the tonic type of response, where the postsynaptic discharge was fairly regular and at a rate that was independent of the input rate. These two types had different frequency response functions - over the frequency range 2-50 Hz the one-to-one type had a frequency response of the form

$$G(j\omega) = \frac{j\omega\tau + 1}{1 + 2j\omega\tau} \exp(-jd\omega) \qquad (11.6.4)$$

where τ is the time constant of a low pass filter and d is a delay time. The coherence function was fairly high up to about 30 Hz: above this the coherence fell but this may have been due to the effects of aliasing produced by the sampling procedure (see French and Holden, 1971). The tonic type of response had a frequency response function similar to that of an underdamped second order system, and a low coherence throughout the frequency range. Thus the postsynaptic response of the tonic type of unit was only weakly related to the input pulse train.

The postsynaptic response of a neurone to a presynaptic, random pulse train has been developed in terms of cross-correlation functions by Knox and Poppele (1975). For a random presynaptic pulse train X(t) and a postsynaptic pulse train Y(t), the first order cross-covariance is

$$\rho_{xy} \ (t,\tau) \quad = \quad E \ \{Y(t + \tau) \ X \ (t)\} \tag{11.6.5}$$

which, if the spikes are treated as Dirac delta functions, can be
expressed in terms of the autocovariance function of the postsynaptic
train ρ_{yy} (τ) and the forward recurrence time distribution (see
section 5.2) V_y (t) of the post synaptic train

$$\rho_{xy} \ (t, \ \tau) \quad = \quad \frac{\lambda_x}{\lambda_y} \int_0^\tau g_{\theta|x}(t,\alpha) \ \{1 - V_y(\alpha)\} \ \rho_{yy}(\tau - \alpha) d\alpha$$

$$\tag{11.6.6}$$

where λ_x and λ_y are the mean input and output spike rates and $g_{\theta|x}(t, \tau)$
is the probability of the postsynaptic potential reaching a threshold
value of θ at a time $(t + \tau)$ given it had an arbitrary value x at
time t. This conditional probability density contains input-produced
changes in postsynaptic potential, and so underlies the contributions
of the primary synaptic and secondary input-periodicity effects
(see section 5.4) to the cross-covariance. Knox (1974) has shown
that when the summation of a large number of inputs is required to
reach threshold, the shape of $\rho_{xy}(\tau)$ is similar to the shape of the
derivative of postsynaptic impulse response. Thus the cross-covariance
is very sensitive to the shape of evoked postsynaptic potentials, even
when their amplitude is small. Knox and Poppele (1975) use this idea
to tentatively identify the location of synapses on Dorsal-spino-
cerebellar tract neurones, a sharp peak in the cross-covariance
function near $\tau = 0$ suggesting a brief postsynaptic potential and
hence a synapse near the spike initiating site.

The use of a random, Poisson distributed pulse train to
investigate the transfer characteristics of synapses has been further
explored by Krausz (1975) and Krausz and Friesen (1975). The
continuous, postsynaptic potential output produced by a Poisson
distributed input train is expanded as series of mutually orthogonal
functionals in the form of a Volterra series, the kernels of which
can be found by cross-correlation methods. This general method is
simplified when the responses to impulses are rapidly rising and
have a similar shape, and so it should be applicable in the analysis

of synaptic transfer dynamics. However a major problem with the
method is the slow convergence of the kernel estimates with length
of pulse train.

Thus it seems that the linear transmission properties of
synapses are similar to those of a low pass filter and a delay, but
that such a linear description misses the interesting effects of the
nonlinearities inherent in the pulse train to postsynaptic potential
transduction. A problem with Krausz's approach is that if the kernels
of the restricted diagonal Volterra series are obtained, and even if
a relatively small number of kernels provide a good characterization
of the input-output relation, the higher order kernels are not readily
interpretable either in terms of possible mechanisms or in terms of
their effects on signal transfer. Knox and Popele (1975) and Bryant,
Marios and Segundo (1973, see also section 5.4) have shown that the
first order, spike to spike cross-covariance functions contain a
wealth of information about the pre- and postsynaptic pulse train
rhythmicity, the shape of the evoked postsynaptic potential and
hence the location of the synapse. It seems likely that higher order
cross-covariances might describe the nonlinearities of synaptic
transfer in a form which could be interpreted in similar terms. This
approach has been developed by Segundo, Bryant and Brillinger (1975)
and Brillinger, Bryant and Segundo (1976).

In these papers the problem of synaptic identification is
approached using the methods of identification of point process
systems developed by Brillinger (1974, 1975). The identification
procedure is based on kernels similar to the Volterra-Wiener kernels,
and these kernels are system invariants which do not change when
the pulse train input to the synapse changes. The model is built up
recursively, incorporating higher order kernels until the synaptic
transfer relation is adequately characterized. The zero-order
kernel is a constant μ reflecting the intensity of the output point
process, and is simply the mean output rate when there is no input.
The first-order kernel, a(u), which is a function of time, is the
best linear predictor of the average change in the instantaneous
rate of the output at a time u given a single input spike occurred
at time 0, and satisfies:

$$\rho_{xy}(u) - \lambda_y = a(u) + \int_{-\infty}^{\infty} a(u - v) \, \rho_{xx}(v) - \lambda_x \, dv \quad (11.6.7)$$

$$-\infty < u < \infty$$

where λ_x and λ_y are the input and output mean rates and ρ_{xy} and ρ_{xx} are the cross- and auto-covariance functions defined in section 5.4. This integral equation assumes that there is no interaction between the input spikes, or that the postsynaptic effects of input spikes add linearly, and so a(u) can be considered the average impulse response function when the input rate is low. In equation 11.6.7 a(u) is defined for negative as well as positive times: this does not imply a reversal of causality, but that the direction of the synapse need not be known, and that in some situations, such as pacemaker discharges, the relationship between the input and output spike trains may appear to be anticipatory.

The coherence function defined by equation 5.4.21 can be used to determine whether or not a first-order model, based on the zero- and first-order kernels, provides an acceptable description of the synaptic relation. When the coherence is close to unity, the linear, first-order model is sufficient; when the coherence is negligible a higher order model is required, incorporating higher order kernels which relate to interactions between the postsynaptic effects of input spikes, such as facilitation or anti-facilitation.

The first order kernel a(u) is obtained from the solution of equation 11.6.7:

$$a(u) \quad = \quad \frac{1}{2\pi} \int_{-\infty}^{\infty} \{ \rho_{xy}(\alpha) / \rho_{xx}(\alpha) \} \exp(ju\alpha) \, d\alpha \qquad (11.6.8)$$

using estimates of the auto- and cross-covariances. A first-order model, based on the zero-order kernel and the first-order kernel a(u):

Prob {output spike in (t, t + u) | input spike at σ_i}

$$\simeq \{ \mu + \sum_{i=-\infty}^{\infty} a(t - \sigma_i) \} \, h \qquad (11.6.9)$$

appears to be adequate when there is a low rate, irregular input: in this case a(u) represents the average impulse response. However, when the input or output are bursting, or regular pacemakers, or when the rates are high, the coherence function is very low and so a higher-order model is required.

Given that, for sinusoidally modulated and random pulse train inputs, a synapse acts as a low pass filter, it may be that some temporal patterns in the input pulse train are more effective than others in producing a postsynaptic pulse. This problem has been investigated by Segundo, Perkel and Moore (1966) by computer simulation and by analysis of intracellularly recorded postsynaptic potentials and spike trains. The problem is really twofold: there is the forward problem of given an input sequence, determining the probability of a postsynaptic discharge, and the backward problem, of given a post-synaptic spike, determining the probability that a certain input sequence has occurred.

A sequence of EPSPs mingled with action potentials could be observed in intracellular recordings from neurones in Aplysia. EPSPs with a uniform amplitude and time course were taken to indicate the postsynaptic responses of the neurone to the same presynaptic input: thus the EPSP times give the times of arrival of presynaptic input spikes and the action potential times give the times of the postsynaptic spikes. The EPSPs in different preparations could be either spontaneous and arriving irregularly or driven by random, Poisson distributed stimulation of a nerve trunk. The EPSPs could be subdivided into two populations; those which were not followed immediately by a postsynaptic spike and those which triggered a postsynaptic spike.

For n EPSPs arriving at times t_1, t_2, ..., t_n the $(n - 1)$ first order intervals of the presynaptic spike train form an ordered set $\{T_2, T_3, ..., T_n\}$, where $T_i = t_i - t_{i-1}$. In any sequence of $(j + 1)$ EPSPs the ordered set of j intervals $\{T_{n-j+1}, T_{n-j+2}, ..., T_n\}$, characterizes the input spike train with a duration $t_n - t_{n-j}$ and mean rate $(j + 1)/(t_n - t_{n-j})$. Any sequence which terminates in a triggering EPSP is a triggering sequence, and a sequence which does not terminate in a triggering EPSP is a non-triggering sequence, even if some EPSPs within the sequence trigger spikes. If the post-synaptic neurone is more sensitive to some input patterns than to others, there should be differences between the patterns of triggering and non-triggering sequences.

Segundo et al defined a pattern as the ordered set of interval ratios:

$$\{ T_{n-j+1}/(t_n - t_{n-j}), \ T_{n-j+2}/(t_n - t_{n-j}), \ \cdots, \ T_n/(t_n - t_{n-j}) \}$$

For a sequence of three EPSPs, the pattern can be fully specified by the position of a point z in the positive quarter-plane with T_n as ordinate and T_{n-1} as abscissa. Any area in this quarter-plane represents a class of input pattern, and the scatter of triggering and non-triggering patterns over this quarter plane were different: thus some input patterns are more effective than others in producing postsynaptic spikes. When the span of the input triplet was short (or the rate was high) the spike-producing probability was independent of pattern. For long-span triplets, or low input rates, no post-synaptic spikes were evoked. However, for triplets with intermediate spans there was a tendency for long-short interval pairs to be more effective in evoking postsynaptic spikes than short-long interval pairs: however, the optimal input pattern tended to be where both T_n and T_{n-1} were of approximately equal length. The effectiveness of different input patterns changed with the triplet span or mean rate. This greater effectiveness of long-short interval pairs was not influenced either by the duration of the preceding interval T_{n-2} or the timing of the previous postsynaptic spike.

The triggering EPSPs patterns can be represented by the compound event (z, s), where a pattern z produced a spike s, with a joint probability density $f_{z,s}$ (z, s)dt. This joint pdf can be expressed as a product of a marginal and a conditional density in two ways: either

$$f_{z,s} \ (z,\ s) \quad = \quad f_z \ (z) \ P_{S|Z} \ (s|z) \qquad\qquad (11.6.10)$$

or

$$f_{z,s} \ (z,\ s) \quad = \quad f_S \ (s) \ P_{Z|S} \ (z|s) \qquad\qquad (11.6.11)$$

Equation 11.6.10 describes the forward probabilities: f_z (z) is the probability of a given input pattern occurring, and so is the product of the mean input rate x and the generating probability $G_{j,k}$ that a particular input pattern $z_{j,k}$ will occur, and $P_{S|Z}$ (s|z) is the probability that this input pattern will evoke a postsynaptic spike. Equation 11.6.11 describes the backward probability, that a given postsynaptic spike s, observed with a probability/unit time f_S (s), was produced by a particular input pattern z_{jk}.

Simple algebraic manipulation gives

$$G_{j,k} \, P_{S|Z} \, (s|z_{j,k}) \quad = \quad P_{Z|S} \, (z_{j,k}|s) E \qquad\qquad (11.6.12)$$

where $E = \sum_{j,k} (G_{j,k} \, P_{S|Z} \, (s|z_{j,k}))$ is the overall spike-evoking probability of the input spike train. All these probabilities can be estimated from the scatter of triggering and non-triggering points on the T_n, T_{n-1} plane. The LHS of 11.6.12 separates the properties of the presynaptic spike train from the postsynaptic properties, while on the RHS both terms reflect the relations between the input and output trains.

This pattern sensitivity of the postsynaptic neurone only extends for small number of intervals, and so only for an integration time given by the average span of the patterns. The computer simulations and Aplysia neurones were specifically chosen so that they required the summation of few presynaptic inputs to reach threshold: when the threshold potential is much greater than the EPSP amplitude it is unlikely that the postsynaptic neurone will exhibit much pattern sensitivity. A functional pattern sensitivity has only been observed at junctions which have large postsynaptic potentials e.g. the neuromuscular junction.

Segundo, Perkel, Wyman, Hegstad and Moore (1968) have examined the sensitivity of a model neurone to the input interval distribution when there are a number of excitatory input lines. When the activity in the input lines was independently and identically distributed, the postsynaptic discharge became more regular as the number of input lines increased while the EPSP amplitude decreased to maintain a constant postsynaptic discharge rate. For a large number of input lines, each of which produces a small EPSP, the postsynaptic interval distribution was independent of the form of the input interval distributions - this is because the superposition theorem (see section 8.2) gives the effective input as Poisson-distributed, irrespective of the distributions of activity in the individual input lines. When the number of input lines was small, and the EPSP amplitude was large, the postsynaptic interval distribution varied with the input distributions. At a given total input rate, the postsynaptic mean interval did not change as the input distribution changed, but the standard deviation changed as the input distribution changed, a Poisson distributed input producing the postsynaptic discharge with the greatest standard deviation. Asymmetric, symmetric and bimodal output distributions could be produced by independently distributed inputs.

When the activity in the input lines was interdependent, the output interval distribution was always dependent on the form of the input distributions, even when the number of input lines was large and the EPSP amplitude was small. Thus, for the postsynaptic cell to be sensitive to the pattern of activity in the input lines, either the inputs must be interdependent or the number of input lines must be small.

11.7. References.

Anderson, C.R. and Stevens, C.F.: Voltage clamp analysis of acetylcholine produced end-plate current fluctuations at frog neuromuscular junction. J. Physiol. 235 655-691 (1973).

Bayly, E.J.: Spectral analysis of pulse frequency modulation in the nervous system. I.E.E.E. Trans. Bio-Med Engng. 15 257-265 (1968).

Bayly, E.J.: Spectral Analysis of pulse frequency modulation. In: Systems analysis in Neurophysiology, ed. C.A.Terzuolo. Univ. of Minnesota 48-60 (1969).

Bennett, M.V.L.: Physiology of Electrotonic junctions. Ann. N.Y. Acad. Sci. 137 509-539 (1966).

Boyd, I.A. and Martin, A.R.: Spontaneous subthreshold activity at mammalian neuromuscular junctions. J. Physiol. 132 61-73 (1956).

Braun, M. and Schmidt, R.F.: Potential changes recorded from the frog motor nerve terminal during its activation. Pflugers Arch. ges. Physiol. 28756-80 (1966).

Brillinger, D.R.: Cross-spectral analysis of processes with stationary increments including the G/G/∞ queue. Annals of Prob. 2 815-827 (1974).

Brillinger, D.R.: The identification of point process sustems. Annals of Prob. 3 909-929 (1975).

Brillinger, D.R., Bryant, H.L. and Segundo, J.P.: Identification of synaptic interactions. Biological Cybernetics (1976)

Bryant, H.L., Marios, A.R. and Segundo, J.P.: Correlations of neuronal spike discharges produced by monosynaptic connections and by common inputs. J. Neurophysiol. 36 205-225 (1973).

Bullock, T.H.: Transfer functions at synaptic junctions. In: Information processing in the nervous system. Proc. I.U.P.S. XXII Congress, Leiden 1962. ed. R.W.Gerard and J.W.Duyff. Excerpta Medica (1964).

Cohen, I., Kita, H. and van der Kloot, W.: Miniature endplate potentials: evidence that the intervals are not fit by a Poisson distribution. Brain Research 54 318-323 (1973).

Cohen, I., Kita, H. and van der Kloot, W.: The intervals between miniature endplate potentials in the frog are unlikely to be exponentially distributed. J. Physiol. 236 327-339 (1974a).

Cohen, I., Kita, H. and van der Kloot, W.: The stochastic properties of spontaneous quantal release of transmitter at the frog neuromuscular junction. J. Physiol. 236 341-361 (1974b).

Cohen, I., Kita, H. and van der Kloot, W.: Stochastic properties of spontaneous release at the crayfish neuromuscular junction. J. Physiol. 236 363-371 (1974c).

Cox, D.R. and Lewis, P.A.W.: The Statistical Analysis of Series of Events. Methuen, London (1966).

Crow, J.F. and Kimura, M.: An Introduction to Population Genetics Theory. Harper and Row, New York. (1970).

del Castillo, J. and Katz, B.: Changes in endplate activity produced by presynaptic polarization. J.Physiol. 124 586-604 (1954).

del Castillo, J. and Katz, B.: Quantal components of the endplate potential. J. Physiol. 124 560-573 (1954).

del Castillo, J. and Katz, B.: Statistical factors involved in neuromuscular facilitation and depression. J. Physiol. 124 574-585 (1954).

del Castillo, J. and Katz, B.: Local activity at a depolarized nerve-muscle junction. J. Physiol. 128 396-411 (1955).

Eccles, J.C.: The Physiology of Synapses. Springer-Verlag, Berlin (1964).

Fatt, P. and Katz, B.: Some observations on biological noise. Nature 166 597-598 (1950).

Fatt, P. and Katz, B.: Spontaneous subthreshold activity at motor nerve endings. J. Physiol. 117 109-128 (1952).

French, A.S. and DiCaprio,R.A.: The dynamic electrical behaviour of the electrotonic junction between Retzius cells in the Leech. Biological Cybernetics 17 129-135 (1975).

French, A.S. and Holden, A.V.: Alias-free sampling of neuronal spike trains. Kybernetik 8 165-171 (1971).

Furshpan, E.J. and Potter, D.D.: Transmission at the giant synapses of the crayfish. J. Physiol. 145 289-325 (1959).

Furukawa, T. and Furshpan, E.J.: Two inhibitory mechanisms in the Mauthner neurons of goldfish. J. Neurophysiol. 26 140-176 (1963).

Hubbard, J.I.: Repetitive stimulation at the mammalian neuromuscular junction, and the mobilization of transmitter. J. Physiol. 169 641-662 (1963).

Jack, J.J.B., Noble, D. and Tsien, R.W.: Electric Current Flow in Excitable Cells. Clarendon Press, Oxford. 502pp. (1975).

Katz, B.: Nerve, Muscle and Synapse. McGraw-Hill, New York. 193pp (1966).

Katz, B. and Miledi, R.: The measurement of synaptic delay, and the time course of acetylcholine release at the neuromuscular junction. Proc. Roy. Soc. (London) B 161 483-495 (1964).

Katz, B. and Miledi, R.: The release of acetylcholine from nerve endings by graded electrical pulses. Proc. Roy. Soc. (London) B 167 23-38 (1967).

Katz, B. and Miledi, R.: Membrane noise produced by acetylcholine. Nature 226 962-963 (1970).

Katz, B. and Miledi, R.: Further observations on acetylcholine noise. Nature, New Biology 232 124-126 (1971).

Katz, B. and Miledi, R.: The statistical nature of the acetylcholine potential and its components. J. Physiol. 224 665-699 (1972).

Katz, B. and Miledi, R.: The characteristics of endplate noise produced by different depolarizing drugs. J. Physiol.230 707-717 (1973a).

Katz, B. and Miledi, R.: The binding of acetylcholine to receptors and its removal from the synaptic cleft. J. Physiol. 231 549-574 (1973b).

Knox, C.K.: Cross-correlation functions for a neuronal model. Biophysical J. 14 567-582 (1974).

Knox, C.K. and Poppele, R.E.: Response of neuronal syatems to random pulse trains. Proc. of the First Symposium on Testing and Identification of Nonlinear Systems. Ed. G.D.McCann and P.Z.Marmarelis. 227-235 (1975).

Krausz, H.I.: Identification of nonlinear systems using random impulse train inputs. Biological Cybernetics 19 217-230 (1975).

Krausz, H.I. and Friesen, W.O.: Identification of discrete input nonlinear systems using Poisson impulse trains. Proc. of the First Symposium on Testing and Identification of Nonlinear Systems, Pasedena, California, March 1975. Ed. G.D.McCann and P.Z.Marmarelis.125-146 (1975)

Kubo, R.: Statistical mechanical theory of irreversible processes. I.General theory and application to magnetic and conduction problems. J. Phys. Soc. Japan 12 570-586 (1957).

Kubo, R.: In: Some Aspects of the Statistical-Mechanical Theory of Irreversible Processes. Lectures in Theoretical Physics Vol I. Ed. Brittin and Dunham. Interscience Publ., New York (1959).

Lewis, P.A.W.: Some results on tests for Poisson processes. Biometrika 52 67-78 (1965).

Liley, A.W.: The effects of presynaptic polarization on the spontaneous activity at the mammalian neuromuscular junction. J. Physiol 134 427-443 (1956).

Liley, A.W.: Spontaneous release of transmitter substance in multiquantal units. J. Physiol. 136 595-605 (1957).

Liley, A.W. and North, K.A.K.: An electrical investigation of the effects of repetitive stimulation on mammalian neuromuscular junctions. J. Neurophysiol. 16 509-527 (1953).

Magleby, K.L. and Stevens, C.F.: The effect of voltage on the time course of endplate currents. J. Physiol. 223 151-171 (1972).

Magleby, K.L. and Stevens, C.F.: A quantitative description of endplate current. J. Physiol. 223 173-197 (1972).

Mallart, A. and Martin, A.R.: An analysis of facilitation of transmitter release at the neuromuscular junction of the frog. J. Physiol. 193 679-694 (1967).

Martin, A.R.: A further study of the statistical composition of the endplate potential. J. Physiol. 130 114-122 (1955).

Martin, A.R.: Quantal nature of synaptic transmission. Physiol. Reviews 46 51-66 (1966).

Perkel, D.H., Schulmam, J., Bullock, T.H., Moore, G.P. and Segundo, J.P.: Pacemaker neurones: effects of regularly spaced synaptic input. Science 145 61-63 (1964).

Segundo, J.P., Bryant, H.L. and Brillinger, D.R.: Identification of synaptic operators. Proc. of the First Symposium on Testing and Identification of Nonlinear Systems. Ed. G.D.McCann and P.Z.Marmarelis. California Institute of Technology, California. 221-227 (1975).

Segundo, J.P., Moore, G.P., Stensaas, J. and Bullock, T.H.: Sensitivity of neurones in Aplysia to temporal pattern of arriving impulses. J. Expl. Biol. $\underline{40}$ 643-667 (1963).

Segundo, J.P., Perkel, D.H. and Moore, G.P.: Spike probability in neurones: influence of temporal structure in the train of synaptic events. Kybernetik $\underline{3}$ 67-82 (1966).

Segundo, J.P., Perkel, D.H., Wyman, H., Hegsted, H. and Moore, G.P.: Input-output relations in computer-simulated nerve cells. Influence of the statistical properties, strength, number and interdependence of excitatory pre-synaptic terminals. Kybernetik $\underline{4}$ 157-171 (1968).

Soucek, B.: Influence of the latency fluctuations and the quantal process of transmitter release on the endplate potentials amplitude distribution. Biophysical J. $\underline{11}$ 127-139 (1971).

Stevens, C.F.: Synaptic physiology. Proc. I.E.E.E. $\underline{56}$ 916-930 (1968).

Stevens, C.F.: Inferences about membrane properties from electrical noise measurements. Biophysical J. $\underline{12}$ 1028-1047 (1972).

Takeuchi, A.: The longlasting depression in neuromuscular transmission of frog. Japan J. Physiol. $\underline{8}$ 102-113 (1958).

Terzuolo, C.A.: and Bayly, E.J.: Data transmission between neurones. Kybernetik $\underline{5}$ 83-84 (1968).

Terzuolo, C.A., McKean, T.A., Poppele, R.E. and Rosenthal, N.P.: Impulse trains, coding and decoding. In: Systems Analysis Approach to Neurophysiological Problems. Ed. C.A.Terzuolo. Univ. of Minnesota, Minneapolis. 86-91 (-969).

Terzuolo, C.A., Purple, R.L., Bayly, E. and Handleman, E.: Post-synaptic inhibition - its action upon the transducer and encoder s systems of neurons.In: Structure and Function of Inhibitory Neuronal Mechanisms : Proc. of the 4th International Meeting of Neurobiologists, Stockholm, Sept. 1961. Ed. C. von Euler, S. Skogland and U.Soderberg. Pergamon Press, Oxford. (1968).

Theis, R.E.: Neuromuscular depression and the apparent depletion of transmitter in mammalian muscle. J. Neurophysiol. $\underline{28}$ 427-442 (1965).

Usherwood, P.N.R.: Transmitter release from insect motor nerve terminals. J. Physiol. 227 527-551 (1972).

van der Kloot, W., Kita, H. and Cohen, I.: The timing of the appearance of miniature endplate potentials. Progress in Neurobiology 4 269-326 (1975).

Verveen, A.A. and DeFelice, L.J.: Membrane noise. Progress in Biophysics and Molecular Biology 28 189-265 (1974).

Watanabe, Y.: Responses of an abdominal ganglion of the crayfish to electrical stimulation with a sinusoidal frequency change. J. Fac. Sci. Hokkaido Univ. Ser VI Zool. 15 93-102 (1962).

Watanabe, Y.: Statistical measurement of signal transmission in the central nervous system of the crayfish. Kybernetik 6 124-130 (1969).

Wernig, A.: Quantum hypothesis of synaptic transmission. J. of Neural Transmission, Suppl. XII 61-74 (1975).

12. MODELS OF THE STOCHASTIC ACTIVITY OF NEURAL AGGREGATES.

The electrical activity of a single neurone varies stochastically
with time, and for neurones which generate action potentials the output from the neurone can be treated as a stochastic point process.
Such a description is incomplete, in that it ignores the spatial and
temporal changes of voltage with distance and time within the neurone,
but this incomplete description appears to be a useful way of characterizing the output from a neurone. The action potentials from a neurone
will disperse spatially and temporally down the axonal branches, and
will form the inputs to synapses on other neurones and effectors.
Thus it might appear reasonable to describe the electrical activity of
neurones forming the nervous system as a multi-variate point process:
however, there are too many neurones for such an approach to be feasible.

In a human nervous system there are approximately 10^{10} neurones
(Blinkov and Glezer, 1968), and even in what is perhaps the simplest
true nervous system, the nerve net of Hydra, there are about 10^3
neurones. If the activity of a component neurone is abstracted as a
function of a single variable, a description of the activity of the
nervous system in terms of the activities of its constituent neurones
would be a trajectory in a space of extremely high dimension. Even
in a micronet of two model neurones with known properties and connections the forward problem of describing the discharge patterns which
can be produced is formidable, and Harmon (1964) suggests that the
inverse problem of identifying the inputs and connectivity of a micronet from the observed discharge pattern may be impossible.

The major impetus behind the development of models of the stochastic activity of neurones has been the extensive analysis of experimentally recorded spike trains, and the problem has been to develop models
which could account for specific experimental findings. However, there
are no techniques available for recording the simultaneous activity of
large numbers of neurones, and so any attempts at modelling the
possible stochastic patterns of activity in the nervous system are

perhaps speculative and premature. However, the behaviour of an
animal, which is in some way a manifestation of the outputs from its
nervous system, is structured and stochastic, and this implies that
the patterns of electrical activity in the nervous system are stochastic,
and that the variability of the activity of single neurones is not
lost by some kind of averaging. The electrical activity of the nervous
system persists throughout the life of an animal, and so whatever the
patterns of electrical activity are, they must be stable. The nervous
system can generate rhythmic discharge patterns, such as the motor
discharges to the respiratory muscles, and so any model of the activity
of the nervous system must be capable of some kind of limit-cycle-like
activity. Thus it is possible to specify some kinds of properties
that models of the activity of the nervous system should possess, even
though the activity of the nervous system as a whole cannot be observed.

Sherrington (1940) described the patterns of electrical activity
within the brain as "the head-mass becomes an enchanted loom where
millions of flashing shuttles weave a dissolving pattern, always a
meaningful pattern though never an abiding one; a shifting harmony of
subpatterns." Although mathematical models of the activity of the
nervous system have not matched the insights of this description, and
have added little to it, in this chapter I will consider some approaches to modelling the nervous system based on assemblies of neurone-like
elements.

The nervous system is not simply a collection of a large number
of neurones, but is an organization of interacting neurones: thus
the patterns of activity in the nervous system may reflect the topology
of the connections between neurones rather than the properties of
individual neurones. In the nervous system a gross organization is
apparent: neurones with similar properties are aggregated into masses,
such as sensory relay nuclei, or sheets, such as the layers of the
cortex, and the axons linking neurones in different structures form
clearly definable tracts and pathways. However, even at this macroscopic level of morphology the structure is not stereotyped; any
given fibre tract almost certainly contains 'aberrant' fibres, and the
gross pattern of the folds and convolutions of the cortex covering the
human brain is almost as individual as a fingerprint. At the microscopic level some parts of the nervous system, such as the cerebellar
cortex, have a regular, lattice-like structure, while other parts,
such as the reticular formation, have little apparent organization.

Given this diversity of the degree and type of topological organization, there can be no general model of the nervous system, only particular models of specific topologies.

If there are only a few elements in a neural net it might be possible to describe the overall behaviour of the net in terms of the properties of its components and the connectivity of the net: this idea underlies the widespread experimental effort directed at the 'simple' nervous systems found in invertebrates. However, it may be that even a micronet of three neurones has one too many for this kind of approach to succeed, and that the behaviour of neural networks cannot be inferred from the behaviour of their components.

Although any given neurone is probably unique in its precise structure, connectivity and behaviour, it is possible to classify neurones consistently into a reasonably small (compared to the number of neurones in the nervous system) number of types. This suggests that instead of emphasizing the diversity of neurones it might be useful to emphasize the similarity of neurones, and to represent a collection of individual neurones by an aggregate of identical model neurones, or an aggregate of similar model neurones whose parameters are drawn from a common distribution function. The problem of describing the overall behaviour of the aggregate then becomes the problem of describing the behaviour of a typical or average component. The behaviour of such a typical component can then be compared with the experimentally observed behaviour of real neurones, as when recordings are made from single neurones it is assumed that the observed neurone is a typical sample from a population of neurones which would all exhibit similar behaviour under similar conditions.

The simplest model neurone considered in chapter 3 is the logical neurone of McCulloch and Pitts (1943): some characteristics of networks of logical neurones are discussed in section 12.1. A major deficiency of the logical neurone as a neural model is that time is quantized - two simple continuous-time model neurones are the perfect and leaky integrators, and the behaviour of homogenous nets of these model neurones are treated in section 12.2. Such homogenous nets exhibit only transient behaviour, as their activity is unstable: for a neural network to generate stable discharge patterns there must be at least two types of component which have antagonistic actions - these represent excitatory and inhibitory neurones. The temporal behaviour of interacting populations of excitatory and inhibitory neurones is treated in section 12.3, and the spatial behaviour of a centre-surround topology of excitatory and inhibitory neurones is considered in section 12.4.

12.1. Nets of logical neurones.

. The logical neurone of McCulloch and Pitts (1943) (see section 3.1) is a simple neural model which retains only the properties of spatial summation and a threshold pulse generating device. If both excitatory and inhibitory connections are permitted, any logical operation which can be specified by a Boolean function can be performed by a small, appropriately connected network of logical neurones, and McCulloch and Pitts suggested that Boolean algebra might provide an adequate language for describing the microscopic behaviour of neural nets. The operations which are performed by a micronet of logical neurones are completely determined by the connectivity of the micronet and the values of the threshold parameters and coupling coefficients - this suggests the problem of designing micronets to perform specific operations. Although this kind of problem is remote from neurophysiology, the idea that the connectivity of a net determines the functions of a net is often implicitly assumed, and underlies the experimental efforts directed at mapping the cell-to-cell connections in simple nervous systems. However, even when the connections of a neural net are known the functional behaviour is often obscure.

Another idea which follows from the emphasis on the connectivity of neural nets is the problem of obtaining reliable behaviour from a net when the connections are unreliable, either because of aberrant connections or because of the random deletion of connections by some kind of injury. Winograd and Cowan (1963) have discussed the problem of synthesizing reliable automata from unreliable components, and in the presence of incorrect connections, and showed that an increase in the redundancy of the automata, either by an increase in the number or the complexity of the components, could permit reliable behaviour in the presence of noise or incorrect connections. The relative insensitivity of the behaviour of the nervous system to neurone loss might suggest that the nervous system has a redundant organization.

The response of a logical neurone h with n input lines and a threshold S_h is governed by:

$$y(m + 1) \quad = \quad 1 \left[\sum_{j=1}^{n} a_j x_j (m) - s_h \right] \qquad (12.1.1)$$

where $1 [x]$ denotes the Heaviside step function

$$1 [x] \quad = \quad \begin{cases} 1 & \text{for } x > 0 \\ 0 & \text{for } x \leqslant 0 \end{cases}$$

and $x_j (m)$ is the state of activity (0 or 1) of the jth input line at a
time m and the coupling coefficients a_j of the input lines are positive
for excitatory inputs and negative for inhibitory inputs, m is an integer:
as in section 3.1 time is quantized. The dynamical behaviour of solutions of this single neuronic equation has been treated in detail by
Kitagawa (1973). In a network of logical neurones, the output from
one unit forms the inputs to other units, and so the microscopic behaviour of a network of N logical neurones is described by

$$y_h (m + 1) \quad = \quad 1 \left[\sum_{k=1}^{N} \sum_{r=0}^{n(h)} a_{hk}^{(r)} y_k (m - r) - s_h \right] \qquad (12.1.2)$$

where for a given neurone h the coefficients $a_{hk}^{(r)}$, for $k \neq h$, represent
the strength of the coupling from neurone k to neurone h at a time r,
and the self-coefficient $a_{hh}^{(r)}$ permits the discharge of the hth neurone
to be influenced by its previous activity. These coefficients are real
numbers which completely map the functional connectivity of the network.
Thus equation 12.1.2 gives a deterministic description of the microscopic activity of the neural net as an automaton. These neuronic or
decision equations were introduced by Caianiello (1961) as part of a
general theory of the brain: His basic assumption is that the time
scale of neuronal activity is much less than, and is clearly separable
from, the time scale of behavioural changes. This 'adiabatic learning
hypothesis' means that the instantaneous behaviour of a neural net can
be described by the neuronic equations with their constant coupling
coefficients, while long term changes in behaviour are produced by slow
changes in the coupling coefficients which are described by separate
mnemonic or evolution equations. In these mnemonic equations the
coupling coefficients $a_{hk}^{(r)}$ change according to some arbitrary rule

if an output from neurone h is preceded in a short time interval by
an input from neurone k.

The form assumed for the mnemonic equations does not matter,
unless the modifiable behaviour of a neural net is to be investigated,
as under the adiabatic learning hypothesis the coupling coefficients
are treated as constants. The neuronic equations appear to be
structurally stable, in that small changes in the thresholds S_h or
coupling coefficients will not change the solution of a system
of neuronic equations.

One property of the solutions of the neuronic equations with
constant coupling coefficients which has been considered by Caianiello
(1966) is the existence of reverberations, or periodic solutions. At
any time the state of activity of the net is completely represented
by an N-element vector, the elements being 0 or 1. If $\vec{v}_m$ is an N-
element column vector denoting the state of activity of the net at a
time m, a periodic solution or reverberation of period R exists if

$$\vec{v}_{m+R} = \vec{v}_m \qquad\qquad (12.1.3)$$

There are two ways of looking at the evolution of the vector $\vec{v}_m$
with time period m: one can either follow the development of
activity (state 1) or of inactivity (state 0) in the net. For a
given unit h to be active at a time (m + 1):

$$\sum_{k=1}^{N} \sum_{r=0}^{n(h)} a_{hk}^{(r)} y_k (m-r) \geq S_h$$

and so it will be inactive if

$$\sum_{k=1}^{N} \sum_{r=0}^{n(h)} a_{hk}^{(r)} y_k (m-r) < S_h$$

or if

$$\sum_{k=1}^{N} \sum_{r=0}^{n(h)} a_{hk}^{(r)} \; y_k (m-r) \leq S_h - 1 \tag{12.1.4}$$

if for convenience the coupling coefficients and threshold only
assume integral values. If one is interested in the development of
inactivity in a network one can either follow the inactivity or set
up a 'dual network' which has the same coupling coefficients but in
which all the thresholds S_h are replaced by

$$\{ \; \sum_{k=1}^{N} a_{hk} \; y(m) \; - \; S_h \; + \; 1 \; \}$$

The activity in this dual network is the same as the inactivity in
the original network. If all the thresholds of a network and its
dual are equal, i.e. if

$$S_h \; = \; \frac{1}{2} \; (\sum_{k=1}^{N} a_{hk} y_k (m) \;)$$

or in matrix form

$$\frac{1}{2} \; \| a_{hk}^{(r)} \| \begin{pmatrix} 1 \\ 1 \\ \cdot \\ \cdot \\ \cdot \\ 1 \end{pmatrix} \; - \; \begin{pmatrix} S_1 \\ S_2 \\ \cdot \\ \cdot \\ \cdot \\ S_N \end{pmatrix} \; = \; 0 \tag{12.1.5}$$

the network is 'self-dual' or normal.
In such a normal or self-dual network every $y_h (m)$ for all m belongs
to a reverberation if equation 12.1.3 holds, and so no parts of the
network are undergoing transients during a reverberation. Caianiello
(1966) and Caianiello, de Luca and Ricciardi (1967, 1968) obtain some
conditions for the existence of reverberations in a neural net and
consider the problem of designing networks with specific reverberation
periods, with reverberations either independent or dependent on the
initial inputs to the network. What is of interest here is the
length of possible transients and the periodicity of possible

reverberations - for a network of N components the maximum transient and the maximum period of a reverberation are both 2^N: this follows from the strong Markov property of the neuronic equations. When N is only reasonably large - say 100 - 2^N is an astronomic number and so, even with a short time unit, activity in a deterministic network undergoing long period reverberations could appear unpredictable or stochastic over any reasonable observation interval.

The approach to neural networks taken by Caianiello et al is deterministic, and so perhaps sheds more light on the behaviour of automata than on the behaviour of real neural nets. Amari (1971, 1974) has introduced a probabilistic element into this approach to neural networks by considering the macroscopic activity of randomly connected networks of McCulloch-Pitts neurones. In order to approach the statistical neurodynamics of randomly connected networks a change in notation is useful: Amari writes the neuronic equation 12.1.2 as:

$$\vec{v}(m+1) \quad = \quad sgn \ (A \ \vec{v}(m) \ - \ \vec{S}) \tag{12.1.6}$$

where $sgn \ (\vec{v})$ is a vector with i^{th} component $sgn \ (v_i)$, where the signum function

$$sgn \ (x) \quad = \quad \begin{cases} 1 & x > 0 \\ -1 & x \le 0 \end{cases}$$

is used instead of the Heaviside step function $1[.]$ defined by equation 3.1.3. This change in notation means that when the threshold of a component neurone is not reached the state 'no action potential' is represented by -1. Equation 12.1.6 gives the instant-to-instant transition of the microstates $\vec{v}(m)$: this microscopic description of the activity in the net uses the connectivity matrix $\underline{A}$ and the threshold vector $\vec{S}$. For a network of N components there are $(N^2 + N)$ parameters in (12.1.6), and these parameters can be denoted by an $(N^2 + N)$-dimensional vector ω:

$$\omega \quad = \quad (A, \vec{S}) \quad \equiv \quad (a_{hk}, S_h) \tag{12.1.7}$$

$$h, k \quad = \quad 1, 2, \dots, N$$

A network is randomly connected when the parameters are associated with some probability distribution $P_N(\omega)$. The transition equation 12.1.6 can be written using the state-transition operator T_ω, which is a function of the parameters ω, as

$$\vec{v}(m+1) \quad = \quad T_\omega \; \vec{v}(m) \tag{12.1.8}$$

Thus a given net, defined by its present microstate $\vec{v}(m)$ and its transition operator T_ω, is considered as a sample drawn from an ensemble or probability space of all possible nets. Assuming that the parameters are all drawn independently from the same probability distribution, if $<\;>$ denotes an ensemble average, for any function $f(\omega)$

$$< f(\omega) > \quad = \quad \int f(\omega) \; P_N(\omega) \, d\omega \tag{12.1.9}$$

The vector $\vec{v}(m)$ defines the microscopic activity of the net at a time m, and the macroscopic activity z can be defined by a function of this vector:

$$z \; = \; \alpha(\vec{v}) \; = \; \frac{1}{N} \sum_{i=1}^{N} \; y_i \tag{12.1.10}$$

$$-1 \leq z \leq 1$$

Thus the macroscopic state activity variable z is a function of $\vec{v}$, which by 12.1.8 is a function of the parameters ω. Since the parameters are random variables, the macroscopic activity z is a random variable dependent on the distribution $P_N(\omega)$. The problem is to obtain the rules for the evolution of the macroscopic state variable z(m) for all members of the ensemble of possible nets. Thus the desired transition equation should be independent of the random variables ω.

As $\vec{v}(m)$ gives rise to $\vec{v}(m+1)$ by the microscopic transition equation, the corresponding macroscopic transition is

$$z(m) \rightarrow z(m+1) \; = \; \alpha(T \; \vec{v}(m) \;) \tag{12.1.11}$$

Assuming that the random variable z(m+1) is closely distributed about a centre dependent on the value of the random variable z(m), Amari proves that the ensemble mean of z(m+1) depends on z(m):

$$< \alpha(T\vec{v}(m) \;) > \; = \; \phi(z(m) \;) \tag{12.1.12}$$

and for large N, use of a central limit theorem argument gives

$$\phi(z) = \Phi(Wz - S) \tag{12.1.13}$$

where

$$W = \frac{N\bar{a}}{\sqrt{\{N\sigma_a{}^2 + \sigma_s{}^2\}}} \qquad\qquad S = \frac{\bar{s}}{\sqrt{\{N\sigma_a{}^2 + \sigma_s{}^2\}}}$$

$$\Phi(x) = 2\int_0^x \frac{1}{\sqrt{2\pi}} \exp(-u^2/2)\,du$$

and $\bar{a}$ and $\sigma_s{}^2$ are the mean and variance of the coupling coefficients a_{hk} and $\bar{s}$ and $\sigma_s{}^2$ the mean and variance of the thresholds S_h.

If a given $z(m + 1)$ in a given net differs from the ensemble average by an amount $\varepsilon_\omega\ (\vec{v}(m)\ )$

$$\tag{12.1.14}$$
$$\alpha(T_\omega\ \vec{v}(m)\ ) = \phi(z(m)\ ) + \varepsilon_\omega\ (\vec{v}(m)\ )$$

it seems reasonable to assume that $\varepsilon_\omega\ (\vec{v})$ is small and

$$< \varepsilon_\omega\ (v)\ > = 0$$

$$< \{\varepsilon_\omega\ (v)\ \}^2 > = 0$$

This has been proved in a special case by Rotenberg (1971), and requires the variance $\mathrm{Var}\ \{\alpha(T_\omega v)\}$ to tend to zero as N tends to infinity.

If $\varepsilon_\omega(v)$ is small enough to be neglected, the macroscopic state transition equation is (from 12.1.11 and 12.1.14)

$$z(m + 1) = \phi(z(m)\ ) = \Phi(Wz(m) - S) \tag{12.1.15}$$

which, since it does not depend on ω, holds for all nets in the ensemble and is what is required. Thus the evolution of macroscopic activity in the net is determined by the weighting or coupling parameter W and the threshold parameter S.

When the macroscopic activity z of a net is conserved, or there is no dissipation or generation of macroscopic activity

$$z = \phi(z) \tag{12.1.16}$$

and the net is not dead, as the microscopic activity can still be
changing, but is at an equilibrium level. Amari considers the number
of equilibrium levels and their stability for a simple net where all
the ω are drawn from the same probability distribution P_N (ω). For
a macrostate to be an equilibrium level equation 12.1.16 must hold,
or the curves y = ϕ(z) and y = z must intersect. From equation
12.1.13 the function ϕ(.) is sigmoid, and so the curves y = ϕ(z) and
y = z can intersect at most at three points. Thus in a simple net
there can be a maximum of three equilibrium levels. Amari proves
that there will be three equilibrium levels if

$$h(W) \quad > \quad |S| \quad , \quad W > \sqrt{ \pi/2 } \} \tag{12.1.17}$$

and only one equilibrium level when $W < \sqrt{ \pi/2 \}$ or

$$h(W) \quad < \quad |S| \quad , \quad W > \sqrt{ \pi/2 } \} \tag{12.1.18}$$

where

$$h(W) \quad = \quad W \, \phi(\sqrt{ \log(2W^2/ \pi) \} }) \quad - \quad \sqrt{ \log(2W^2/\pi) \}}$$

Having found the number of possible equilibrium levels in a
given type of net, the next problem is to investigate their stability.
In this context a level is stable if the activity returns to that level
after a small disturbance. For a level z to be stable:

$$\phi'(z) < 1 \tag{12.1.19}$$

and for z to be unstable

$$\phi'(z) \quad > \quad 1$$

where ϕ' is the derivative of the function ϕ given by equation 12.1.13.

Using equations 12.1.17-18 and the stability criterion 12.1.19,
Amari shows:
a) if a net has three equilibrium levels (12.1.17 holds), the
upper and lower levels are stable and the middle level is unstable.
Thus the net is bistable, and the middle equilibrium level is a
threshold:
b) if there is only one equilibrium level (i.e. 12.1.18 holds),

then if

$$-h(W) \quad > \quad |S| \quad , \quad W \; < \; - \sqrt{ \{ \; \pi/2 \} }$$

this level is unstable;   otherwise the net is monostable.

c)   when

$$h(W) \quad > \quad |S| \quad , \quad W \quad > \quad \{ \; \pi/2 \}$$

the net is bistable, and when

$$-h(W) \quad > \quad |S| \quad , \quad W \; < \quad \sqrt{ \{ \; \pi/2 \} }$$

the net has no stable levels, but oscillates, with a period of twice the quantal time period.

Thus the two lines $h(W) = \pm |S|$ divide the S-W plane into three regions of behaviour, an oscillatory region, a monostable region where whatever the initial level the activity converges to a single stable level, and a bistable region. When the parameters W and S fall on the lines $h(W) = \pm |S|$ the net is structurally unstable, in that small changes in the parameters produce large or qualitative changes in the dynamical behaviour. Such qualitative changes in behaviour are catastrophes in the terminology of Thom (1972): and the line $h(W) = + |S|$ represents a cusp catastrophe.

The patterns of dynamic behaviour of a simple net are relatively restricted, and are similar to the behaviour of a single formal neurone. This emphasizes the flip-flop nature of the formal neurone: a Mc-Culloch-Pitts neurone can represent an enzymic switch, a simple net of such enzymic switches, or a neurone as a molecular automaton, or a simple net of neurones. In each case the behaviour is dominated by the threshold properties, and can either be periodic, with a period of twice the discrete time unit, or monostable, when whatever the inputs the activity tends to a single stable level, or bistable, when the output is controllable by the inputs. The stability can be changed by changes in the threshold parameters - this fits with the idea of changes in behaviour being produced by changes in threshold as a possible model of learning.

More complex dynamic behaviour can be produced by systems of

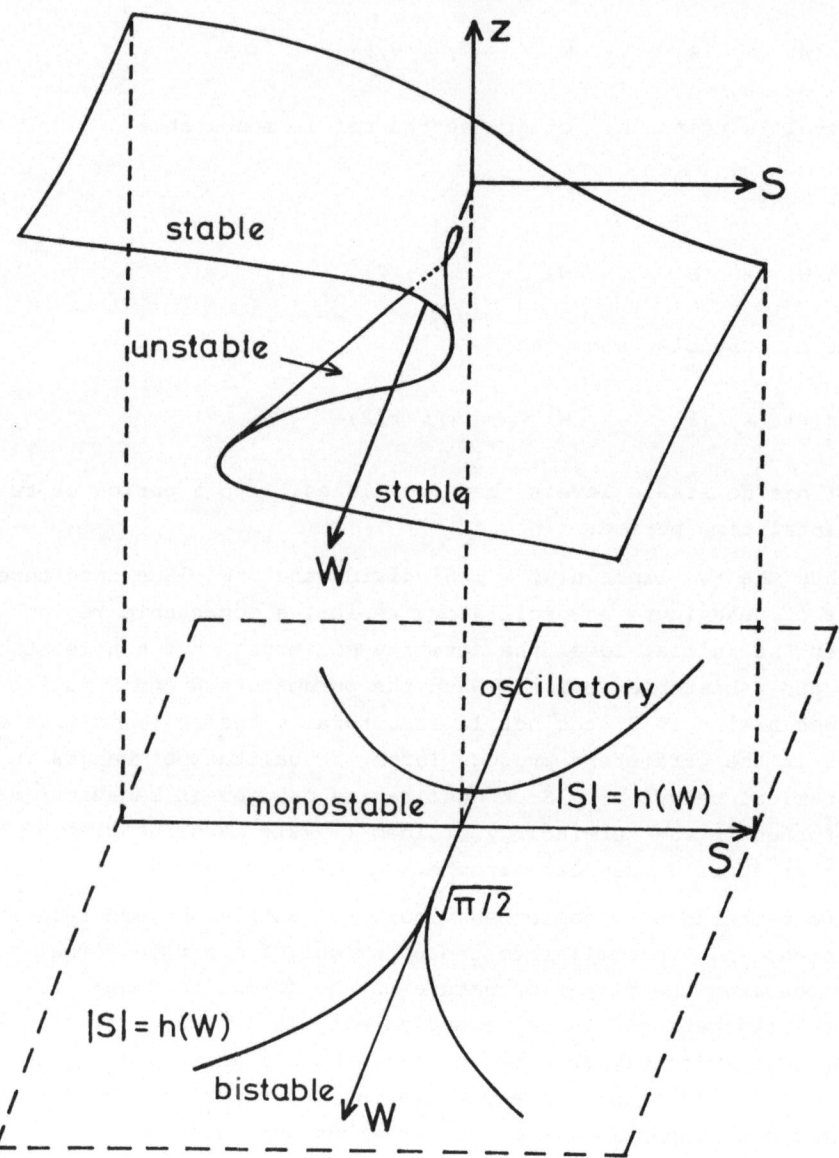

Figure 12.1. Equilibrium activity levels z (the macroscopic activity, defined by
12.1.10) as a function of the coupling parameter W and the threshold parameter S
for a simple net. The parameters W and S are defined by 12.1.13. For W > 0 the
behaviour is that of a cusp catastrophe.

interacting simple nets: in this way the simple net is viewed as an intermediate between single units and an organized network. If there are m simple nets in a connected, complex net, and each simple net has a macroscopic activity z_i, i = 1, 2, $\cdots$, m, the macroscopic activity of the complex net will be an m-element vector $\vec{z}$ = $(z_1, z_2, \cdots z_m)$. The transition equations will be analogous to those of a simple net, except in place of z and ϕ there will be vectors $\vec{z}$ and $\vec{\phi}$.

For a complex net of two connected simple nets, there will be coupling parameters W_{11} and W_{22} for connections within each simple net, and coupling parameters W_{12} and W_{21} for connections between the two nets. The effect of these cross-connections can be considered as equivalent to a change in the threshold parameters S_1 and S_2 of the two simple nets, as

$$z_1 \ (m + 1) \quad = \quad \phi_1 \ (z_1 \ (m), \ z_2 \ (m) \ )$$

$$= \quad \phi \ (W_{11}z_1 + W_{12}z_2 - S_1)$$

$$= \quad \phi \ (W_{11}z_1 - \tilde{S}_1) \tag{12.1.20}$$

where $\tilde{S} = S_1 - W_{12}z_2$

represents the effective threshold parameter of the first net for activity originating within the first net: the effect of the second net has been to add an average excitatory or inhibitory level. Thus the equilibrium states of the complex net will be at the intersections of:

$$z_1 = (W_{11}z_1 - \tilde{S}_1)$$
$$\tag{12.1.21}$$
$$z_2 = (W_{22}z_2 - \tilde{S}_2)$$

Since, for an isolated, bistable simple net, the equilibrium level z_{eq} as a function of S at a fixed W is a section through the cliff of a cusp catastrophe (see Figure 12.1), each equation of (12.1.21) is a Z-shaped curve, and so the two macrostate curves will intersect at several (no more than 9) points, which are the equilibrium levels of the complex net. These equilibrium levels $\vec{z}_{eq}$ are stable if the matrix:

$$\frac{\partial \phi_1(\vec{z}_{eq})}{\partial z_2}$$

has no eigenvalues with absolute magnitudes greater than one. If
there are no stable equilibrium levels the activity of the complex
net will oscillate with an arbitrary period.

The range of possible dynamic behaviour in a complex net of
only two simple nets is much richer than the dynamic behaviour of a
simple net: it seems reasonable to assume that as the number of
simple nets in a complex net increase there is a corresponding increase
in the range of dynamic behaviours. The types of dynamic behaviour
will be dependent on the connectivity between the simple nets, or the
average cross-coupling terms W_{ij}. Thus this digression into randomly
assembled automata returns to the idea of the connectivity, in this
case between simple random nets, determining the behaviour.

12.2 Networks Continuous in Time and Space: a Field Approach.

A problem with the approach using McCulloch-Pitts formal
neurones is that there are a large number of discrete units, each
operating in discrete, synchronized time steps. A logical neurone
has no memory of its inputs, and so the behaviour of the neural nets
of section 12.1, although interesting, is a poor representation of
what might be occurring in parts of a real nervous system. The perfect and leaky integrator models, introduced in sections 3.2 and 3.3,
are simple but plausible models of neurones, and so a network of such
continuous time model neurones might provide a more convincing model
of a real neural net.

A difficulty of such an approach is that each model neurone is
represented by an integral equation, or by an integro-differential
equation if adaptation and accommodation are introduced into the
model neurones. For the simplest case of a net of N perfect integrator
models, there will be N coupled integral equations (see section 10.4).
A way out of this difficulty is to replace the discrete collection

of N neurones by a continuum, and to describe the activity in such a
continuous system as a field quantity which satisfies some kind of
field equations. This approach was started by Beurle (1956), and
the first stage is to arrange the conditions of the network so that
a field description provides a good approximation.

The field quantity chosen by Beurle was the proportion of cells
becoming active in a small volume/unit time: this activity F(.)
forms the sources of the field and varies in three spatial dimensions
and with time, but for simplicity the activity can be considered to
vary only in one spatial dimension and so can be represented as F(x,t).

The activity F(x, t) will spread with distance x in the continuous system by some law reflecting the patterns of connectivity
between the discrete neurones. Beurle assumed that excitatory connections are formed whenever a dendrite and an axon come into close
proximity, and the dendritic and axonal structures were determined
probabilistically and independently of factors outside the cell:
there is no tendency for axons to grow towards dendrites. Thus the
connections between neurones are random. The dendrites of a single
neurone spread out in a localized, tree-like structure, and there is
a tendency for the density of dendrites to decrease with distance
r from the soma. A simple model is that the dendritic density D
(which is the total length of dendrites in a unit volume) decreases
exponentially with distance from the soma:

$$D \;=\; c(r_0/r) \, \exp \, (-r/r_0) \tag{12.2.1}$$

where r_0 and c are constants. This model is not purely arbitrary,
but is compatible with the data of Scholl (1953, 1956) on the dendritic arborizations of cortical neurones. The way in which dendritic
density changes with distance in real neurones obviously depends on
the histological type of neurone - see Braitenberg & Lauria (1960),
and ten Hoopen and Reuver (1970, 1971) have discussed dendritic
branching patterns in terms of a probabilistic model of dendrite
growth. Networks with different connectivity patterns will be
approximated by different field equations, but the choice of equation
12.2.1 to represent dendritic branching gives a field equation which
is isotropic and of only second order. Since the problem is to see
if a field theoretic approach is useful it seems reasonable to choose
a dendritic branching pattern which gives a reasonably simple field
equation.

Uttley (1955) has considered the probability of connections between axons and dendrites, when a connection is formed whenever an axon and dendrite are in close proximity, for a variety of axonal distributions interweaving an exponential dendritic field. The richness of connectivity can be given by a function $\xi(x)$, which is the mean number of connections from all cells in an infinite plane of unit thickness on to a given cell at distance x, and for a compact axonal system passing through an exponential dendritic field is given by

$$\xi(x) \quad = \quad b \exp (- |x| /r_o) \tag{12.2.2}$$

where the constant b depends on the axonal and dendritic densities and the size of the critical distance within which an excitatory connection is formed. $\xi(x)$ is the mean number of connections: in any small volume δV the probability that a given neurone has k connections is given by the Poisson distribution.

$$P(q) \quad = \quad \exp (-\xi_o) \; \xi_o^{\;k} / k! \tag{12.2.3}$$

where ξ_o is the value of $\xi(x)$ at the centre of the small volume δV.

When a given neurone is active it excites, after a delay τ, all the neurones on which it has connections: the delay τ represents the conduction time and synaptic delay. After generating an action potential the neurone is refractory, or insensitive to any inputs, for a time θ representing the absolute and relative refractory period. The effect of a subthreshold excitatory input decays with time: this can be represented by a function h(t) which gives the time course of the average post-synaptic potential: for simplicity Beurle took h(t) to be a rectangular function of duration s.

If cells are becoming active at a mean rate F within a plane of thickness dX at X, the mean rate of arrival of impulses at cells in the plane x will be $F(x, t - \frac{x-X}{v}) \; \xi \; (x-X) dX$, where v is the conduction velocity. The excitation $\psi(x,t)$ is given by

$$\psi(x, t) \quad = \quad \int_{-\infty}^{\infty} F(x, t- \frac{x-X}{v}) \; \xi \; (x-X) dX \tag{12.2.4}$$

which represents the total mean rate of arrival of impulses at cells
in the plane x from all other cells.   The mean excitation of all cells
in the plane x will be the double convolution:

$$\bar{\psi}(x, t) \;=\; \int_{-\infty}^{t} \int_{-\infty}^{\infty} F(X, T)\,\xi\,(x - X)\; h\;(t - T)\; dXdT \qquad (12.2.5)$$

Suppose the activity F(x, t) is uniform with distance x: can this
uniform level of activity remain constant with time?   For the activity
to remain constant,

$$F\,(t + \tau) \;=\; F(t)$$

and the rate of recovery of refractory cells must equal the rate of
activation of cells.   If the fraction of cells which are not refractory,
or are sensitive to excitation, is R, this means

$$(1 - R) \;=\; F\theta \qquad (12.2.6)$$

For a uniform activity, the convolution 12.2.4 and equation 12.2.2
give the mean rate of arrival of impulses on a cell as

$$\bar{\psi}(x, t) \;=\; F \int_{-\infty}^{\infty} \xi(x)\,dx \;=\; 2Fbr_o \qquad (12.2.7)$$

For a rectangular h(t), with a duration s, equations 12.2.5 and
12.2.7 give the mean integrated excitation $\bar{\psi}$ as:

$$\bar{\psi} \;=\; 2Fbr_o s \qquad (12.2.8)$$

Since the distribution of excitation over the cells obeys a Poisson
law, the fraction of cells which receive an excitation of (q - 1)
will be:

$$P(q - 1) \;=\; \exp(-\bar{\psi})\; \frac{\bar{\psi}^{(q - 1)}}{(q - 1)!}$$

$$=\; \exp(-2Fbr_o s)\; \frac{(2Fbr_o s)^{(q - 1)}}{(q - 1)!} \qquad (12.2.9)$$

and of this fraction only RP(q - 1) are sensitive to excitation.
Suppose that q represents the threshold of the cells: excitation
is arriving at a rate $2Fbr_o$, and so the rate at which the just
subthreshold cells become active is

$$F(t + \tau) = 2R\ P(q - 1)\ F(t)\ br_o \qquad (12.2.10)$$

and so for $F(t + \tau)$ to be equal to $F(t)$

$$2R\ P(q - 1)\ br_o = 1 \qquad (12.2.11)$$

From equations 12.2.6, 12.2.9 and 12.2.11, if $F(x, t)$ is to be
constant in space and time

$$\exp(-2Fbr_o)\ \frac{(2Fbr_o)^{(q - 1)}}{(q - 1)!}\ (2br_o - 2Fbr_o\theta) = 1 \qquad (12.2.12)$$

When values of $(2br_o)$ from Scholl's data and a reasonable ratio θ:s
are substituted into (12.2.12) there are no solutions for F unless
the threshold q is small. There are no stable levels of uniformly
distributed activity which are self-maintaining, and even for the
case with a small threshold q when there is a solution this uniform
state of activity is unstable. This lack of any stable, uniformly
distributed activity is also seen in computer simulations of random
nets of McCulloch-Pitts neurones (Griffith, 1971): the activity
rapidly tends to either a quiescent state or a state of total activity.
This means that the only type of activity this model can support is
in some way nonuniform, or organized in space and time i.e. some
kind of propagating waves. As an example Beurle considers the
propagation of plane waves: such a plane wave will propagate as
the region ahead of the wavefront will contain a smaller fraction of
refractory cells than the region behind, and as the refractory cells
recover there is the possibility of a series of waves or a circulating
wave. The kinds of propagating waves will depend on the form of the
connectivity and time course functions $\xi(x)$ and $h(t)$.

The types of propagating wave activity will be given by solutions
of the appropriate field equations, and so the problem is to specify
an appropriate form of the field equations, which in general will
be nonlinear partial differential equations of at least second order.

Beurle obtained the differential equation for a plane wave:

$$\frac{d^2}{dt^2}(R) \quad + \quad 1 - \frac{R\phi}{F} \quad \frac{d}{dt}(R) \quad = \quad 0 \tag{12.2.13}$$

where : R is the fraction of sensitive cells

$\phi\delta t$ denotes the fraction of sensitive cells which reach
threshold in an interval δt, and is a function of $F(x, t)$, $\xi(x)$ and
$h(t)$. Equation 12.2.13 is really a partial differential equation,
as differentials wrt x are included in the ratio ϕ/F, but when
$\xi(x)$ is exponential (equation 12.2.2) ϕ/F is a constant. For
$\phi/F = m$, a constant, equation 12.2.13 becomes

$$\frac{d^2}{dt^2} \quad + \quad 1 - Rm \quad \frac{d}{dt} \quad (R) \quad = \quad 0$$

and substituting the solution of this nonlinear differential equation

$$R \quad = \quad \frac{1}{m} - \frac{2\beta}{m} \quad \tanh \beta(t - t_o)$$

into 12.2.5 gives a propagating wave of activity:

$$F \quad = \quad \frac{2(m - 1)}{m} \quad \frac{\beta}{\cosh^2 \beta(t - t_o)} \tag{12.2.14}$$

with an amplitude $2(m - 1)/m$ and rate of change determined by the
constant β.

Thus Beurle's approach leads to a field equation which cannot
support uniform activity, but can generate propagating plane waves.
Other propagative disturbances can be produced by different forms of
$\xi(x)$ and $h(t)$ e.g. circular and spherical waves, and vortex effects.
The equations describe transformation (reaction) and spread or transport of a scalar quantity: if the activity is taken as analogous to
a concentration of a single substance these reaction-diffusion
equations are similar to those of the chemical model for morphogenesis
introduced by Turing (1952), in which a homogenous equilibrium is
unstable, and small disturbances give rise to a spatial morphogenesis.

ten Hoopen (1965) has outlined a field theoretic approach

similar to that of Beurle: the field quantity is the activity
$F(\vec{r}, t)$, the expected fraction of neurones becoming active at the
position vector $\vec{r} \equiv (x, y, z)$. The excitation $\psi(\vec{r}, t)$ depends on
the effective integrated impulses from external inputs $E(\vec{r}, t)dt$ and
from interaction within the net $K(\vec{r}, t)dt$:

$$\psi(r, t) = \int_{t-\tau-s}^{t-s} \{K(\vec{r}, u) + E(\vec{r}, u)\}du \qquad (12.2.15)$$

with the interaction component given by

$$K(\vec{r}, t) = \xi(\vec{r}) F(\vec{r}, t) \qquad (12.2.16)$$

as an approximation of

$$K(\vec{r}, t) = \iiint \xi(\vec{r}, \vec{u}) F(\vec{u}, t) d^3\vec{u}$$

analogous to equation 12.2.4.

The activity $F(\vec{r}, t)$ depends on:

(a) the expected fraction of sensitive cells becoming active.
The expected fraction of sensitive cells is

$$1 - \int_{t-\theta}^{t} F(\vec{r}, u) du$$

and the probability that a sensitive cell becomes active is

$$\frac{1}{s} S\{\psi(\vec{r}, t), q\} dt$$

with q the threshold, as in (12.2.9)

(b) the expected fraction of refractory cells becoming active
as soon as they become sensitive. The expected fraction of cells
leaving the refractory state is

$$F(r, t-\theta)dt$$

and the probability that a refractory cell becomes active as soon as
it becomes sensitive is

$$S\{\psi(r, t), q\}$$

Combining these four terms gives $F(\vec{r}, t)$ as:

$$F(\vec{r}, t) = \frac{1}{s} S\{\psi(\vec{r}, t), q\} \quad \{1 - \int_{t-\theta}^{t} F(\vec{r}, u)du\}$$

$$+ S\{\psi(\vec{r}, t), q\} F(\vec{r}, t-\theta) \qquad (12.2.17)$$

where $S\{.\}$ is some nonlinear functional of the sources $F(.)$ and excitation $\psi(.)$, and represents the population response function. If the period of latent addition of the cell, s, is small, the activity will be proportional to the rate at which sensitive cells reach threshold, and this will not depend very much on whether or not these cells have previously reached threshold i.e. the second term of equation 12.2.17 can be neglected, and

$$F(\vec{r}, t) = \frac{1}{s} S\{\psi(\vec{r}, t), q\} \quad \{1 - \int_{t-\theta}^{t} F(\vec{r}, u)du\} \quad (12.2.18)$$

The first term of (12.2.18) is essentially the event density, and for the nonlinear functional $S\{.\}$ ten Hoopen uses

$$S\{\psi(\vec{r}, t), q\} = 1 - \Phi\{\frac{(q - \psi(\vec{r}, t))}{\sqrt{\psi(\vec{r}, t)}}\}$$

where $\Phi(y) = \frac{1}{\sqrt{(2\pi)}} \int_{-\infty}^{x} \exp(-t/2)^2 dt$

Numerical solution of equation 12.2.17 in response to a rectangular input of excitation $E(t)$ showed a threshold kind of behaviour, the activity $F(t)$ tending to either quiescence or maximal activity, with a 'threshold' determined by the integrated external excitatory input. The activity tended towards these two stable levels in an oscillatory manner: the oscillations are produced by the reactivation of recovered refractory cells.

Griffith (1963a, 1965) also considers a field theoretic approach to a net of excitatory neurones, and has shown that for the field equation to be isotropic (invariant under translation or rotation) and of second order, the connectivity function must decrease exponentially with distance, as in equation 12.2.2. The behaviour of his model is similar to that obtained by Beurle: the important point is that of the three equilibrium states of uniform activity, the two stable states are those of inactivity and maximal activity, and the central equilibrium state is unstable. Thus an excitatory net cannot maintain a stable, uniform level of activity, but can only act as a switch or discriminator with only two stable output levels. Griffith (1963b) suggests two possible ways of extending this limited range of stable behaviour: in an extended net with incomplete connectivity, different parts may have different activities, or the introduction of inhibitory neurones can permit stable inter-mediate levels of activity. Nets with inhibitory as well as excitatory neurones are considered in sections 12.3 and 12.4.

12.3 Temporal Behaviour of Nets of Excitatory and Inhibitory Neurones

The inability of a random, excitatory net to sustain a uniform level of activity, and the obvious dynamic stability of real nervous systems, were considered paraxoxical by Ashby, von Foerster and Walker (1963): the introduction of inhibition by Griffith (1963) removes this paradox. Retrospectively, it is difficult to see why early models of neural nets neglected inhibition, as Sherrington (1906) emphasized the role of reciprocal inhibition in the activity of the nervous system. It is possible to define a nervous system as an organization of excitatory and inhibitory neurones: the only exceptions to such a definition would be some primitive coelenterate nerve nets. Thus it is not altogether surprising that nets without inhibition present a poor model of the nervous system.

Cowan (1970) has outlined a statistical mechanical approach to nets of excitatory and inhibitory neurones in terms of the time averaged behaviour of individual cells. The activity of a cell is given by its 'sensitivity' α, which is the fraction of the time the cell is not refractory and is sensitive to inputs (this is the single cell, time average equivalent of (1-R) of equation 12.2.6). If, as in section 12.2, the refractory period is θ,

$$\alpha = 1 - \theta/\bar{T}$$

where $\bar{T}$ is the mean interspike interval. With $\vec{\alpha}_i$ and $\vec{\alpha}_e$ representing the activities of inhibitory and excitatory cells, the general kinetic equation for slowly varying inputs is:

$$\left(\frac{\bar{T}d}{dt} + 1\right)f(\vec{\alpha}_i) = B_i\vec{u}_i + A_{ii}\vec{\alpha}_i - A_{ie}\vec{\alpha}_e$$

$$\left(\frac{\bar{T}d}{dt} + 1\right)f(\vec{\alpha}_e) = B_e\vec{u}_e + A_{ei}\vec{\alpha}_i - A_{ee}\vec{\alpha}_e \qquad (12.3.1)$$

where the B are control matrices giving the strengths of specific inputs $\vec{u}_i$ and $\vec{u}_e$ to inhibitory and excitatory cells, the A are non-negative, random coupling matrices and f(.) is a nonlinear function of the form $\ln(\alpha/(1 - \alpha))$.

Cowan considers the statistical behaviour of this model for the case of negligible damping and no self-excitation or self-inhibition, with symmetric connectivity i.e. for the connectivities between cells j and k, $a_{jk} + a_{kj} = 0$, where a_{jk} is an element of the coupling matrices. The existence and stability of solutions of this system of ordinary differential equations are discussed in Orguztoreli (1972) and Leung et al (1974): see also section 10.4.

An interesting aspect of this general network equation is its close similarity to the kinetic equations for interacting populations of predators (excitatory cells) and prey (inhibitory cells) developed by Volterra (1931) and explored by Kerner (1957, 1959, 1961) and Leigh (1968): see the review by Goel, Maitra and Montroll (1971), and the monograph by Kerner (1971). This analogy with predator-prey dynamics immediately suggests that as well as the stable solution corresponding to mutual extinction (total inactivity) there will be, under appropriate conditions, stable, uniform solutions and the possibility of oscillatory solutions.

To demonstrate the existence of multiple stable levels of uniform activity, periodic activity and perhaps even hysteresis, one could apply Beurle's method to a net of excitatory and inhibitory cells. Wilson and Cowan (1972) pursue this approach, assuming that the neural net is locally redundant and is made up of homogenous subpopulations of excitatory and inhibitory cells. In place of the activity F(x, t) of Beurle, two variables are needed: E(t) and I(t) which represent the proportion of excitatory and inhibitory cells which are firing/ unit time. Under the assumption of a dense connectivity and an isotropic net, all spatial changes will be ignored. Since I(t) and I(t) are proportions, they can vary from 0 to 1: this produces a

problem when a subpopulation of cells with a low level of activity
is strongly inhibited. To avoid this problem of a hard nonlinearity
when the activity is zero, small negative values of E(t) and I(t)
should be acceptable; an interpretation is that E(t) = 0, I(t) = 0
represents a resting or background level of activity analogous to the
intrinsic background activity of the nervous system. This device presupposes the existence of stable, uniform levels of activity.

As in Beurle's model, the refractory period is a constant θ and
conduction times and synaptic delays are lumped as a constant τ:
E(t + τ) and I(t + τ) will be the proportions of excitatory and
inhibitory cells which are active at a time (t + τ); this means that
at a time t they were sensitive or non-refractory and received a
suprathreshold excitation. The proportion of excitatory and inhibitory
cells which are sensitive will be:

$$1 - \int_{t-\theta}^{t} E(t')dt' \qquad \text{and} \qquad 1 - \int_{t-\theta}^{t} I(t')dt' \qquad (12.3.2)$$

Of these sensitive cells only a proportion will have received a suprathreshold excitation: let $S_e(\alpha)$ and $S_i(\alpha)$ be the expected fraction of
cells in the excitatory and inhibitory subpopulations which would
respond to a level of excitation α if all the cells were sensitive.
The problem is to choose an appropriate form for these subpopulation
response functions.

In early work on the reflexes of the spinal cord the dorsal and
ventral root compound action potentials were taken as measures of the
number of active input and output fibres. Input-output curves of
the monosynaptic reflex obtained in this way were continuous sigmoid
functions. In order to model these input-output curves Rosenbleuth,
Wiener, Pitts and Ramos (1949) proposed that afferent fibres branch
randomly to give synaptic knobs distributed over the motorneurone pool.
Assuming that each afferent gave the same number of synaptic knobs,
and that each motorneurone had the same number of synaptic knobs and
required the same number of active input fibres to reach threshold,
the propability that a given motorneurone discharges has a Bernouilli
distribution. Since this does not fit the experimentally obtained
sigmoid input-output curves Rosenbleuth et al relaxed the assumption
that the motorneurones had a constant threshold, and were able to fit
the experimental input-output curves and obtain theoretical threshold
distributions. However, the theoretical threshold distribution

functions which were required were peculiar, multimodal curves
requiring up to thirty arbitrary constants. Rall (1955) has
proposed a simpler model which assumes Gaussian distributed thresholds
and can generate a sigmoid input-output relation.

Whatever the form of the threshold distribution, the required
population response function should be monotone increasing with
asymptotic values of 0 and 1 as α approaches $-\infty$ and ∞. If the
threshold distribution is $D(T)$, and all cells in the population
receive the same (large) number of input fibres, the population
response function $S(\alpha)$ will be

$$S(\alpha) = \int_{0}^{\alpha(t)} D(t')dt' \tag{12.3.3}$$

Alternatively, if all cells have the same threshold T and the synaptic
distribution function is $C(w)$ with each synapse delivering an average
excitation $\bar{\alpha}(t)$, the population response function will be:

$$S(\alpha) = \int_{T/\bar{\alpha}(t)}^{\infty} C(w)dw \tag{12.3.4}$$

For each of these models, $S(\alpha)$ will be sigmoid if $D(T)$ or $C(w)$ are
unimodal. In fact one can define these distributions as unimodal as
if a neural population had a clear bimodal or multimodal threshold or
synaptic distribution it would be natural to treat it as a mixture of
two or more homogenous populations with unimodal distributions.

Given that $S(\alpha)$ is to be sigmoid, Wilson and Cowan chooses the
logistic curve

$$S(\alpha) = \frac{1}{1 - \exp\{-k(\alpha - T)\}} \tag{12.3.5}$$

as a simple, arbitrary sigmoid function. The constants k and T
control the maximal slope and population threshold (value of α at
which $S(\alpha) = 0.5$) of the population response curve. Other arbitrary
sigmoid curves could be used, or $S(\alpha)$ could be related to the threshold or synaptic distributions by (12.3.3-4): in this case the
smaller the variance of the threshold or synaptic distributions, the
greater the maximal slope of $S(\alpha)$.

If the postsynaptic effects of an input sum linearly and decay
with a time course h(t), the average level of excitation of a cell
in the excitatory subpopulation will be

$$\int_{-\infty}^{t} h(t - t') \{c_1 E(t') - c_2 I(t') + u_e(t')\} dt \qquad (12.3.6)$$

where c_1 and c_2 are positive coefficients representing the number of
excitatory and inhibitory synapses per cell, and $u_e(t)$ is the external input specific to excitatory cells. Thus from equations
12.3.2, 12.3.5 and 12.3.6:

$$E(t+\tau) = \{1 - \int_{t-\theta}^{t} E(t')dt'\} S_e (\int_{-\infty}^{t} h(t-t') \{c_1 E(t') - c_2 I(t') + u_e(t')\} dt')$$

$$(12.3.7)$$

and similarly for the inhibitory subpopulation, with coefficients
c_3 and c_4 and external input $u_i(t)$

$$I(t+\tau) = \{1 - \int_{t-\theta}^{t} I(t')dt'\} S_i (\int_{-\infty}^{t} h(t-t') \{c_3 E(t') - c_4 I(t') + u_i(t')\} dt')$$

$$(12.3.8)$$

This assumes that any correlation between excitation and the probability that a cell is sensitive may be neglected.

Although all the terms of these equations correspond to simple,
clearly defined physiological concepts, it is necessary to drastically
simplify the equations in order to permit a qualitative analysis of
the types of behaviour exhibited by these coupled, nonlinear integral
equations. Wilson and Cowan achieve this by replacing the temporal
integrals by a local time average: this temporal coarse-graining
gives the replacements:

$$\int_{t-\theta}^{t} E(t')dt' \rightarrow \theta \bar{E}(t)$$

$$\int_{t-\theta}^{t} I(t')dt' \rightarrow \theta \bar{I}(t)$$

$$\int_{-\infty}^{t} h(t - t')E(t')dt' \rightarrow k\bar{E}(t)$$

$$\int_{-\infty}^{t} h(t - t')I(t')dt' \rightarrow k'\bar{I}(t)$$

where k and k' are constants. Replacing $E(t + \tau)$ and $I(t + \tau)$ by Taylor expansions around $\tau = 0$ gives the simplified analogs of equations 12.3.7-8 as:

$$\tau \frac{d\bar{E}}{dt} = -\bar{E} + (1 - \theta\bar{E}) S_e \{kc_1\bar{E} - kc_2\bar{I} + ku_e(t)\}$$

$$(12.3.9)$$

$$\tau \frac{d\bar{I}}{dt} = -\bar{I} + (1 - \theta\bar{I}) S_i \{k'c_3\bar{E} - k'c_4\bar{I} + k'u_i(t)\}$$

This system of equations retains the interesting behaviour of the full equations: all that appears to be lost is a tendency for damped oscillations with a periodicity of twice the refractory period θ.

A phase plane analysis of the simplified equations is facilitated by having $\bar{E}(t) = \bar{I}(t) = 0$ as a steady state solution when $\bar{u}_e(t) = \bar{u}_i(t) = 0$. This can be readily achieved by subtracting $S(0)$ from $S_e(\alpha)$ and $S_i(\alpha)$: the effect is that the maximal value of the subpopulation response curves is now a constant less than one. Let these new maximum levels of the subpopulation response functions by K_e and K_i: the isoclines of (12.3.9) are than

$$c_2\bar{I} = c_1\bar{E} - S_e^{-1} \{ \frac{\bar{E}}{K_e - \theta\bar{E}} \} + u_e \quad \text{for} \quad d\bar{E}/dt = 0$$

$$(12.3.10)$$

$$c_3\bar{E} = c_4\bar{I} + S_i^{-1} \{ \frac{\bar{I}}{K_i - \theta\bar{I}} \} + u_i \quad \text{for} \quad d\bar{I}/dt = 0$$

$$(12.3.11)$$

$S^{-1}(\alpha)$ is the inverse of the population response functions; whatever the detailed form taken for the population response functions $S(\alpha)$, they are monotone increasing sigmoid curves, and so have unique inverses which are monotone increasing functions of α from $-\infty$ to ∞. Thus $\bar{E}$, as given by equation 12.3.11, will increase monotonically with $\bar{I}$; however, because of the minus sign before $S^{-1}(\alpha)$ in equation 12.3.10, $\bar{I}$ will tend to decrease with $\bar{E}$ except over a restricted range where it will increase. It is this kink in the

isocline for $d\bar{E}/dt = 0$ which permits interesting behaviour of the
solutions, and this will arise when there is negative feedback between
the two subpopulations (c_2, $c_3 \neq 0$) and reflects the antisymmetric
nature of the system (12.3.9). Thus the problem is to obtain
conditions, in terms of the coefficients c_j, and population response
curve parameters k and T, for interesting solutions. The behaviour
is going to be strongly determined by the strengths of the connectivity
parameters c_j: essentially, the net is composed of models with a
single time delay, or a single pole in their open loop transfer
functions, and to generate oscillations there must be appropriate
coupling coefficients. A higher order approximation of the dynamics
of the constituent neurones could permit limit cycle behaviour with
less restriction on the coefficients c_j - see section 10.4.

Wilson and Cowan obtain a necessary and sufficient condition
for there to be a kink in the $d\bar{E}/dt = 0$ isocline:

$$c_1 > 9/k_e \qquad\qquad\qquad (12.3.12)$$

where k_e is the slope parameter of $S_e(\alpha)$ and depends on the variance
of the threshold or connectivity distributions (equations 12.3.3-4),
and c_i is the average number of connections between excitatory
neurones. This numerical relationship assumes that the refractory
period has a numerical value of one.

If equation 12.3.12 holds, there will be a class of input
levels (u_i, u_e) such that the isoclines of equations 12.3.10 and
12.3.11 will intersect at a minimum of three points: these intersections are steady state-solutions. Not all the steady state-
solutions need be stable, and the stability of specific solutions
can be investigated by standard methods. From the results of section
12.1, one kind of behaviour that might be expected is two stable
steady-state solutions flanking a central, unstable steady-state
solution: Wilson and Cowan give a numerical example of this. In
such a case, with a Z-shaped $d\bar{E}/dt = 0$ isocline intersecting a monotone increasing $d\bar{I}/dt = 0$ isocline, since the effect of external
input levels u_e and u_i is simply to translate the isoclines parallel
to the I and E axes, hysteresis should result if the input level u_e
(or u_i) is changed while the level u_i (or u_e) remains constant.
Since the condition for such simple hysteresis behaviour with two
stable states depends only on the properties of the excitatory cell

population (c_1 and k_e of 12.3.12), such simple hysteresis should
also be found in predominantly excitatory populations. Harth et al
(1970) and Anninos et al (1970) have found simple hysteresis under
such conditions.

The system of equations 12.3.10-11 can exhibit complex hysteresis, with multiple stable steady-states, as well as simple hysteresis with only two stable steady-states. Wilson and Cowan obtain
a necessary but not sufficient condition for such multiple hysteresis
in terms of the negative feedback between the subpopulations being
stronger than the interactions within the subpopulations. Such
multiple hysteresis can only be obtained when there is inhibition as
well as excitation, and so will not be obtained in a single variable
field-theoretic model. Such multiple hysteresis can be considered
as a model of short-term memory, and Wilson and Cowan point out
that the population threshold-behaviour of the switching between two
stable states will obey a strength-duration curve similar to that
of the leaky integrator model (see equation 3.4.1), in that the
threshold will be greater for short duration transient increases
in the input level. This means that the switching between two
stable states will be relatively insensitive to noise in the input.

Wilson and Cowan consider two aspects of the temporal behaviour
of the system of equations 12.3.10-11. The impulse response of
($E(t) - I(t)$) is a damped oscillation, similar to average evoked
potential. More interestingly, the condition for limit cycle
behaviour is that the two isoclines intersect at only one steady
state point, and this point is unstable: this occurs when interactions within the excitatory subpopulation are stronger than the
interactions within the inhibitory subpopulation. As the point
($E = 0$, $I = 0$) is the stable, resting point of the system when
$u_e = u_i = 0$, limit cycles will only be obtained when there is a
constant input bias. Keeping $u_i = 0$, as u_e is increased above a
threshold value, the frequency of the limit cycles and the average
value of $E(t)$ increase monotonically until $E(t)$ saturates. Wilson
and Cowan suggest that this might be analogous to the experimental
results of Poggio and Viernstein, (1964), who showed an increase in
the mean firing rate and frequency of oscillation of the expectation
density of single thalamic neurones as the level of joint flexion
was increased.

In spite of the complexity of the full system of coupled,

nonlinear integral equations of 12.3.7-8, all the terms of the equations have a simple physiological or anatomical interpretation. Thus although the behaviour of the model is similar to that discussed for a more abstract model in section 12.1, the model of Wilson and Cowan has some kind of plausibility. This might be mistaken, as although the coupling coefficients c_j are defined in an anatomical way, they are used functionally as parameters which can be adjusted to determine types of behaviour. Thus histological estimates of the coupling densities c_j could not be used to test the model. If this model is just considered as a general model for the behaviour of coupled excitatory and inhibitory neural masses, the results are remarkable enough: application of the model to specific neurophysiological or even psychophysical results is nice, but adds little apart from confusing details. What is important is the range of behaviour that can be obtained from interacting excitatory and inhibitory subpopulations, and the crucial role of a sigmoid-shaped population response function. Furthermore, the equation 12.3.7-8 are structurally stable, in that small changes in parameters do not necessarily produce large, qualitative changes in the dynamics.

A limitation of this model is the neglect of spatial variations: these are considered in section 12.4.

12.4 Spatially Distributed Excitatory and Inhibitory Subpopulations

Wilson and Cowan (1973) have extended their model for two types of interacting neural subpopulations to include spatial variations in activity in a one dimensional isotropic sheet. As before, the connectivity is locally redundant, and following Buerle (1956) the connectivity function $\xi(x)$ decays exponentially with distance x

$$\xi_{jj'}(x) = b_{jj'} \exp(-|x|/r_{jj'}) \qquad (12.4.1)$$

where $\xi_{jj}(x)$ is a measure of the probability that neurones of type j' receive connections from neurones of type j at a distance $|x|$, and $r_{jj'}$ is the space constant for the connectivity from neurones of type j to type j'. The constant $b_{jj'}$ depends on the axonal and dendritic densities of the cell populations. In order to localize excitation $r_{ei} > r_{ee}$: this will prevent the propagation of active waves found by Beurle, and gives a centre-surround topology.

The postsynaptic response of any cell is taken to be the linear sum of the postsynaptic responses produced by excitatory and inhibitory inputs: the net excitation of a cell is the difference between the excitation and inhibition. Thus the effect of inhibition is subtractive: a different approach would be a multiplicative or shunting inhibition, and this is considered by Grossman - see below. All possible connectivity patterns are permitted: $E \rightarrow E$, $E \rightarrow I$, $I \rightarrow E$, $I \rightarrow I$, thus the interconnections are recurrent and either local or lateral.

Following the derivation of equation 12.2.5, the mean value of the integrated excitation $\bar{\psi}_e$ of excitatory neurones at x will be:

$$\bar{\psi}_e (x, t) = \int_{-\infty}^{t} \{ \int_{-\infty}^{\infty} \rho_e E(X, T - \frac{x - X}{v_e}) \; \xi_{ee}(x - X) dX$$

$$- \int_{-\infty}^{\infty} \rho_i I(X, T - \frac{x - X}{v_e}) \; \xi_{ie}(x - X) \pm u_e(x, T) \} \; h(t - T) dT$$

$$(12.4.2)$$

where, as in section 12.2, $h(t)$ is the time course of the average postsynaptic potential, v_e and v_i are the conduction velocities of excitatory and inhibitory axons, and as in section 12.3, $E(x, t)$ and $I(x, t)$ are the excitatory and inhibitory activities, which are the proportion of cells becoming active/unit time at a point x and instant t, and $u_e(x, t)$ represents external inputs, ρ_e and ρ_i are the cell densities. An analogous equation can be written for the mean value of the integrated excitation $\bar{\psi}_i$ of inhibitory neurones.

As in the derivation of equation 12.2.17, the expected number of excitatory cells activated during the interval δt at a time $(t + \tau)$ is obtained from the fraction of excitatory cells which are sensitive, and the sigmoid subpopulation response function $S_e(\bar{\psi}_e)$:

$$E(x, t + \tau) \rho_e \delta x \delta t = \{ 1 - \int_{t-\theta}^{t} E(x, T) dT \} \rho_e \delta x \; S_e(\bar{\psi}_e) \delta t$$

$$(12.4.3)$$

A similar equation can be derived for $I(x, t + \tau) \delta x \delta t$. Writing these equations in full for the excitatory and inhibitory neuronal activities:

$$E(x,\ t+\tau)\ \rho_e\, \delta x\delta t\ =\ \{1 - \int_{t-\theta}^{t} E(x,\ T)\, dT\}\, \rho_e \delta x\ .$$

$$S_e\{\ \int_{-\infty}^{t}\ \{\int_{-\infty}^{\infty}\rho_e\ E(X,\ T - \frac{x - X}{v_e})\ \xi_{ee}\ (x - X)\, dX$$

$$-\ \int_{-\infty}^{\infty}\rho_i\ I(X,\ T - \frac{x - X}{v_i})\ \xi_{ie}\ (x - X)\, dX\ \pm\ u_e(x,\ T)\}\quad h(t - T)\, dT\}\delta t$$

$$(12.4.4)$$

and for inhibition:

$$I(x,\ t+\tau)\ \rho_i\ \delta x\delta t\ =\ \{1 - \int_{t-\theta}^{t} I(x,\ T)\, dT\}\ \rho_i\ \delta x\ .$$

$$S_i\{\ \int_{-\infty}^{t}\ \{\int_{-\infty}^{\infty}\rho_e\ E(X,\ T\ -\ \frac{x - X}{v_e})\ \xi_{ei}\ (x - X)\ dX$$

$$-\ \int_{-\infty}^{\infty}\rho_i\ I(X,\ T\ -\ \frac{x - X}{v_i})\ \xi_{ii}\ (x - X)\ \pm\ u_i(x,\ T)\}\quad h(t - T)\, dT\}\ \delta t$$

$$(12.4.5)$$

These cumbersome equations are analogous to equations 12.3.7-8, and
can be simplified by assuming that the distances $|x - X|$ are small
and so the conduction time lag is negligible, and by using the method
of time coarse-graining introduced in section 12.3. These simplific-
ations give the equations analogous to equations 12.3.9 as:

$$\mu\frac{\partial}{\partial t}\ \bar{E}(x,\ t)\ =\ -\bar{E}(x,\ t) + \{1 - \theta\bar{E}(x,\ t)\}\ .$$

$$S_e\{\alpha\mu\{\ \int_{-\infty}^{\infty}\rho_e\ \bar{E}(X,\ t)\ \xi_{ee}(x - X)\, dX - \int_{-\infty}^{\infty}\rho_i\ \bar{I}(X,\ t)\ \xi_{ie}(x - X)\, dX \pm\bar{u}(x,\ t)\}\}$$

$$(12.4.6)$$

$$\mu\frac{\partial}{\partial t}\ \bar{I}(x,\ t)\ =\ -\bar{I}(x,\ t)+1\ \{1 - \bar{I}(x,\ t)\}\ .$$

$$S_i\{\alpha\mu\ \{\ \int_{-\infty}^{\infty}\rho_e\ \bar{E}(X,\ t)\ \xi_{ei}(x - X)\, dX - \int_{-\infty}^{\infty}\rho_i\ \bar{I}(X,\ t)\ \xi_{ii}(x - X)\, dX \pm\ \bar{u}_i(x,\ t)\}\}$$

$$(12.4.7)$$

where μ is the time constant of the function $h(t)$. These partial-
differential integral equations are far more complicated than the
analogous equations of section 12. 3 and so a qualitative, topological
analysis is not possible. Wilson and Cowan investigate the behaviour
of these simplified, time coarse-grained equations by numerical methods

using a logistic function (equation 12.3.5) for the subpopulation response functions. The values of the parameters were chosen to ensure that the state $E = I = 0$ is a stable solution, and that no uniform level of excitation is a stable solution in the absence of inputs. This is an interesting condition in that inhibition was introduced to neural field models by Griffith (1963) as a means of permitting uniform levels of activity to be stable.

Wilson and Cowan obtain three different modes of behaviour when different sets of parameters are used: active transients, spatially restricted limit cycles and spatially nonuniform stable steady states. The suggestion is that these different types of behaviour correspond to the behaviour of histologically specialized neural sheets occurring in different parts of the cortex or thalamus.

The active transient response to a rectangular spatio-temporal input excitation has a threshold and shows temporal and spatial summation. After the cessation of the input, the excitatory activity continues to rise rapidly, and then decays with a slower time course to the resting level. For suprathreshold inputs, the latency of the peak of the response decreases as the stimulus intensity increases. This kind of behaviour is similar to that seen in excitable membranes, where in place of recurrent excitation there is a regenerative current-voltage relation. This behaviour will arise in any system which has the dynamic properties of a stable equilibrium, a threshold for triggering an active response and a smooth return to equilibrium. These are all local properties, as the recurrent lateral inhibition limits the spread of excitation. Zeeman (1972) has discussed these dynamic properties with reference to the action potential in terms of Thom's catastrophe theory, and points out that the simplest possible model for these dynamics is a cusp catastrophe.

Perhaps the dynamic behaviour of the system of equations 12.4.4-5 is simpler than the complexity of the equations suggest. This idea is reinforced by the limit cycle mode of behaviour - in response to a constant, localized excitatory input there can be localized limit cycle activity in which the frequency of the limit cycles increases with the intensity of the excitatory input until $\bar{E}$ saturates. This is qualitatively similar to rate coding in a single model neurone, and is produced by the recurrent nature of the excitatory and inhibitory connections. Further, the response to a localized pulse train input can show frequency division effects similar to those of the simple K-scaler model of section 2.2; instead of generating an action potential every few input pulses the continuum generates a wave of activity every few input pulses. These simple dynamics were all obtained from numerical solutions of the coarse-grained equations: either the range of dynamic behaviour of the equations is in fact simple, or the simplicity of the behaviour results from qualitatively classifying the numerical solutions into familiar modes of behaviour. It seems reasonable to accept that the range of behaviour is similar to that of the models of section 12.1, and to that of a single neurone: essentially the recurrent excitation acts as a regenerative system, and the recurrent inhibition acts as a restoring or damping system. This idea suggests that the dynamics might be simpler than the mechanics, and if there was an experimental situation where the overall spatio-temporal pattern of activity of part of the nervous system was observable, the dynamic behaviour might be relatively simple to model. Activity in a homogenous motorneurone pool seems a suitable case for treatment - see below.

There are also interesting spatial behaviours: edge enhancement occurs when there is a spatially extensive excitatory input to the system, the maximum edge enhancement occurring at the peak of the local active response. Thus this edge enhancement is produced by recurrent lateral inhibition rather than by lateral inhibition. The latency of the maximal edge enhancement decreases as the excitatory input intensity increases, and the disparity of the two points with maximum edge enhancement increases as the excitatory input intensity increases. Wilson and Cowan point out the similarity of these effects to some phenomena in visual psychophysics.

A further aspect of the spatial behaviour is that in response to spatially nonuniform input patterns, a spatial pattern of activity is set up and this spatial pattern can persist after the cessation of the input pattern. Thus there are stable, spatially nonuniform steady state solutions: these are generated and maintained by localized recurrent activity, and can be erased by a strong inhibitory input, or a strong excitatory input to the inhibitory subpopulation. Somewhat speculatively, these persistent nonuniform steady states might represent a mechanism for short term memory.

A problem with this approach is that since we do not know what spatio-temporal patterns of activity are to be found in the nervous system, there is very little experimental information which can be compared with the solutions of the equations. The modes of behaviour obtained by Wilson and Cowan all seem reasonable, and are the kinds of behaviour any model of the nervous system should be capable of generating. Since there is little relevant electrophysiological data, it is tempting to pursue analogies between the solutions and psychophysical results. Although entertaining, this is rather speculative in that it assumes a one-to-one correspondence between the local spatiotemporal pattern of the mean excitatory activity and whatever psychophysical observations measure.

However, the important point from the Wilson and Cowan spatially extensive model is the method of progressing from a model neurone to a two-variable field model which generates plausible behaviour. Different neurone models and connectivity patterns will generate a different field model, but perhaps the overall qualitative behaviour of such different field models will not be all that different. As long as there are both excitatory and inhibitory cells, there is the possibility of stable steady states. Increasing the complexity of the neurone properties will give rise to more complex time behaviour, and altering the connectivity functions so that excitation spreads further than inhibition will give propagative waves.

Rather than extending this approach by looking at detailed variants of the model, it might be more useful to consider the behaviour of this kind of system in more detail: perhaps by considering only the expected values of the excitatory and inhibitory activities interesting behaviour is being missed. In particular, under conditions which generate active transients there might be a large increase in the variance of the excitatory activity as the threshold for generating an active response is approached. Cowan and Wilson suggest that in general E and I will be closely distributed about their mean values - this follows from the dense connectivity required for local redundancy, and the noise insensitivity of the sigmoid subpopulation response functions. However, the cooperative nature of the recurrent excitation generating the active transient response suggests that fluctuations might be enhanced near the threshold. Such fluctuations in the excitatory activity would be smoothed out by the time coarse-graining approximation. Rall (1955) considered both the mean value and variance of excitation

in his model of motoneurone input-output relations, but this was
only for steady state relations. The behaviour of the variance of
the activities, and perhaps higher moments, seems to be a difficult
problem which might be worth exploring.

The different modes of behaviour are found for different sets
of the connectivity parameters, which control the anatomy of the net:
thus the suggestion is that the different types of behaviour are
associated with parts of the nervous system with different net anatomies - the active transient response mode with the sensory cortex, the
limit cycle mode with the thalamus and the spatially nonuniform
stable steady states with short term memory in the pre-frontal cortex.
Without taking these suggestions too seriously, the equations might
be taken to represent the general local, recurrent, redundant nature
of cortical interactions, and the fact that different modes of
behaviour are obtained by quantitative changes in the coupling
parameters might suggest that quantitative cytoarchitectonic changes
of the same, basic cortical structure might produce different
behaviours in different parts of the cortex.

Feldman and Cowan (1975a, b) have applied this field theoretic
approach to model the cyclic discharge patterns found in the respiratory system. Here the overall periodic behaviour is well known,
and the problem is to find appropriate parameters which can fit the
experimental results. Rather than considering spatio-temporal patterns,
the method is to take as an element a homogenous, totally connected
net represented by a field equation similar to (12.3.7-8), and then
to connect a small number of these nets into a supernet by a vector
differential equation similar to (12.3.1). Each totally connected
net represents a homogenous group of respiratory-related units. The
details of the model depend on accepting the classification of six
types of respiratory-related units arranged in an appropriate anatomy:
this is a fuzzy area of neurophysiology, but perhaps the success of
the model tends to favour the particular classification and anatomy
chosen. However, simpler models using different anatomies are also
fairly successful (see Rubio, 1967, 1972) and periodic behaviour is
readily obtainable from equations of the Lotka-Volterra type. What
is interesting is the method of obtaining the population response
function using the distribution of axonal diameters as an index of
α-motoneurone cell size, and hence threshold.

In Cowan's approach to neural nets the main ideas are sub-

populations of excitatory and inhibitory cells, local redundancy
permitting a field theoretic rather than matrix formulation, simple
but plausible cell properties with linear summation representing the
interaction of excitation and inhibition, and sigmoid population
response functions. The behaviours generated by the models are nice
but not all that surprising, which is just as well as there is so
little experimental evidence which can be compared to the predictions
of the model. Grossberg has developed an alternative approach to on-
center, off-surround neural networks in which the inhibition is shunting rather than a simple subtractive inhibition, and obtains a
sigmoid population response curve by considering the desirable
properties the population response curve should have; this work is
reviewed in Grossberg (1973) and Ellias and Grossberg (1975). A
problem with Grossberg's work is that he tends to use his model to
account for psychological phenomena, and this detracts from the
seriousness of the mathematics.

A different kind of approach has been developed by Freeman, and
this has been reviewed in his monograph (1975). Rather than starting
with a collection of plausible model neurones, organizing the connections in some way and exploring the behaviour of the field
approximations, Freeman starts with his experimental results on
averaged evoked potentials in the olfactory system of the cat and
analyses these results in terms of linear systems theory.

Freeman considers the behaviour of a homogenous neural mass of
some $10^4 - 10^7$ neurones to be characterized by the averaged evoked
potential. This index of activity is **not** simply related to the
activities of the constituent cells, but at least it is experimentally
observable and the input (compound action potentials in fibre tracts)
is experimentally controllable. The amplitude of the evoked potential
is continuously graded, and twin input pulses can be used to obtain
an input intensity range over which superposition holds, and where
the amplitude of the averaged evoked potential is proportional to the
input. This linear range is assumed to be part of the physiological
range, and suggests that the dynamics can be modelled by linear
differential equations.

However, the shape of the average evoked potential changes
with the input amplitude; the major cause of this static nonlinearity
is the sigmoid population response function. Thus in place of
simple linear differential equations there will be linear differential

equations with state dependent coefficients. The problem is to obtain these equations and investigate their behaviour.

The olfactory system consists of a layer of receptors whose axons project to the olfactory bulb and terminate on periglomerular cells on the surface of the bulb. The bulb is a 3-layered structure, the outermost layer of periglomerular cells feeding the underlying layer of mitral and tufted cells, which in turn feed underlying granule cells. The output from the bulb (the lateral olfactory tract) are the axons of the mitral cells, whose discharges are influenced by the other cell populations of the bulb. The lateral olfactory tract projects to the 3-layered prepyriform cortex, terminating on the superficial layer of pyramidal cells. These superficial pyramidal cells project onto underlying granule cells, which in turn project onto deep pyramidal cells. Thus the olfactory system is a spatially organized structure of two layered sheets of interacting subpopulations, where each subpopulation is a set of interconnected neurones with a common afferent and efferent projection. The dense, recurrent interactions between the cells forming a subpopulation mean that the subpopulation can be lumped as a positive feedback loop: the problem is to characterize the dynamics of the system by means of the closed loop transfer function.

The open loop transfer function of any linear system can be obtained as the Laplace transform of its impulse response: the impulse response of a homogenous neural mass can be obtained by stimulating a homogenous afferent tract to the mass while synaptic transmission, either within the mass or by feedback loops external to the mass, is depressed by deep anaesthesia. Freeman takes the time-averaged evoked potential as the response: the assumption of linearity permits this to be identified with the impulse response. There is a hidden assumption of redundancy: the expected number of cells becoming active per unit time is equal to the mean firing rate of any cell only in a redundant net, and Freeman is assuming that the time-averaged evoked potential is a monotone function of the expected number of cells which are becoming active (the activity of section 12.2), which in turn is equal to the mean firing rate of any cell. Thus post-stimulus time histograms of single unit spike discharges and averaged evoked potentials should obey the same dynamics: the success of the analysis appears to justify this implicit assumption of ergodicity.

If the open-loop, time-averaged evoked potential is fitted by some simple arbitrary function r(t) with a Laplace transform R*(s), the closed loop transfer function P*(s) for a positive feedback loop with a gain K_e is

$$P*(s) \quad = \quad R*(s)/(1 - K_e R*^2(s))$$

where the gain K_e is dependent on the amplitude of the stimulus. If the impulse response r(t) is the sum of two exponentials, the inverse Laplace transform of P*(s) (the closed loop impulse response) p(t) will be of the form

$$p(t) \quad = \quad p_1 \exp(-p_1 t) + p_2 \exp(-p_2 t) + p_3 \cos(\omega t + \phi) \exp(-p_3 t)$$

which will rise rapidly and decay slowly if the gain of the positive feedback loop is less than one. This closed loop impulse response can be estimated experimentally by a post-stimulus time histogram. Thus from the open loop response and the gain the closed loop response can be obtained theoretically, and compared to experimental results, either for positive or negative feedback loops, and the stability and behaviour of the whole system investigated by root locus methods. The advantage of the root locus display method is that the linear, frequency dependent part of the behaviour (determined by the rate constants) and the nonlinear, amplitude dependent part of the behaviour (the variable gain coefficients) are clearly separable.

12.5   Some qualitative speculations

The types of model I have considered in sections 12.1 - 12.4 all attempt to describe the types of behaviour generated by assemblies of neurone-like elements, in terms of the properties of the components and the interactions between them. The results of these approaches - steady states, limit cycles and propagating waves -- seem trivial compared to Sherrington's idea of 'meaningful patterns', and since there is so little relevant experimental evidence there is a danger that further development of these approaches might become lost in their plausibility and progressively less and less relevant to

neurophysiology. In the problem of obtaining the behaviour of a net
the more plausible the model, the less tractable the system of
equations, while the behaviours obtained by numerical solutions still
appear simple. The simplicity of the behaviours obtained suggests
that one might forget the mechanics, and just explore the possible
dynamics of neural activity: by deliberately taking a more abstract
approach there is less danger in believing in a model because it can
fit some experimental data, and perhaps one might gain some insights
into the possible dynamics of meaningful patterns of neural activity.

A crucial feature of the activity of the nervous system is
that it is persistent throughout the life of an animal, and so is in
some sense stable, and that it appears to be stochastic, but not chaotic:
the activity is patterned but is not stereotyped. Perhaps the stochastic nature of the activity is not some kind of noise which can be
ignored, but is an essential aspect of the neural control of behaviour:
a nervous system with a stereotyped, deterministic behaviour would
not exhibit adaptive behaviour in a changing environment. Essentially,
one cannot learn without making mistakes. In this sense the stochastic
nature of the activity, together with interaction with the environment
by the motor and sensory systems, are analogous to random mutation
and natural selection in the evolutionary development of species.
This may be a non-trivial analogy, as in both cases one is interested
in the development of patterns or structures. This suggests that one
could approach the modelling of the stochastic patterns of activity of
the nervous system not by reference to networks of nerve cells but
in terms of simpler examples of dynamic pattern formation.

There are many examples of the formation of static spatial
patterns or evolving, spatio-temporal patterned structures in physico-
chemical systems. Static spatial structures, such as the honeycomb
pattern obtained when a thin film of varnish dries, can arise as
equilibrium solutions of some kind of interference between reaction
and diffusion or transport processes. However, evolutionary, spatio-
temporal patterned structures arise in dissipative systems which are
far from equilibrium, and in which there is a high density of interacting components, the interactions between components being nonlinear.
The idea is that an initially chaotic or homogenous state is moved
by random fluctuations towards a structured, stable state: Turing
(1952) uses this as the basis for a reaction-diffusion theory for
morphogenesis. A thermodynamic theory for such far from equilibrium,

dissipative systems has been developed by Glansdorff and Prigogine (1971), and related ideas relevant to neural dynamic patterns are discussed in Katchalsky, Rowland and Blumenthal (1974). A dynamic pattern will be established as random fluctuations take the system over a jump transition between two stable states: far from equilibrium there can be many critical points where such jump transitions can occur, and so there can be a richness in the evolving, dynamic pattern. These physical arguments suggest that the thermodynamics of dissipative structures may provide insights into dynamic patterns in the nervous system, where there is a high density of nonlinearly interacting components. Cowan (1971) emphasizes the dissipative nature of his neural field models (see section 12.3 and 12.4), and that the dissipative dynamics underly the structural stability of the field equations when they are subject to small perturbations of parameters. However, dissipative structures are in general just as intractable as nonlinear field equations, and so although the thermodynamics of dissipative structures might tell us how dynamic patterns can be formed, it cannot tell us what sort of dynamic structures correspond to Sherrington's meaningful patterns.

A more abstract approach to pattern formation, or morphogenesis, is found in Thom's catastrophe theory (Thom, 1970, 1972), and this might be used to classify what sort of dynamic patterns could occur. Although Thom's remarkable ideas have been applied to embryological development (Zeeman, 1974) and evolutionary development (Waddington, 1974; Dodson, 1976) there has been little application of his ideas to neurophysiology. Essentially, it is reasonable to assume that the activity in the nervous system obeys some dynamic which is structurally stable: a small change in the parameters will produce a small change in behaviour. In section 5.1 I have argued that the stationary activity of neurones is interesting, as changes in activity may be by abrupt changes between stationary, locally stable patterns of activity; further, the neuronic equations of section 12.1 and the field equations of sections 12.3 and 12.4 are structurally stable. A dynamic pattern involves a qualitative change in activity, or a small perturbation produces a large change in activity, or a jump transition. This is the critical point, or catastrophe set, where the equations are no longer structurally stable. The patterns of activity are spatio-temporal, and so only require a four-dimensional control space. In such a control space there are seven, and only seven, elementary catastrophes, or types of transition between stable points or attractors.

If the only attractors of the neural dynamic were point attractors, Sherrington's meaningful patterns could only be the seven elementary catastrophes. Since the attractors of the neural dynamic are also cyclic or cyclic in higher dimensions there will also be other general catastrophes forming the dynamic pattern of activity. However, among the dynamic patterns will be the seven elementary catastrophes, and so one can identify these elementary catastrophes with part of Sherrington's meaningful patterns. The unique two-dimensional elementary catastrophe is the cusp catastrophe illustrated in Figure 12.1: this simple catastrophe underlies the threshold behaviour of the action potential (Zeeman, 1972), the bistable behaviour of a simple net of formal neurones (Amari, 1974; section 12.1) and the active transient response of Wilson and Cowan's field equations (section 12.4). If one could record the spatio-temporal pattern of activity in parts of the nervous system, one would expect the other elementary catastrophes could model the experimentally observed dynamics. One system where the spatio-temporal pattern is observable is motorneurone activity during the swimming escape sequence in Tritonia: this has been modelled by Harth, Lewis and Csermely (1975) using the dynamics of the cusp catastrophe.

Thom's catastrophe theory generates deterministic dynamic patterns, and so in place of the clockwork discrete automata of section 12.1 there is a system of continuous, deterministic equations. The observed activity of the nervous system is apparently stochastic: this could be the result of so many bifurcations that the activity is nonpredictable because it is beyond computation, although deterministic; alternatively the activity might be truly stochastic. These two alternatives are indistinguishable: however I tend to favour the view that the activity is stochastic, and that both the dynamic patterns and fluctuations in activity ultimately derive from random events at the membrane or molecular level.

12.6. References.

Amari, S.: Characteristics of randomly connected threshold-element networks and network synthesis. Proc. I.E.E.E. $\underline{59}$ 35-47 (1971).

Amari, S.: A method of statistical neurodynamics. Kybernetik $\underline{14}$ 201-215 (1974).

Amari, S.: A mathematical theory of nerve nets. Advances in Biophysics $\underline{6}$ 75-120 (1974).

Anninos, P.A., Beek, B., Csermely, T.J., Harth, E.M. and Pertile, G.: Dynamics of neural structures. J. Theoretical Biol. $\underline{26}$ 121-148 (1970).

Ashby, W.R., von Foerster, H and Walker, C.C.: Instability of pulse activity in a net with threshold. Nature $\underline{196}$ 561-2 (1962).

Beurle, R.L.: Properties of a mass of cells capable of regenerating pulses. Phil. Trans Roy. Soc. (London) B $\underline{240}$ 55-94 (1956).

Blinkov, S.M. and Glezer, I.T.: The Human Brain in Figures and Tables: a quantitative handbook. Plenum, New York (1968).

Braitenberg, V. and Lauria, F.: Toward a mathematical description of the grey substance of nervous systems. Il Nuovo Cimento series X $\underline{18}$ Suppl. 2 149-165 (1960).

Caianiello, E.R.: Outline of a theory of thought processes and thinking machines. J. Theoretical Biology $\underline{1}$ 204-235 (1961).

Caianiello, E.R.: Decision equations and reverberations. Kybernetik $\underline{3}$ 98-100 (1966).

Caianiello, E.R. and de Luca, A.: Decision equation for binary systems: application to neuronal behaviour. Kybernetik $\underline{3}$ 33-40 (1966).

Caianiello, E.R., de Luca, A. and Ricciardi, L.M.: Reverberations and control of neural networks. Kybernetik $\underline{4}$ 10-18 (1967).

Caianiello, E.R., de Luca, A. and Ricciardi, L.M.: Neural networks: reverberations, constants of motion and general behaviour. Neural Networks:ed. E.R.Caianiello,p.92-99. Springer-Verlag, Berlin (1968).

Cowan, J.D.: A statistical mechanics of nervous activity. Lectures on Mathematics in the Life Sciences II: Some Mathematical Questions in Biology 2-57. A.M.S., Providence, Rhode Island (1970).

Cowan, J.D.: Stochastic models of neuroelectric activity. Proc. 6th. I.U.P.A.P. Conference on Statistical Mechanics.ed. S.A. Rice, K.F.Freed and J.C.Light. Univ. of Chicago Press, Chicago. p. 109-129 (1971).

Dodson, M.M.: Darwin's law of natural selection and Thom's theory of catastrophes. Math. Biosci. 28 243-274 (1976).

Ellias, S.A. and Grossberg, S.: Pattern formation, contrast control and oscillations in the short term memory of shunting on-center, off-surround networks. Biological Cybernetics 20 69-98 (1975).

Feldman, J.L. and Cowan, J.D.: Large scale activity in neural nets I: theory with applications to motoneurone pool responses. Biological Cybernetics 17 29-38 (1975a).

Feldman, J.L. and Cowan, J.D.: Large scale activity in neural nets II: a model for the brainstem respiratory oscillator. Biological Cybernetics 17 39-51 (1975b).

Freeman, W.J.: Mass Action in the Nervous System. Academic Press, New York (1975).

Glansdorff, P. and Prigogine, I.: Thermodynamic Theory of Structure, Stability and Fluctuations. Wiley, New York (1971).

Goel, N.S., Maitra, S.C. and Montroll, E.W.: On the Volterra and other nonlinear models of interacting populations. Rev. Modern Physics 43 231-276 (1971).

Griffith, J.S.: On the stability of brain-like structures. Biophys. J. 3 299-308 (1963).

Griffith, J.S.: A field theory of neural nets I: derivation of field equations. Bull. Math. Biophys. 25 111-120 (1963).

Griffith, J.S.: A field theory of neural nets II: properties of the field equations. Bull. Math. Biophys. 27 187-195 (1965).

Griffith, J.S.: A View of the Brain. Oxford Univ. Press, London (1967).

Griffith, J.S.: Mathematical Neurobiology. Academic Press, New York (1971)

Grossberg, S.: Contour enhancement, short term memory and constancies in reverberating neural networks. Studies in Applied Maths. 52 213-257 (1973).

Harmon, L.D.: Neuromimes: action of a reciprocally inhibitory pair. Science 146 1323-1325 (1964).

Harth, E.M., Csermely, T.J., Beek, B. and Lindsay, R.: Brain functions and neural dynamics. J. Theoretical Biology 26 93-120 (1970).

Harth, E., Lewis, N.S., Csermely, T.J.: The escape of Tritonia: dynamics of a neuromuscular control mechanism. J. Theoretical

Biology 55 201-228 (1975).

ten Hoopen, M.: Mathematical model of cooperative processes in neural populations. In: Cybernetics of Neural Processes. ed. E.R.Caianiello p.279-286. C.N.R., Rome (1965).

ten Hoopen, M. and Reuver, H.A.: Probabilistic analysis of dendritic branching patterns of cortical neurones. Kybernetik 6 176-188 (1970)

ten Hoopen, M. and Reuver, H.A.: Growth patterns of neuronal dendrites-an attempted probabilistic description. Kybernetik 8 234-239 (1971).

Katchalsky, A.K., Rowland, V. and Blumenthal, R.: Dynamic Patterns of Brain Cell Assemblies. Neuroscience Research Program Bulletin 12 (1974)

Kerner, E.H.: A statistical mechanics of interacting biological species. Bull. Math. Biophys. 19 121-146 (1957).

Kerner, E.H.: Further considerations on the statistical mechanics of biological associations. Bull. Math. Biophys. 21 217-255 (1959).

Kerner, E.H.: On the Volterra-Lotka principle. Bull. Math. Biophys. 23 141-157 (1961).

Kerner, E.H.: Gibbs Ensemble; Biological Ensemble. Gordon and Breach, New York (1971).

Kitagawa, T.: Dynamical systems and operators associated with a single neuronic equation. Math. Biosci. 18 191-244 (1973).

Leigh, E.G.: The ecological role of Volterra's equations. Lectures on Mathematics in the Life Sciences I 1-62 A.M.S. Providence, R.I. (1968).

Leung, K.V., Mangeron, D., Oguztoreli, M.N. and Stein, R.B.: On the stability and numerical solutions of two neural models. Utilitas Mathematica 5 167-212 (1974).

McCulloch, W.S. and Pitts, W.H.: A logical calculus of ideas immanent in nervous activity. Bull. Math. Biophys. 5 115-133 (1943).

Oguztoreli, M.N.: On the neural equations of Cowan and Stein. Utilitas Mathematica 2 305-317 (1972).

Poggio, G.F. and Viernstein, L.J.: Time series analysis of impulse sequences of thalamic somatic sensory neurons. J. Neurophysiology 27 517-545 (1964).

Rall,W.: A statistical theory of monosynaptic input-output relations. J. Cellular and Comparative Physiol. 46 373-411 (1955).

Rosenblueth, A., Wiener, N., Pitts, W. and Garcia-Ramos, J.: A statistical analysis of synaptic excitation. J. Cellular and Comparative Physiol 34 173-205 (1949).

Rotenberg, A.R.: Behaviour of homogenous statistical ensemble of finite automata.Automat. Telemekh. 9 84-92 (1971).

Rubio, J.E.: A mathematical model of the respiratory center. Bull. Math. Biophys. 29 719-736 (1967).

Rubio, J.E.: A new mathematical model of the respiratory center.Bull. Math. Biophys. 34 467-481 (1972).

Scholl, D.A.: Dendritic organization in the neurones of the visual and motor cortex of the cat. J. Anatomy (London) 87 387-406 (1953).

Scholl, D.A.: The Organization of the Cerebral Cortex. Methuen , London (1956).

Sherrington, C.S.: The Integrative Action of the Nervous System. Constable, London (1906).

Sherrington, C.S.: Man and his Nature. (1946).

Thom, R.: Topological models in biology. Towards a theoretical biology 3 drafts.p.89-116 . ed. C.H.Waddington. Edinburgh Univ. Press (1970).

Thom, R.: Stabilite Structurelle et Morphogenese. Benjamin, New York (1972). English translation by D.H.Fowler:Addison-Wesley, New York(1975)

Turing, A.M.: The chemical basis of morphogenesis. Phil. Trans. Roy. Soc. (London) B 237 37-72 (1952).

Uttley, A.M.: The probability of neural connections. Proc. Roy. Soc. (London) B 144 229-240 (1955).

Volterra, V.: Lecons sur la theorie mathematique de la lutte pour la vie. Gauthier-Villars, Paris (1931).

Waddington, C.H.: A catastrophe theory of evolution. Ann. N.Y. Acad. Sci. 231 32-42 (1974).

Wilson, H.R. and Cowan, J.D.: Excitatory and inhibitory interactions in localized populations of model neurones. Biophys. J.12 1-24 (1972)

Wilson, H.R. and Cowan, J.D.: A mathematical theory of the functional dynamics of cortical and thalamic nervous tissue. Kybernetik 13 55-80 (1973).

Winograd, S. and Cowan, J.D.: Reliable Computation in the Presence of Noise. M.I.T.Press, Cambridge, Mass. (1963).

Zeeman, E.C.: Differential equations for the heartbeat and nerve impulse. Towards a theoetical biology 4 essays 8-67. ed. C.H. Waddington. Edinburgh Univ. Press (1972).

Zeeman, E.C.: Primary and secondary waves in developmental biology. Some Mathematical Questions in Biology, A.A.A.S. Meeting, San Francisco, Feb. 1974. Lectures on Mathematics in the Life Sciences 7 A.M.A. Providence, Rhode Island (1976)

13. INFORMATION TRANSMISSION BY MODEL NEURONES.

In the models discussed in sections 1 - 12, the idea has been that the experimentally recorded electrical activity of neurones, whether fluctuations in membrane potential, spike discharges or some measure of the activity of neural aggregates, is stochastic, and the problem has been to obtain models which generate a stochastic process with properties similar to the observed samples of activity. These models may be abstract, or may contain parameters and variables which correspond to biophysically measurable quantities, and so the models might give insights into possible mechanisms generating the stochastic activity. In this chapter I will consider a different kind of problem: does the stochastic nature of neural activity have any functional implications?

The nervous system does not have a single, clearly defined function: rather it has a series of functions, some of which can be considered to be concerned with control and with communication. These are fuzzy concepts, and are difficult to quantitate: however, a simple function of some neurones and neural systems is the trans- mission of information. In particular, the regenerative, non-linear current voltage relation of axonal membrane can be considered as a specialization which permits the transmission of electrical signals, as a train of action potentials, over a long distance (see section 5.0). This suggests that it might be useful to use the methods of information theory to investigate the effect of neuronal variability on information transmission by neurones.

In Shannon's mathematical theory of communication (Shannon, 1948, 1949) a communication system is modelled by a series arrangement of:

    a) a source, which generates the information to be transmitted,
    b) a source encoder, and

c) a channel encoder, both of which transform the output of the source into a signal which can be transmitted through

d) the channel, which represents the physical mechanism of information transmission;

e) a channel decoder, and

f) a source decoder, both of which recover the signal from the output of the channel and present it to

g) the receiver.

This is an abstract sequence of operations, and in a man-made communication system it is easy to identify the separate steps with separate mechanisms. This is not so straightforward when a neurone is analyzed as a communication system, as the idea behind the encoder and decoder operations is that the signal is transmitted as a sequence of symbols, and the relationship between the signal and the symbols, which need not be a one-to-one relation, is the code. This code, or the properties of the encoder and decoder, needs to be known. In neurophysiology the code is not known, in fact there is not a code, but different neurones use different codes. A number of possible, candidate codes are available - these have been reviewed by Perkel and Bullock (1968). Part of the problem is that the set of symbols itself is not known, and so in order to apply information theory to analyse the properties of neurones one must choose a set of symbols and a code. The results of such an analysis may have little relevance to the biology of the neurone; however, if the source and receiver are related to the input and output of the neurone such an analysis does provide a method of input-output analysis based on conditional probabilities, and so is a useful alternative to the methods based on linear systems theory (see section 5.4).

If the source is generating as its output a random variable X which assumes values x_i, $i = 1, 2, \cdots$, the x_i form the symbols and the symbol x_i occurs with some probability, $p(x_i)$. A measure, in bits, of the information content of a symbol x_i is

$$I_i \equiv \log_2 \frac{1}{p(x_i)} \qquad (13.0.1)$$

and so, obviously, the more unlikely the occurrence of a symbol the higher its information content. The average of I_i taken over all the symbols is the uncertainty of the source, H(X), or the source entropy, given by:

$$H(X) = - \sum_{i=1}^{\infty} p(x_i) \log_2 p(x_i) \quad \text{bits/symbol} \quad (13.0.2)$$

Similarly, if the output of the channel to the receiver is a random variable Y, with values y_j, j = 1, 2, $\cdots$, there will be an uncertainty for the output distribution

$$H(Y) = - \sum_{j=1}^{\infty} p(y_j) \log_2 p(y_j) \quad (13.0.3)$$

The output to the receiver will depend on the input to the channel, and so there is a conditional uncertainty

$$H(Y|x_i) = - \sum_{j=1}^{\infty} p(y_j|x_i) \log_2 p(y_j|x_i) \quad (13.0.4)$$

and an average conditional uncertainty

$$H(Y|X) = \int_{min(x)}^{max(x)} p(x) H(Y|x) \, dx \quad (13.0.5)$$

The information transmitted, I, is the reduction in the uncertainty which results from the dependence of the output of the channel Y on the source X, and so is the difference

$$I \equiv H(Y) - H(Y|X) \quad (13.0.6)$$

$$= \sum_{i,j} p(y_j, x_i) \log_2 \frac{p(y_j, x_i)}{p(y_j) p(x_i)}$$

and the channel capacity, C, is the maximum of I over all possible distributions p(x) of the source

$$C \equiv \max_{p(x)} I \quad (13.0.7)$$

These definitions do not take account of any correlations between

adjacent symbols, which would tend to reduce the information transmitted. If

$$p(x_1, x_2, \cdots, x_n) \ne p(x_1)\, p(x_2) \cdots p(x_n)$$

the uncertainty of the source can be defined as the uncertainty of
a particular symbol given that all previous symbols have been observed:

$$H(X) = -\lim_{n \to \infty} \sum_{i_0, \ldots, i_n} p(x_{i_0}, \ldots, x_{i_n})\, \log_2 p(x_{i_0} | x_{i_1}, \ldots x_{i_n})$$

$$\le -\sum_i p(x_i)\, \log_2 p(x_i) \tag{13.0.8}$$

The equality sign of equation 13.0.8 holds if there is no correlation
between the symbols, and so any correlation among the signals reduces
the uncertainty. As it stands, equation 13.0.8 is not very useful,
as it requires knowledge of all (as $n \to \infty$) previous symbols: however,
in practice, any source will have a finite memory, or the statistical
dependencies will wear off in some time. This is the correlation
time, T_c, which can be estimated from the autocorrelation function of
the source: for times $> T_c$ the autocorrelation function will be flat.

Correlation between the source symbols means that the output
symbols will be correlated with more than one input symbol and with
previous output symbols. The transinformation will be reduced, and in
place of equation 13.0.6 the transinformation will be given by:

$$I = \lim_{n \to \infty} H(Y_0 | Y_1, \ldots, Y_n) -$$

$$\lim_{n \to \infty} H(Y_0 | Y_1, \ldots, Y_n, X_0, \ldots, X_n)$$

$$= \lim_{n \to \infty} \sum_{\substack{j_0, \ldots, j_n \\ i_0, \ldots, i_n}} p(y_{j_0}, \ldots, y_{j_n}, x_{i_0}, \ldots, x_{i_n}) \, . \tag{13.0.9}$$

$$. \log_2 \frac{p(y_{j_0} | y_{j_1}, \ldots, y_{j_n}, x_{i_0}, \ldots, x_{i_n})}{p(y_{j_0} | y_{j_1}, \ldots, y_{j_n})}$$

The problem is to apply these equations to obtain the information transmission rate T' of neurones and model neurones: not only is there a choice of model and code, but also a choice of symbols.

13.1 The axon as a channel

The output of most long-axoned neurones is a sequence of action potentials which propagate along the axon to the axon terminals, and so the axon could be considered to physically represent a communication channel through which the action potentials are transmitted. If the axonal membrane had a linear current-voltage relation, any steady electrical signals would attenuate with distance, and transients would decay more rapidly with distance as the current would be shunted through the membrane capacity. The regenerative membrane current-voltage relation permits the propagation, without decrement, and at a constant conduction velocity, of action potentials. Thus information is transmitted as a series of all/none events, and since the equations describing the propagation of action potentials (the Hodgkin-Huxley and cable equations) are deterministic, the axon could be thought of as a noise-free channel. If information is to be transmitted through such a noise-free channel as a series of all/none events, the only restriction on the channel capacity would be the number of events which could be transmitted/unit time. Thus the channel capacity would be limited by the absolute refractory period and the channel is using a binary pulse code, where the symbols are the occurrence or non-occurrence of an action potential in a discrete time interval. A neurone can have an interspike interval as short as one msec.; even though it is unlikely that a neurone would discharge at 1000 impulses/sec. such a maximal discharge rate with a binary pulse code would give a capacity of 1000 bits/ sec. If the interspike intervals were continuously graded even higher capacities could be found using an interval code: up to 4000 bits/sec. (Mackay and McCulloch, 1952; Rapoport and Horvarth, 1960; Färber, 1968).

These very high estimates of the channel capacity do not appear to be all that meaningful: perhaps one should think of the all/none nature of the action potential in terms of reliable propagation of the action potential rather than imposing an upper limit on the

channel capacity. Further, neurones do not appear to use either a
binary pulse code or an interval code, and these estimates do not
consider the stochastic nature of neural activity.

Although the Hodgkin-Huxley equations are deterministic, they
can be interpreted probabilistically, (see section 4.3) and so the
Hodgkin-Huxley axon is not a noise-free channel. Further the
fluctuations in membrane potential considered in section 1 will
also act as an intrinsic noise source. Lass and Abeles (1975) have
investigated the characteristics of an axon as a channel by taking
the interval between twin stimuli applied at one end of the axon as
the input and the interval between the two responses recorded at
the other end of the axon as the output. The histogram of the output intervals had a standard deviation of about 3-5 μsec and was
Gaussian: this suggests treating the axon as continuous constant
channel with an additive Gaussian noise. Information transmission
in such a channel is optimal when the input intervals are Poisson
distributed (Abeles and Lass, 1975). The channel capacity determined
in this way is limited by the refractory period and the noise: since
the intrinsic fluctuations in membrane potential considered in
section 1 are physiologically almost negligible, and the resultant
fluctuations in output interval of only a few μsec, the channel
capacity of several hundred bits/sec is not very meaningful. Perhaps
it is a mistake to consider the axon as a channel, and an action
potential, or the interval between two action potentials, as a
signal. Neurones transmit information by changes in their discharge
of action potentials - perhaps spike trains form the channel and
some change in the statistical characteristics of a spike train
forms the message.

13.2. Information preservation in a neural chain.

If information is being transmitted as a series of stereotyped,
all/none action potentials, where the action potentials are only
distinguishable by their times of occurrence, the information content
of the signal will depend only on the patterns of times of occurrence
of the action potentials. Thus, as far as information transmission
is concerned, the spike train is equivalent to a (stochastic) point
process. This simplification ignores the information given by which

cell is active: in fact the identity of the active cell may be more
significant than the details of its activity. If the statistical
properties of the spike train do not change with time the discharge
can be represented by a realization of a stationary stochastic point
process, and if the intervals between successive action potentials are
independent, by a stationary renewal process (see section 5).

The statistical properties of such a stationary, renewal point
process can be completely characterized by the interval probability
density function, $f(t)dt$, which contains the same amount of information
as the original process. $f(t)dt$ is the probability that the intervals
between successive action potentials lie between t and (t + dt), as
dt → 0, and can be estimated from the interspike interval histogram.

A spike train in an axon is of no direct biological significance:
it is effective only when it produces some change in the activity of
a post-synaptic cell or effector. Thus the axonal spike train
represents the signal in the channel, and information will only be
successfully transmitted when the signal reaches the receiver. As a
linear approximation, the subthreshold post-synaptic response of the
neurone receiving the spike train as an input will be the convolution
of the input train $x(t) = \Sigma \delta(\tau - \tau_i)$ with the impulse response of
the neurone. If the post-synaptic cell is modelled by a perfect
integrator with constant threshold and instantaneous reset (see sections
3.2 and 7.6) and if the post-synaptic response to a single input has
an amplitude h, the membrane potential of the postsynaptic neurone
will be given by:

$$V(t) = h \int_0^t \Sigma_i \delta(\tau - \tau_i) \tag{13.2.1}$$

for $V(t) < V_o$, the threshold. For excitatory inputs, when $V(t)$ first
reaches V_o at a time t' a post-synaptic output will be generated, and
the membrane potential instantaneously reset to its resting value.

The output intervals $(t'_n - t'_{n-1})$ are random variables given by
the sum of those input intervals $(\tau_i - \tau_{i-1})$ where $t'_n = \tau_i$ and
$t'_{n-1} = \tau_{i-k}$ and where k is the integral part of $(1 + V_o/h)$. Thus the
probability density function of the intervals $(t'_n - t'_{n-1})$ will be the
convolution of the probability density functions of the intervals
$(\tau_i - \tau_{i-1})$. Since the input is stationary and renewal the probability
density of the output train will be given by the k-fold convolution
of the input density $f(t)$ with itself.

In any excitatory, feed-forward pathway in the nervous system information will be transmitted through a cascade of neurones, $(j-1)$, j, $(j+1),\ldots,$. If each neurone is modelled by a perfect integrator with instantaneous reset the interval density function $f_j(t)$ of the output of the jth neurone, which forms the input to the $(j+1)^{th}$ neurone, will be the k-fold self-convolution of the interval density of its input, $f_{j-1}(t)$. This assumes that the spike trains are stationary and renewal, and that all the neurones in the chain have the same threshold k, and ignores the effects of convergence. Since the information content of the stationary, renewal spike train is equivalent to the information content of its interval density function, this information will be preserved as the signal passes up the neural chain only if $f_j(t)$ has a form which is invariant (apart from a linear scale change) under the operation of self-convolution. A probability density function which has this property of invariance of form under self-convolution is said to obey a stable law. This suggests as a hypothesis (Holden, 1975,1976) that if action potential trains in a cascade of sensory neurones have interval densities which are stable, then the sensory pathway is preserving information, and the converse, that if the action potential trains do not have stable interval distributions information is not being preserved, and so the pathway is processing rather than transmitting information. It may not be very sensible to estimate the information transmission rates of pathways which are processing information.

The properties of the stable distributions have been extensively investigated (Levy, 1940; Gnedenko and Kolmogorov, 1968). The interest in the stable distributions stems from the fact that the sum of n independent random variables with probability density functions $f_1(t)$, $f_2(t),\ldots,f_n(t)$ has a density function $f_s(t)$ given by the n-fold convolution

$$f_s(t) \quad = \quad f_1(t)*f_2(t)*\ldots*f_n(t) \tag{13.2.2}$$

where $*$ denotes the convolution operation. Khintchine and Levy (1936) have given the theorem:

A necessary and sufficient condition for a distribution function $F(x)$ to be stable is that the logarithm of its characteristic function $Q(t)$ can be represented by

$$\log Q(t) \quad = \quad i\gamma t - c|t|^\alpha \{1 + i\beta \frac{t}{|t|} w(t,\alpha)\} \tag{13.2.3}$$

where $0 < \alpha \leq 2$

                        $-1 \leq \beta \leq 1$

                        $c \geq 0$

                        γ is any real number

and

$$w(t,\alpha) \;=\; \tan(\pi/2)\alpha \quad \text{if } \alpha \neq 1$$
$$w(t,\alpha) \;=\; (2/\pi)\log|t| \quad \text{if } \alpha = 1$$

and the characteristic function $Q(t)$ is the Fourier transform of the distribution function $F(x)$.

There are only three stable distributions which are available in closed form; these are the Gaussian distribution ($\alpha = 2$), the Cauchy distribution ($\alpha = 1$) and the stable distribution of order one half ($\alpha = 1/2$). Of these distributions, the Cauchy distribution cannot be considered a suitable candidate for an interspike interval distribution as it exists for negative values of time and the interspike intervals are non-negative by definition. Thus there are only two stable distributions available in closed form which may be considered as possible candidates for interspike intervals in a pathway which is preserving information: these are the Gaussian distribution with a density

$$f(x) \;=\; \frac{1}{\sqrt{2\pi}} \; \exp\{-(1/2)x^2\}$$

and the stable distribution of order one half, with a density

$$f(x) \;=\; \frac{1}{\sqrt{2\pi}} \; \exp\{-1/(2x)\}\, x^{-3/2} \quad \text{for } x > 0$$

$$0 \quad \text{for } x \leq 0$$

which is tabulated in Holt and Crow (1973) and occurs as the interspike interval density in random walk and diffusion neural models (sections 6 and 7). Since

$$\Gamma(x) \;=\; (x-1)! \quad \text{for } x \text{ integeral}$$
$$= \int_0^\infty t^{x-1}\exp(-x)\,dt \;\simeq\; \sqrt{2\pi}\,\exp(-x)\,x^{x-1/2}$$

the stable distribution of order one half contains the gamma distribution defined by equation 5.2.6. Thus, for the interval density of an action potential train to be stable and available in closed form, it must be unimodal and either symmetric and Gaussian or positively

skewed and approximately a gamma density. It is interesting to note
that the interspike interval histograms of a large number of first,
second and third order sensory neurones have been fitted by a Gaussian
or gamma density function. This would suggest that information is being
preserved in these pathways. Further, the interval histograms of
more central sensory neurones, believed to be involved in processing
information by acting as feature detectors, are multimodal and are
obviously not stable. Thus perhaps one should estimate information transmission rates only in neurones with stable interval distributions.
However, only Gerstein and Mandlebrot (1964) have explicitly commented
on the stability of interval distributions, and one can only assume that
spike trains which have been fitted by Gaussian or gamma densities
are in fact stable.

13.3. Signal detection.

One simple problem in modelling the transmission of information
by neurones is to determine whether or not there is a signal. This
corresponds to the experimental situation of presenting a controlled
stimulus and attempting to decide whether or not there is a response:
this might be to estimate a threshold, map a pathway or determine
a receptive field. The widespread use of post-stimulus time
histograms emphasizes that it is not necessarily easy to detect a
stimulus produced change in neuronal activity.

Stein (1967) has considered a binary channel, for which there
are two possible responses y_1 (no response) and y_2 (one or more action
potentials) which may be produced by two possible input signals x_1
and x_2, which have probabilities a and b of producing a response,
$b > a$. Thus if $a = 0$ and $b = 1$, the transinformation/trial is one
bit; for other conditional probabilities a and b the transinformation
is obtained as:

$$I = (-bH_a + aH_b)/(b - a) + \log_2 \{ 1 + 2^{(H_a-H_b)/(b-a)} \}$$

$$(13.3.1)$$

where $H_a = H(Y| x_1) = -(1 - a)\log_2(1 - b) - a.\log_2 a$

$H_b = H(Y| x_2) = -(1 - b)\log_2(1 - b) - b.\log_2 b$

and so when there is no spontaneous activity ($H_a = 0$)

$$I = \log_2 \{ 1 + b(1 - b)^{(1-b)/b} \}$$

A common experimental situation is to determine the threshold; in this case $b = \frac{1}{2}$ and for a binary channel with no spontaneous activity the traninformation is 0.32 bits/trial. This analysis of the binary channel serves to emphasize the dependence of the transinformation on the choice of the input signals. If the function of a neurone was purely to transmit information, the stimulus distribution which maximized the transinformation might correspond to the 'natural' stimulus distribution.

Fitzhugh (1957) has analyzed the responses of retinal ganglion cells to brief flashes of light using a similar approach. For a particular flash intensity, the two possible inputs are either a flash or no flash: these are equiprobable. The response index was the number of impulses occuring in a specified time after the input. Since the ganglion cells are spontaneously active there can be a response even when the input is no flash. The transinformation was computed using 13.0.6 where the conditional probability of obtaining a response index of y given a flash intensity x was estimated from the histograms of the response index . The trans- information increased with increasing flash intensity, approaching the capacity of one bit/trial at intensities 3-4 times greater than threshold.

13.4 Transmission of steady signals using a rate code.

Adrian (1928) demonstrated that the number of action potentials occuring in a period, or the average discharge rate, is simply related to the stimulus intensity in sensory neurones - this generaliz- ation followed from Adrian and Zotterman's 1926 experiments on frog muscle receptors. Such a rate code is assumed to be used by many neurones, particularly in the peripheral nervous system, and this is a very attractive assumption as it provides a simple measure of action potential activity as the rate in impulses/sec. This rate code is often called a frequency code, but this introduces confusion when periodic signals are considered. A simple model neurone which has a rate code is the perfect integrator with instantaneous reset, where

the discharge rate is proportional to the stimulus intensity when
steady inputs are used. Such a rate code need not imply that there
is a continuous range of biologically significant signals - Bittner
(1968) has shown that the discharge rate of excitatory motoneurones
to the claw opener muscle of the crayfish acts as the control signal
to a switch: the motoneurones innervate a number of muscle fibres,
and the different neuromuscular junctions are effective at different
discharge rates. Thus a change in the discharge rate switches the
response between different groups of muscle fibres.

Expressions for the information capacity of a regular neurone
using a rate code have been obtained by Grusser (1962), corrected by
Barlow (1963) and extended by Stein (1967). Grusser obtains the
capacity of a regular neurone using a rate code as

$$C = \log_2 (r_{max} t) \text{ bits} \tag{13.4.1}$$

where r_{max} is the maximum discharge rate and t is the period for which
the signal is applied, which Grusser considers as the integration
time of the neurone, and so the neurone is looking at the input
signal through a rectangular weighting window. Barlow notes that
13.4.1. gives a capacity of zero when $t = 1/r_{max}$, which is not
appropriate. Essentially, a discharge rate of zero is as just as much
a signal as any other discharge rate, and so the corrected expression
is

$$C = \log_2 (r_{max} t + 1) \text{ bits} \tag{13.4.2}$$

Stein obtains the capacity as lying between the limits:

$$\log_2 (Rt + d) \leq C \leq \log_2 (Rt + 1) \tag{13.4.3}$$

where $R = r_{max} - r_{min}$ is the range of possible discharge rates, and
$d < 1$ is a constant whose value depends on whether or not the neurone
is spontaneously active. For a given range of discharge rates the
capacity can be calculated precisely. For small duration signals the
number of possible numbers of action potentials in the response will
be a (small) integer n, and so an n x n stimulus-response matrix
can be constructed where there are distinct subsets of stimuli which
cannot be confused. For such a matrix the capacity can be calculated
precisely from the capacities of the distinct subsets. For long

duration signals the capacity can be calculated accurately from
equation 13.4.3 as the limits converge as $1/t$. The slope of the
capacity - $\log_2$ stimulus duration curves increases to one and then
slowly decreases to its asymptotic value of $\frac{1}{2}$.

The transinformation will approach the capacity only if the input
distribution is optimal. Since the channel matrix is square the
optimal input distribution for a given signal duration can be calculated.
The optimal input distributions for different signal durations for the
completely regular neurone are discrete, with equiprobable, equispaced
stimuli. These are peculiar distributions and are not physiological:
a more plausible input distribution would be a rectangular, continuous
distribution, which would give a transinformation about a bit less
than the capacity.

The completely regular neurone represents one extreme case: the
other extreme is a random neurone, where the discharge has a Poisson
distribution. For such a random neurone the capacity is always less
than the capacity of the regular neurone with the same parameters.
The capacities of these two extreme cases diverge with increasing
signal durations. For large signal durations the capacity - $\log_2$
stimulus duration curve of the random neurone approaches an asymptotic
slope of $\frac{1}{2}$.

The behaviour of the stochastic spike trains generated by the models
of chapters 6 - 10 will be between these two extremes of the regular
and random discharge. Stein approaches the problem of information
transmission by stochastic spike trains by obtaining asymptotic
formulae for the capacities as a function of signal duration for
long duration signals. These formulae assume that the discharge is
stationary and renewal - there are no significant correlations between
adjacent interspike intervals. Further, it is necessary to assume that
the variability in the numbers of impulses occuring in a fixed period
is large enough to have a Gaussian distribution, but much less than the
total number of impulses occuring in that period, so that the response
distribution is close to the stimulus distribution. These last two
assumptions will be exact as $t \to \infty$. For a stationary renewal process,
with a mean interval μ and variance σ^2, the number of events occuring
in $(0,t]$ has an asymptotically Gaussian distribution as $t \to \infty$, with
a mean

$$x \simeq t/\mu \qquad\qquad\qquad (13.4.4)$$

and a variance

$$s^2 \approx \sigma^2 t/\mu^3 \qquad (13.4.5)$$

The uncertainty H of a Gaussian distribution depends only on its standard deviation s (Shannon, 1948):

$$H = \log_2\{ \sqrt{(2\pi e)} s\} \qquad (13.4.6)$$

Stein obtains the optimal input distribution p(x) as

$$p(x) = \frac{1}{s(x) \int_{x_{min}}^{x_{max}} \frac{dx}{s(x)}} \qquad (13.4.7)$$

and so the probability density for the occurence of a particular stimulus varies inversely with s(x), the standard deviation in the number of impulses, which may vary with the mean number of impulses. His derivation shows that for a finite stimulus duration, the optimal stimulus distribution is a discrete distribution of equiprobable stimuli with a spacing proportional to s(x). As t → ∞ the spacing becomes negligible and the distribution continuous, with the density inversely proportional to s(x).

To obtain an asymptotic formula for the capacity:

$$H(Y) \approx H(X) = -\int p(x) \log_2 p(x)\, dx \qquad (13.4.8)$$

$$H(Y|X) \equiv \int p(x) H(Y|x)\, dx \approx \int p(x) \log_2\{ \sqrt{(2\pi e)} s(x)\}\, dx \qquad (13.4.9)$$

$$I \approx -\int p(x) \log_2 \{ \sqrt{(2\pi e)} p(x) s(x)\}\, dx \qquad (13.4.10)$$

and so using 13.4.7 to obtain the maximum of 13.4.10:

$$C \approx \log_2 \{ \int_{x_{min}}^{x_{max}} \frac{dx}{\sqrt{(2\pi e)}\ s(x)} \} \qquad (13.4.11)$$

From 13.4.4 and 13.4.5 the capacity can be written in terms of the maximum μ_1, minimum μ_0 and mean μ intervals and the standard deviation σ of the interval distribution of the renewal process:

$$C \simeq \log_2 \left\{ \sqrt{(t/2\pi e)} \int_{\mu_o}^{\mu_1} \frac{d\mu}{\sqrt{\mu\sigma}} \right\} \quad \text{bits} \qquad (13.4.12)$$

Thus if a sensory spike train is treated as a realization of a stationary renewal process the information capacity for steady signals can be directly estimated from the statistical properties of the spike train by using 13.4.12.

However, spike trains often show significant correlations between adjacent intervals, and so 13.4.12 would overestimate the capacity. If the serial correlation coefficients decrease rapidly with increasing order, which will be true for virtually any spike train, the corrected expression for the capacity is:

$$C \simeq \log_2 \left\{ \sqrt{(t/2\pi e)} \int_{\mu_o}^{\mu_1} \frac{d\mu}{\sqrt{(\mu a)}\sigma} \right\} \qquad (13.4.13)$$

where $\quad a = 1 + 2 \sum_{i=1}^{\infty} \rho_i$

and ρ_i is the serial correlation coefficient estimated by equation 5.2.18.

These expressions have been used to estimate the information capacity of muscle spindle afferents by Matthews and Stein (1969) and their second-order sensory neurones, the dorsal spino-cerebellar tract neurones, by Walloe (1968, 1970). Walloe also used two other methods for estimating the capacity of the DSCT neurones - a method based on the convergence of the mean firing rate, and a method using a stimulus-response matrix.

The method based on the convergence of the mean firing rate obtains the number of distinguishable lengths (and hence the capacity) as a function of observation time. For a stationary sequence of intervals t_i an estimator f_n of the mean discharge rate f_∞ is given by

$$f_n = n / \sum_{i=1}^{n} t_i \qquad (13.4.14)$$

which will converge to f_∞ as $n \to \infty$. Experimentally, it was found that f_n converged faster when the intervals were in their original order than when they were shuffled: this supports the idea that the convergence of the mean rate provides a biologically plausible index of the response. Treating the output spike trains in blocks of 150 intervals, the rate of convergence was measured by the smallest number N_{min} such that:

$$|f_n - f_{150}| < \Delta f \quad \text{for all } n \geq N \tag{13.4.15}$$

and so for any positive Δf a single value N_{min} can be calculated: the smaller the value of N_{min}, the faster the convergence. A time for convergence T, or an observation time, is then defined by

$$T = \sum_{i=1}^{N_{min}} t_i \tag{13.4.16}$$

The mean observation time $\bar{T}$ was found to be independent of the firing rate and hence the level of stretch applied to the muscle. This suggests that this mean time for convergence might be identified with Grusser's (1962) integration time. A plot of $\bar{T}$ against $1/\Delta f$ was linear

$$\bar{T} = a/\Delta f \tag{13.4.17}$$

and the parameter a varied between different cells and was larger when the intervals were read in the correct rather than shuffled order. If R is the range of discharge rates between zero and full stretch of the muscle, the number d of discriminable lengths is

$$d = R/2\Delta f \tag{13.4.18}$$

For a given observation time, Δf can be calculated using the experimentally determined value of a and 13.4.17. The transinformation is then $\log_2 d$. If all muscle lengths are equiprobable, the capacity is

$$C = \log_2 d = \log_2 R\bar{T}/2a \tag{13.4.19}$$

where R and a are experimentally determined constants for a given cell.

The second method used by Walloe was to set up a stimulus-response matrix. A not too small number of equiprobable muscle lengths was taken to be the the input signals: these should be equidistant and span the physiological length range. Any one trial will give a response with a mean discharge rate measured over some observation time. The mean output rates obtained on different trials were divided into m equispaced groups of discharge rate. From this stimulus-response array an n x m matrix can be constructed with an element r_{jk} giving the probability that a discharge rate falling in the jth discharge rate group was produced by a muscle length falling in the kth length group, j=1,2,..,m, k=1,2,..,n. For the matrix

$$\begin{bmatrix} r_{11} & r_{12} & \cdot & \cdot & \cdot & r_{1m} \\ r_{21} & r_{22} & \cdot & \cdot & \cdot & r_{2m} \\ \cdot & \cdot & \cdot & r_{jk} & \cdot & \cdot \\ \cdot & \cdot & \cdot & \cdot & \cdot & \cdot \\ \cdot & \cdot & \cdot & \cdot & \cdot & \cdot \\ r_{n1} & r_{n2} & \cdot & \cdot & \cdot & r_{nm} \end{bmatrix}$$

since the elements r_{jk} are probabilities

$$\sum_{j=1}^{n} \sum_{k=1}^{m} r_{jk} = 1 \qquad (13.4.20)$$

and since the input lengths are equiprobable

$$\sum_{k=1}^{m} r_{jk} = p_j = 1/n \qquad (13.4.21)$$

and the other marginal probability is

$$\sum_{j=1}^{n} r_{jk} = q_k \qquad (13.4.22)$$

Thus the entropies are:

$$H(X) = -\sum_{j=1}^{n} p_j \log_2 p_j \qquad (13.4.23)$$

$$H(Y) = -\sum_{k=1}^{m} q_k \log_2 q_k \qquad (13.4.24)$$

$$H(X,Y) = -\sum_{j=1}^{n} \sum_{k=1}^{m} r_{jk} \log_2 r_{jk} \qquad (13.4.25)$$

and the transinformation is

$$I = H(X) + H(Y) - H(X,Y) \qquad (13.4.26)$$

Application of these three methods - equations 13.4.13, 13.4.19 and
13.4.26 - gave similar results for the DSCT neurones. For short
observation times, less than 30msecs., the transinformation was very
small. With longer observation times the transinformation increases
steeply to about 2 bits at about 150 msec., with a maximum transinformation rate at about 100 msec. of 20 bits/sec. With observation
times of about 1 sec. the transinformation rate is about 4 bits/sec.
Thus an optimal integration time is about 100 msec.

The DSCT neurone has inputs from about 15 primary muscle spindle
afferents, and so if the activities of the afferents were mutually
independent the maximal information input to the DSCT neurone would
be 15 times the capacity of an afferent. Matthews and Stein (1969)
estimate that a single, de-efferented primary muscle spindle afferent
can transmit about 4.8 bits in a second, and so 15 could transmit
about 6.6 bits in 200 msec. while each DSCT neurone is transmitting
about 3 bits. Thus , given that the afferents excite a number of DSCT
neurones, a surprising high proportion of the information is transmitted
across the primary muscle spindle-DSCT synapse. The system is
preserving information, and even though there are appreciable serial
dependencies between the intervals the interval histograms of both
primary muscle spindle afferents and DSCT cells are unimodal and appear
as if they might be stable under self-convolution when the muscle is
held at a constant length.

This analysis of information tranmission using a rate code assumes
that the nervous system is interested in the levels of steady signals,
and Walloe avoided the effects of adaptation by ignoring the first
½ second of the response. Werner and Mountcastle (1963, 1965) and
Mountcastle, Poggio and Werner (1962) have adopted a similar kind of
approach to estimate transinformations using an interval code in the
somatic afferent system.It seems likely that the nervous system might
be interested in changing signals, and so it might be more sensible
to estimate the capacity when the signals are varying with time.

13.5. Information transmission of periodic signals.

The response of the perfect integrator model to a sinusoidally
modulated input is a pulse train whose density is sinusoidally
modulated with no change of phase and a frequency-independent gain:
see section 3.2. If the density is taken as the output, there is no
distortion of the signal; however, the spectral density will have

components at the carrier frequency, integer multiples of the carrier
frequency, a series of sidebands at integer multiples of the modulating
frequency above and below each carrier band, as well as at the modulating frequency (see equation 5.3.16). The pulse density of the
sinusoidally modulated spike train only has a steady level and a
component at the modulating frequency, as other frequency components
do not have a constant phase relation to the input cycle and are
averaged out. The response of sensory neurones to a cyclic input is
often obtained as a cycle histogram, which estimates the density of
action potentials as a function of phase angle of the cyclic input,
and so the smooth, sinusoidal cycle histograms or pulse densities
obtained when a perfect integrator is driven by a sinusoidal input
suggest that the sinusoidal signal is being transmitted without
distortion (Stein, French and Holden, 1972). However, the modulation
of the intervals between pulses, or the instantaneous rate, will
show distortions in both gain and phase (Partridge, 1966; McKean,Poppele
Rosenthal and Terzuolo, 1970; Holden, 1976), and when the modulation
depth of the input is greater than one there will be distortions
similar to rectification. These distortions appear as the impulse
density by definition cannot become negative, and as the input modulation depth increases these distortions grow, giving increasing
harmonics of the input frequency. Cycle histograms showing such
rectification have been observed in a variety of sensory neurones,
and are to be expected in rapidly adapting neurones or neurones
which do not have a background discharge rate. This rectification
can be considered as a distortion of the sinusoidal signal.

However, for modulation depths less than one the impulse density
averaged over a number of cycles does not show any distortion. This
idea of distortion-free transmission is reinforced when the sinusoidally
modulated output of a perfect integrator model is the input signal to
a second perfect integrator model: if

$$h_1(\phi) = a.\exp(j\phi) + c$$

is the pulse density of the output of the first perfect integrator,
the pulse density of the second is simply

$$h_2(\phi) = \{ a.\exp(j\phi) + c \}/r$$

when a/c is the modulation depth and r is the number of impulses

required to reach threshold. Thus the frequency response function is
simply a frequency independent constant, and there are no changes in
phase.

The perfect integrator model is a rather artificial model: a
more plausible model is the leaky integrator discussed in section 3.3.
The response of the leaky integrator model to a sinusoidally modulated
input is a series of pulses occurring at phase angles fixed with
respect to the driving cycle: Rescigno, Stein, Purple, Poppele, 1970;
Knight, 1972; Stein, French, Holden 1972; Ascoli, Barbi, Chillemi and
Petracchi, 1976). Thus the cycle histogram is not a smooth sinusoid,
but a series of discrete peaks: these represent distortions of the
sinusoidal signal (French and Stein, 1970). This phase-locked
behaviour is found in a variety of sensory neurones which have fairly
regular discharges, and has been obtained in numerical solutions of
the Hodgkin-Huxley membrane equations (Holden, 1976).

In both these cases the input and output are deterministic: the
output spike trains can be made to be stochastic by introducing an
internal noise source, which acts as an auxillary signal adding to
the sinusoidal input signal. The introduction of such an internal
noise source tends to reduce the distortions in the pulse density
produced by rectification (Spekreijse, 1969; Spekreijse and Oostings,
1970; French, Holden and Stein, 1972) and the distortions due to phase
locking (Stein and French, 1970; Stein, 1970; Stein, French and Holden,
1972)

Thus, if the impulse density averaged over a number of cycles is
taken to be the output of the neurone, variability produced by an
auxillary or internal noise source appears to enhance the ability of
neurones to transmit information about periodic signals for a range
of amplitudes and frequencies where rectification and phase-locking
would occur. This is a curious idea, as variability or noise generally
impairs the ability of a system to transmit information. However, the
arguement suggests that cells responding to rapidly changing signals
would require a higher degree of variability than cells responding
to static or slowly changing signals. There are a number of experimental
examples which support this idea - primary muscle spindle afferents
have a more variable discharge than secondary muscle spindle afferents,
cochlear neurones are extremely irregular while slowly adapting skin
receptor neurones are very regular.

If the spectral density of the output trains are considered, the
responses of both the perfect and leaky integrator models to a cyclic

input have components at frequencies other than the modulation frequency: these other components may be considered to represent distortions of the signal. If a synapse is taken to act as a low pass filter (see section 11.6), the synapse will filter out the higher frequencies components of the spectral density, and so the sinusoidal signal could be recovered even from a phase locked spike train. The addition of noise as an auxillary signal, although reducing the distortions of the pulse density, would give a continuous output spectral density and so would leave some noise power in the filtered response. Thus this frequency domain arguement gives the opposite conclusion - when the response is viewed through a low pass synaptic filter noise cannot enhance the ability of the leaky integrator model to transmit information about periodic signals.

This paradox is somewhat artificial, as to obtain the impulse density one has to average over a number of cycles. Such temporal averaging cannot occur in the nervous system, which operates in real time. However, averaging over a spatially distributed population of similar neurones can occur in the nervous system, and so if a population of cells represents an ergodic process, the behaviour of a population average would be equivalent to the behaviour of a temporal average. This would require the activities of the cells in the population to be uncorrelated and the cell parameters the same, and so at best is only approximately true. Knight (1972) has investigated the behaviour of populations of perfect and leaky integrators, and has shown that the presense of internal noise improves the behaviour of a heterogenous population of leaky integrator models, in that the population firing rate can be made to follow the stimulus waveform. The source of this noise was assumed to be additive to the input: Gestri (1971, see section 4.1) considers random fluctuations in threshold as a possible noise source. Poggio and Torre (1975) point out that an internal noise source is not additive when nonlinear neural models are considered. Milgram and Inbar (1976) apply these ideas to a model of the stretch reflex, and show that fluctuating thresholds appear to be the most effective source for noise to reduce distortions.

Thus neuronal variability appears to improve the ability of populations of neurones to follow periodic signals, and can provide a carrier rate which prevents the distortions produced by rectification. However, periodic signals form an unlikely input to the nervous system, and so these apparently advantageous effects of neuronal variability may have no biological significance.

13.6. Transmission of stochastic, time-varying signals.

The response of the perfect and leaky integrator models to a
white noise input has been derived in sections 7.6 and 7.7. Such a
white noise input is completely random and contains power distributed
equally at all frequencies. The natural input to a receptor will be
some stochastic process, and so rather than using the deterministic
steady and periodic input signals of sections 13.4 and 13.5 it might be
appropriate to consider the ability of a neurone to transmit information
when the input signal is a sample of a stochastic process, with a
continuous spectral density. White noise provides a mathematically
convenient stochastic process; even though it might be thought to
arise as the limiting process obtained by the superposition of a large
number of excitatory and inhibitory pulse trains (see section 7.1),
a delta-correlated white noise is just as artificial an input as an
infinitely long sinusoidal process, as the input to any real sensory
neurone will be restricted to a band of frequencies and will have
some spectral shape.

When a neurone is subject to a realization of a band-limited
white noise process as its input signal, it is possible to obtain
estimates of the input, output and cross-spectra defined in section 5.4.
Computational methods for obtaining these spectral estimates, when the
input is a continuous stochastic signal and the output is a spike train
treated as a stochastic point process, are treated in French and
Holden (1971a,b,c). From these estimates an estimate of the coherence
function defined by 5.4.21 can be obtained:

$$\gamma^2(f) \quad = \quad \frac{|S_{xy}(f)|^2}{S_{xx}(f) \; S_{yy}(f)}$$

and the coherence function of diffusion process model neurones can be found
analytically (section 7.6) or numerically (section 7.7). The coherence
function $\gamma^2(f)$ is a normalized measure of the linearity of the relation
between the input and the output: it is equal to one for a linear,
noise-free relation and can be reduced by nonlinearities or when part of
the output is produced by a signal, which may be stochastic, which is
not related to the input. The coherence function represents that
proportion of the output spectral density which can be accounted for by

linear regression on the input spectral density. Such a measure of correlation can be related to a measure of the information capacity (Stein and French, 1970) as if there is a continuous signal $x(t)$ additively contaminated by a noise signal $n(t)$, with spectral densities $X(f)$ and $N(f)$, the capacity is related to the signal-to-noise ratio by (Shannon, 1948):

$$C = \int_o^W \log_2 \{ 1 + \frac{X(f)}{N(f)} \} \ df \qquad (13.6.1)$$

where the signals and noise have a bandwidth w, and the coherence is given by

$$\gamma^2(f) = \{ 1 + \frac{N(f)}{X(f)} \}^{-1} \qquad (13.6.2)$$

and so the capacity may be estimated by:

$$C \approx - \int_o^W \log_2 \{ 1 - \gamma^2(f) \} \ df \qquad (13.6.3)$$

This estimate of the capacity assumes that it is the linear correlation between the input and output which is transmitting information; any nonlinear relations are treated as noise.

The coherence functions computed for a perfect integrator model when the input is a Poisson distributed train of impulses and a white noise are shown in Figures 3.1. and 7.1. The striking characteristic of these curves is the very low value of the coherence function for frequencies above the carrier rate. Such low values of the coherence found even in the absence of an intrinsic noise source suggest that the linear frequency response function is a poor description of the input-output relation, and that higher terms of a Volterra series representation of the input-output relation will contibute appreciably to the output. Such a Volterra series representation has been obtained by Poggio and Torre (1975) for the perfect and leaky integrator models. Similarly low values of the coherence function have been found for an insect mechanoreceptor (French, Holden and Stein 1972).

The effect of neuronal variability produced by an internal noise source on information transmission of stochastic signals can be found adding a second, independent noise signal to the input. The effect of this internal noise is to reduce the coherence at all frequencies, except close to the carrier rate: thus the capacity is

reduced as the small increase in coherence for frequencies close to
the carrier rate will not be balanced by the decrease in coherence for
frequencies above and below the carrier rate : see Figure 13.1.

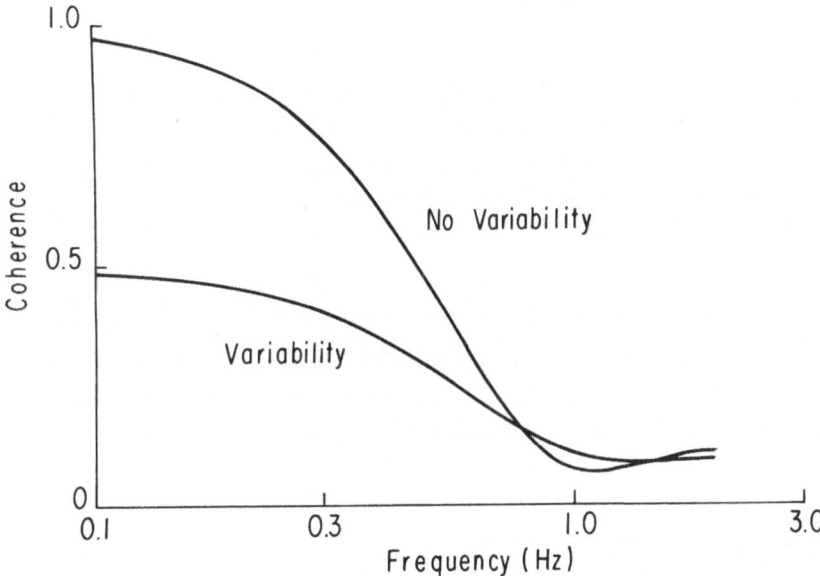

Figure 13.1. The decrease in the coherence function of a perfect integrator model
produced when n = a (variability) and n = 0 (no variability). a^2 = 0.1, c = 1.

For the perfect integrator model with an intrinsic noise source
subject to a white noise input, if a and n are the root mean square
fluctuations produced by the noise signal and the intrinsic noise, the
coherence will be (Stein, French and Holden, 1972):

$$\gamma^2(f) = \begin{cases} a^2/(n^2 + a^2) & \text{for} \quad f \ll c \\[2ex] \dfrac{a^2 c \pi}{(a^2 + n^2)8f} & \text{for} \quad f \gg c \end{cases} \qquad (13.6.4)$$

where c is the carrier rate. From 13.6.3 and 13.6.4, the transinformation for low frequency signals will be

$$I \simeq w\log_2 \{1 + (a/n)^2\} \qquad (13.6.5)$$

and so the transinformation increases linearly with the upper band

limit w when w << c. For rapidly changing signals, with w>> c, the
transinformation will increase with the logarithm of w:

$$I \;\simeq\; \frac{c\,a^2}{16\,(a^2 + n^2)}\; \log_2 w \qquad\qquad (13.6.6)$$

These transinformations will approach the capacity when the input has
a spectral shape which maximizes the transinformation: for w << c the
optimal input is a white noise and so 13.6.5 estimates the capacity,
which is limited only by the internal noise. For w >> c, the optimal
input would have a spectral shape which was related to the shape of
the coherence function, and so concentrated signal power at low
frequencies: thus 13.6.6 is less than the capacity.

A similar reduction of the transinformation by internal noise was
found numerically for the leaky integrator model. Thus internal
noise reduces the ability of a neurone to linearly transmit information
about broad-band, stochastic signals.

13.7. Transinformations of different codes.

In the approaches used in sections 13.1 - 13.6 it has been necessary
to make a choice of what constitutes the symbols and of what kind of
code is used. For steady signals, a rate code seems a reasonable
candidate code: however, use of such a rate code implicitly treats
the neurone as a perfect integrator model with a fixed integration
time, similar to the KC-scaler of section 2.2. Estimating the rate as
a number of spikes in some fixed interval views the spike train
through a rectangular window: Grusser (1962) considers this integration
time to represent the time constant of the receiving cell, say some
15 - 80 msec., Walloe (1968,1970) suggests that an integration time
of 100 msec. will maximize the transinformation rate for DSCT cells.
Whatever the postsynaptic response to a spike train, it is likely that
presynaptic spikes which arrive recently will have more effect on the
postsynaptic response than earlier spikes, and so some weighted
average may be more suitable, with the rectangular window replaced by
some decaying, weighting function (Barlow, 1962; Perkel, 1971;
Segundo, 1971).

For stochastic signals, the transinformation rate can be calculated from the spectral density functions, which are estimated on the
assumption that the spike train is a linearly filtered train of Dirac

delta functions (French and Holden, 1971a). This does not make any explicit assumptions about the code; however, obtaining the trans-information from the coherence assumes that information is transmitted only by the linear part of the relation between the input and the output. Thus the effects of any nonlinear operations are treated as equivalent to distortions or noise.

Ekhorn and Popel (1974) have proposed a method for estimating the transinformation rate without making any assumptions about the code. The output signal is the spike train, and since the action potentials are indistinguishable apart from their times of occurrence, this signal is completely characterized either by the ordered sequence of intervals or the ordered sequence of occurrence times. These representations are equivalent, as one can be obtained from the other. The times of occurrence are chosen for convenience. Time is divided into small sections of duration Δt, where Δt is sufficiently small so that during one section there can be at most one action potential. If there is an action potential, the section is given a value of 1, if there is no action potential, a value of 0. These valued sections form the symbols and the signal is an ordered sequence of these symbols. If the uncertainty of the spike train is given by H(Y) defined by 13.0.2 , which neglects serial correlations between symbols, or by an equation analogous to 13.0.8, which considers the reduction of uncertainty produced by serial correlations, the inform-ation rate H'(Y) is

$$H'(Y) \;=\; \frac{H(Y)}{\Delta t} \tag{13.7.1}$$

Any serial correlations will vanish for times greater than the correlation time T_c defined in section 13.0: this correlation time will correspond to a fixed number of symbols n_o, and so the trans-information can be found from 13.0.9 when the summmations are taken to n_o. Thus the transinformation rate for $n \geq n_o$ is found as:

$$I' \;=\; \frac{1}{\Delta t} \sum_{\substack{i \\ j_0,\ldots,j_n}} p(y_{j_0},\ldots,y_{j_n},x_i) \;\cdot \tag{13.7.2}$$

$$\cdot \left\{ \log_2 \frac{p(y_{j_0},\ldots,y_{j_n},x_i)}{p(y_{j_0},\ldots,y_{j_n})p(x_i)} \;-\; \log_2 \frac{p(y_{j_0},\ldots,y_{j_{n-1}},x_i)}{p(y_{j_0},\ldots,y_{j_{n-1}})p(x_i)} \right\}$$

This method of obtaining the transinformation does not make any assumptions about the code, but treats the spike train as the signal: thus it can be considered to be a signal code. If the action potentials are treated as point events, the ordered sequence of intervals between events contains the same information as the ordered sequence of symbols. The transinformation could be calculated either from the sequence of symbols or the sequence of intervals; with the exception of regular spike trains, it is computationally more efficient to use the ordered sequence of symbols as the correlation time corresponds to a fixed number of symbols. A further advantage of the signal code is that the time course of information transfer can be obtained: instead of obtaining the expectation of the transinformation in bits/symbol the information transmitted by each symbol can be obtained. This has been applied to obtain the time course of information transfer between retinal ganglion cell axons and principal cells of the lateral geniculate nucleus of the cat, when the input to the visual system was a series of flashes of light.(Eckhorn and Popel, 1975).

Since the signal code does not make any restrictive assumptions about the form of the relationship between the input and output, the transinformation obtained using the signal code will be the maximal transinformation possible for the input signal used. Thus, in any comparison of the transinformations achieved by different putative codes, the transinformation obtained from the signal code will give the upper bound. Eckhorn, Grusser, Kroller, Pellnitz and Popel (1976) have estimated the transinformations of three different neuronal systems using different assumptions about the code. The transinformations obtained using a rate code could be much less (down to 5%) of the transinformations obtained with the signal code. However, a weighted average code, where the shape of the weighting function was obtained from the linear correlation between the input and output signals, gave transinformations which were 90 - 95% of those given using the signal code. Thus most of the information is transfered by the linear relation between the input and output. This apparently contradicts the results discussed in section 13.6, where the coherence function was very low for frequencies above the carrier rate. However, the systems considered by Eckhorn et al. were all spike to spike transmission across a synapse : retinal ganglion cell to lateral geniculate cell, γ-efferents to muscle spindle afferent, and muscle spindle afferent to DSCT cell, whereas the systems considered in section 13.6 were continuous signal to spike train signal transmission.

13.8. References.

Abeles, M. and Lass, Y.: Transmission of information by the axon II. The channel capacity. Biol. Cybernetics $\underline{19}$ 121-125 (1975).

Adrian, E.D.: The Basis of Sensation. W.W.Norton: New York (1928).

Adrian, E.D. and Zotterman, Y.: The impulses produced by sensory endings 2: The response of a single end organ. J. Physiol. $\underline{61}$ 151 (1926).

Ascoli, C., Barbi, M., Chillemi, S. and Petracchi, D.: Phase locking of the neural discharge to periodic stimuli. Proc. 3rd. European Meeting on Cybernetics and Systems Research, Vienna April 1976. Transcripta Press (1976).

Barlow, H.B.: The information capacity of nervous transmission. Kybernetik $\underline{2}$ 1 (1963).

Bittner, G.D.: Differentiation of nerve terminals in the crayfish opener muscle and its functional significance. J. Gen. Physiol. $\underline{51}$ 731-758 (1968).

Eckhorn, R., Grusser, O.J., Kroller, J., Pellnitz, K. and Popel, B.: Efficiency of different neuronal codes: information transfer calculations for three different neuronal sustems. Biol. Cybernetics $\underline{22}$ 49-60 (1976).

Eckhorn, R. and Popel, B.: Rigorous and extended application of information theory to the afferent visual s stem of the cat.
    I. Basic concepts. Kybernetik $\underline{16}$ 191-200 (1974).
    II. Experimental results. Biol. Cybernetics $\underline{17}$ 7-17 (1975).

Farber, G.: Berchnung und Messung des Informationsflusses der Nervenfaser. Kybernetik $\underline{5}$ 17-29 (1968).

Fitzhugh, R.: The statistical detection of threshold signals in the retina. J. Gen. Physiol. $\underline{40}$ 925-948 (1957).

French, A.S. and Holden, A.V.: Alias-free sampling of neuronal spike trains. Kybernetik $\underline{8}$ 165-171 (1971a).

French, A.S. and Holden, A.V.: Frequency domain analysis of neuro-physiological data. Computer Programs in Biomedicine $\underline{1}$ 219-234 (1971b).

French, A.S. and Holden, A.V.: Semi-on-line implementation of an alias-free sampling system for neuronal signals. Computer Programs in Biomedicine $\underline{2}$ 1 (1971c).

French, A.S., Holden, A.V. and Stein, R.B.: The estimation of the frequency response function of a mechanreceptor. Kybernetik $\underline{11}$ 15-23 (1972).

Gerstein, G.L. and Mandlebrot, B.: Random walk models for the activity of a single neurone. Biophys. J. $\underline{4}$ 41-68 (1964).

Gestri, G.: Pulse frequency modulation in neural systems: a random model. Biophys. J. $\underline{11}$ 98-109 (1971).

Gnedenko, B.V. and Kolmogorov, A.N.: Limit Distributions for Sums of Independent Random Variables. Trans. K.L.Chung. Addison- Wesley Publ. Co, Reading, Mass. (1968).

Grusser, O.J.: Die Informationskapazitat einelner Nervenzellen fur die signalubermittlung im Zentralnervensystem. Kybernetik $\underline{1}$ 209-11(1962).

Grusser, O.J., Hellner, K.A. and Grusser-Cornehls, U.: Die Inforation-subertrangung im afferenten visuellen system. Kybernetik $\underline{1}$ 175 (1962).

Holden, A.V.: A note on convolution and stable distributions in the nervous system. Biol. Cybernetics 20 171-173 (1975).

Holden, A.V.: The response of excitable membrane models to a cyclic input. Biol. Cybernetics 21 1-8 (1976).

Holden, A.V.: Information transfer in a chain of model neurones. Proc. 3rd European Meeting on Cybernetics and Systems Research, Vienna, April 1976. Transcripta Press, London (1976).

Holt D.R. and Crow, E.L.: Tables and graphs of the stable probability density functions. J. Res. Nat. Bur. Stds. 77B 143 (1973).

Khintchine, A. Ya. and Levy, P.: Sur les lois stable. C.R. Acad. Sci. (Paris) 202 374 (1936).

Knight, B.W.: Dynamics of encoding in a population of neurones. J. Gen. Physiol. 59 734-766 (1972).

Knight, B.W.: The relationship between the firing rate of a single neuron and the level of activity in a population of neurons. J. Gen. Physiol. 59 767-778 (1972).

Lass, Y. and Abeles, M.: Transmission of information by the axon.I: Noise and memory in the myelinated nerve fibre of the frog. Biol. Cybernetics 19 61-67 (1975).

Levy, P.: Sur certain processus stochastiques homogenes. Comp. Math. 7 283 (1940).

MacKay, D.M. and McCulloch, W.S.: The limiting information capacity of a neuronal link. Bull. Math. Biophys. 14 127-135 (1952).

McKean, T.A., Poppele, R.E., Rosenthal, N.P. and Terzuolo, C.A.: The biologically relevant parameter in nerve impulse trains, Kybernetik 6 168-170 (1970).

Matthews, P.B.C. and Stein, R.B.: The regularity of primary and secondary muscle spindle afferent discharges. J. Physiol 202 59-82 (1969)

Milgram, P. and Inbar, G.F.: Distortion suppression in neuromuscular information transmission due to interchannel dispersion in muscle spindle firing thresholds. I.E.E.E. Trans B-M.E. 23 1-15 (1976).

Mountcastle, V.B., Poggio, G.F. and Werner, G.: The neural transformation of the sensory stimulus at the cortical input level of the somatic afferent system. In : Information Processing in the Nervous System. 196-217. Proc. I.U.P.S. Vol. III XXII Int. Congress, Leiden 1962. Ed. R.W. Gerard and J.W.Duyff. Excerpta Medica, Amsterdam (1962).

Partridge, L.D.: A possible source of nerve signal distortion arising in pulse rate encoding of signals. J. Theoretical Biol. 11 257-281 (1966)

Perkel, D.H.: Spike trains as carriers of information. In: The Neurosciences: second study program. Ed. F.o.Schmitt. 587-597. Rockefeller U.P.,N.Y. (1971).

Perkel, D.H. and Bullock, T.H.: Neural Coding. N.R.P. Bulletin 6 221-348 (1967).

Poggio, T. and Torre, V.: A nonlinear transfer function for some neuron models. Proc. First Symposium on Testing and Identification of Nonlinear Systems, Pasedena, March 1975. Ed. G.D.McCann and P.Z. Marmarelis. 292-300. (1975).

Rapoport, A. and Horvarth, W.J.: The theoretical channel capacity of a single neuron as determined by various coding schemes. Information and Control 3 335-350 (1960).

Rescigno, A., Stein, R.B.,Puple, R.L. and Poppele, R.E.: A neuronal model for the discharge patterns produced by cyclic inputs. Bull. Math. Biophys. 32 337-353 (1970).

Segundo, J.P.: Communication and coding by nerve cells. In: The Neurosciences- second study program.Ed. F.O.Schmitt. 569-585. Rockefeller Univ. Press, N.Y. (1971).

Shannon, C.E.: A mathematical theory of commumication. Bell Systems Techn. J. 7 379-423,623-653 (1948).

Shannon, C.E.: Communication in the presence of noise. Proc. I.R.E. 37 10 (1949).

Spekreijse, H.: Rectification in the goldfish retins. Analysis by sinusoidal and auxillary stimulation. Vision Research 9 1461-72 (1969)

Spekreijse, H. and Oostings, H.: Linearizing: a method for analyzing and synthesizing nonlinear systems. Kybernetik 7 22-31 (1970).

Stein, R.B.: The information capacity of nerve cells using a frequency code. Biophys. J. 7 797-826 (1967).

Stein, R.B.: The role of spike trains in transmitting and distorting sensory signals. in : The Neurosciences- second study program. Ed. F.O.Schmitt. 597-604. Rockefeller Univ. Press, N.Y. (1971)

Stein, R.B. and French, A.S.: Models for the transmission of information by nerve cells. In: Excitatory Synaptic Mechanisms. Proc. 5th Int. Meeting of Neurobiologists. Ed. P. Andersen and J.K.S.Jansen Oslo Univ. Press (1970).

Stein, R.B., French, A.S. and Holden, A.V.: The frequency response, coherence and information capacity of two neuronal models. Biophys. J. 12 295-322 (1972).

Walloe, L.: Transfer of signals through a second order sensory neurone. Thesis, Univ. of Oslo. (1968).

Walloe, L.: On the transmission of information through sensory neurones. Biophys. J. 10 745-763 (1970).

Werner, G and Mountcastle, V.B.: The variability of central neural activity in a sensory system, and its implications for the central reflection of sensory events. J. Neurophysiol. 26 958-977 (1963).

Werner, G. and Mountcastle, V.B.: Neural activity in mechanoreceptor afferents. Stimulus-response relation, Weber functions and information transmission. J. Neurophysiol. 28 359-397 (1965).

INDEX.

Editors: K. Krickeberg;
S. Levin; R. C. Lewontin;
J. Neyman; M. Schreiber

Biomathematics

Springer-Verlag
Berlin
Heidelberg
New York